S0-BMA-446

AEROBIOLOGY

US/IBP SYNTHESIS SERIES

This volume is a contribution to the International Biological Program. The United States effort was sponsored by the National Academy of Sciences through the National Committee for the IBP. The lead federal agency in providing support for IBP has been the National Science Foundation.

Views expressed in this volume do not necessarily represent those of the National Academy of Sciences or the National Science Foundation.

Volume

1 MAN IN THE ANDES: A Multidisciplinary Study of High-Altitude Quechua/ *Paul T. Baker and Michael A. Little*

2 CHILE-CALIFORNIA MEDITERRANEAN SCRUB ATLAS: A Comparative Analysis/*Norman J. W. Thrower and David E. Bradbury*

3 CONVERGENT EVOLUTION IN WARM DESERTS: An Examination of Strategies and Patterns in Deserts of Argentina and the United States/ *Gordon H. Orians and Otto T. Solbrig*

4 MESQUITE: Its Biology in Two Desert Scrub Ecosystems/*B. B. Simpson*

5 CONVERGENT EVOLUTION IN CHILE AND CALIFORNIA: Mediterranean Climate Ecosystems/*Harold A. Mooney*

6 CREOSOTE BUSH: Biology and Chemistry of *Larrea* in New World Deserts/ *T. J. Mabry, J. H. Hunziker, and D. R. DiFeo, Jr.*

7 BIG BIOLOGY: The US/IBP/*W. Frank Blair*

8 ESKIMOS OF NORTHWESTERN ALASKA: A Biological Perspective/ *Paul L. Jamison, Stephen L. Zegura, and Frederick A. Milan*

9 NITROGEN IN DESERT ECOSYSTEMS/*N. E. West and John Skujiņš*

10 AEROBIOLOGY: The Ecological Systems Approach/*Robert L. Edmonds*

Additional volumes in preparation

US/IBP SYNTHESIS SERIES 10

AEROBIOLOGY
The Ecological Systems Approach

Edited by

Robert L. Edmonds
University of Washington

Dowden, Hutchinson & Ross, Inc.
Stroudsburg Pennsylvania

Library of Congress Catalog Card Number: 78-23769
ISBN: 0-87933-346-4

81 80 79 1 2 3 4 5
Manufactured in the United States of America

Library of Congress Cataloging in Publication Data

Main entry under title:
Aerobiology: the ecological systems approach.
(US/IBP synthesis series; 10)
Includes index.
1. Air-Microbiology. 2. Airborne infection. 3. Micro-organisms-Dispersal. 4. Air quality. I. Edmonds, Robert L. II. Series.
QR101.A37 574.909'6 78-23769
ISBN 0-87933-346-4

Distributed world wide by Academic Press,
a subsidiary of Harcourt Brace Jovanovich,
Publishers

FOREWORD

This book is one of a series of volumes reporting results of research by U.S. scientists participating in the International Biological Program (IBP). As one of the 58 nations taking part in the IBP during the period July 1967 to June 1974, the United States organized a number of large, multidisciplinary studies pertinent to the central IBP theme of "the biological basis of productivity and human welfare."

These multidisciplinary studies (integrated research programs), directed toward an understanding of the structure and function of major ecological or human systems, have been a distinctive feature of the United States' participation in the IBP. Many of the detailed investigations that represent individual contributions to the overall objectives of each integrated research program have been published in the journal literature. The main purpose of this series of books is to accomplish a synthesis of the many contributions for each principal program and thus answer the larger questions pertinent to the structure and function of the major systems that have been studied.

Publications Committee: US/IBP
Gabriel Lasker
Robert B. Platt
Frederick E. Smith
W. Frank Blair, Chairman

PREFACE

Airborne diseases threaten the lives and productivity of humans, animals, and plants. Despite recent advances in medical, veterinary, and plant sciences, many airborne diseases remain as problems. Aerobiology is a discipline focused on the atmospheric transport of the microorganisms causing these diseases, with the ultimate objective of providing insights into disease control.

Aerobiology, however, includes more than just the study of airborne diseases. Many airborne particles have allergenic properties and are important even though they may be dead or incapable of causing disease, e.g., pollen. Plant distributions are closely related to pollen dispersal. The field of palynology, as a result, is included under the umbrella of aerobiology, as is entomology, since many insects are wind dispersed. Because microorganisms are transported in the atmosphere the principles of atmospheric transport and deposition in both indoor and outdoor environments are included. Recently the impact of biogenic particles on precipitation processes has also emerged as a subject matter of interest to aerobiologists.

In the last few decades air pollution has emerged as a major problem and, while air pollution is not strictly considered in the domain of aerobiology, microorganisms are impacted by air pollutants while airborne, perhaps affecting their viability and their ultimate ability to cause disease.

Before the inception of the US/IBP Aerobiology Program, aerobiology tended to be fragmented and had little cohesiveness as an integrated field of study. The US/IBP Aerobiology Program provided a cohesive framework by placing aerobiology in a systems analyses framework, where each particle is considered with respect to its source, release into the atmosphere, dispersion, deposition, and impact.

The material in this volume consists of contributions from 24 scientists who helped shape aerobiology during the tenure of the International Biological Program. The systems analysis framework was largely developed at a series of workshop/conferences held from 1970 to 1973, which were conceived by Dr. William S. Benninghoff of the University of Michigan, the initial director of the US/IBP Aerobiology Program. Dr. Benninghoff worked diligently from the inception of the International Biological Program in 1964 until 1969, when the Aerobiology Program received its first funding, to ensure that aerobiology was included as a component of the United States and international efforts. His work reached culmination in 1974 when the International Association for Aerobiology was formed under the auspices of the International Union of Biological Sciences. In this volume we hope to represent the current state of the art of aerobiology and establish it as a true but broad interdisciplinary area.

Financial support for this volume was provided by the National Science Foundation (grant no. DEB 7000968) and by many universities and research institutes. Research by Mr. Gilbert Raynor was carried out under the auspices of the New York State Museum and Science Service and the U.S. Atomic Energy Commission (now Energy Research and Development Administration), and was partially supported by research grant no. R-800677 from the Division of Meteorology, U.S. Environmental Protection Agency, and previous grants. Dr. Eugene C. Ogden, New York State Botanist, and Janet V. Hayes collaborated in most of the research drawn upon for his chapter. Appreciation is expressed to Dr. Allen Solomon, Dr. Jack Wallin, Dr. Leonard Bernstein, Dr. C. L. Kramer, and Dr. H. E. Schlichting, Jr., who also supplied information.

Drs. R. L. Dimmick and M. A. Chatigny conducted research funded under the auspices of the Office of Naval Research, Department of the Navy, contract no. N00014-75-C-0157. These authors wish to express their appreciation to Muriel Johnson for her helpful assistance.

Dr. W. R. Solomon's work was supported in part by research grant AI-10181 from the National Institute of Allergy and Infectious Diseases, NIH, USPHS; and Dr. I. L. Bernstein's work was supported in part by training grant AM-05509, National Institute of Arthritis and Metabolic Diseases, NIH, USPHS. Dr. J. M. Leedom wishes to express his appreciation to the Hastings Foundation Fund for partial support of the work done by him and Dr. C. G. Loosli. Finally, I wish to thank Mrs. Martha G. Ellis for assisting in the editing of this volume.

Robert L. Edmonds
Seattle, Washington

TABLE OF CONTENTS

Thomas G. Akers
School of Public Health
and Tropical Medicine
Tulane University
New Orleans, LA 70118

William S. Benninghoff
Department of Botany
University of Michigan
Ann Arbor, MI 48104

I. Leonard Bernstein
Division of Immunology
Department of Internal Medicine
University of Cincinnati Medical Center
Cincinnati, OH 45229

James R. Burleigh
Plant Science Department
Chico State College
Chico, CA 95926

Lucas Calpouzos
Department of Plant Sciences
University of Idaho
Moscow, ID 83843

Mark A. Chatigny
Naval Biosciences Laboratory
Naval Supply Center
Oakland, CA 94625

Robert L. Dimmick
Naval Biosciences Laboratory
Naval Supply Center
Oakland, CA 94625

Robert L. Edmonds
University of Washington
College of Forest Resources
Seattle, WA 98195

James B. Harrington, Jr.
Forest Fire Research Institute
Canadian Forestry Service
Ottawa, Ontario, CANADA

Herbert F. Hasenclever
U.S. Department of Health,
Education and Welfare
Rocky Mountain Laboratory
Hamilton, MT 59840

Charles L. Kramer
Division of Biology
Kansas State University
Manhattan, KS 66506

John M. Leedom
Departments of Medicine
and Pathology
University of Southern California
Los Angeles, CA 90033

Bruce Lighthart
Corvallis Environmental Research Center
U.S. Environmental Protection Agency
Corvallis, OR 97330

Clayton G. Loosli*
Los Angeles County–University
of Southern California
Medical Center
Los Angeles, CA 90033

Michael L. McManus
USDA Forest Service
105 Sanford Street
Hamden, CT 06514

Randolph E. McCoy
Agricultural Research
Center
University of Florida
Ft. Lauderdale, FL 33314

Conrad J. Mason
Department of Atmospheric and
Oceanic Sciences
University of Michigan
Ann Arbor, MI 48104

George H. Quentin
College of Science and
Engineering
University of Texas of the
Permian Basin
Odessa, TX 79762

Gilbert S. Raynor
Meteorology Division
Brookhaven National Laboratory
Upton, NY 11973

*Deceased

Robert S. Safferman
Environmental Protection Agency
National Environmental Research Center
Cincinnati, OH 45268

Harold E. Schlichting, Jr.
President
Bio Control Company
Port Sanilac, MI 48649

Allen M. Solomon
Environmental Sciences Division
Oak Ridge National Laboratory
Oak Ridge, TN 37830

William R. Solomon
Division of Allergy
Department of Internal Medicine
University of Michigan Medical School
Ann Arbor, MI 48104

J. Clifton Spendlove
Environment and Life Science Division
Department of the Army
U.S. Army Dugway Proving Ground
Dugway, UT 80422

Gabor Vali
Department of Atmospheric Science
University of Wyoming
Laramie, WY 82070

Jack R. Wallin
Agricultural Research Service
U.S. Department of Agriculture
Columbia, MO 65201

1. INTRODUCTION

R. L. Edmonds

DEFINITION OF AEROBIOLOGY

It has long been known that biological objects, nonliving particulates, and gases are present in the atmosphere. Aerobiology, a term coined in the 1930s, is a scientific discipline focused on the transport of airborne organisms in both outdoor (extramural) and indoor (intramural) environments. A great variety of fields are represented under the umbrella of aerobiology, including plant, human, and animal pathology; entomology; allergology; air pollution effects; palynology; phytogeography; and meteorology.

Aerobiology, in its original definition, was restricted to studies of airborne living organisms. Man-made and natural air pollutants affect the viability of organisms in the atmosphere and on the ground, however, and for this reason pollutant gases and particulates are included when such biological effects are demonstrated. The more important gases known to affect microorganisms adversely are sulfur dioxide, oxides of nitrogen, carbon monoxide, and ozone, but many more are capable of impact. The particulate materials considered important in aerobiology include smokes and dusts, radionuclides, pesticides, and the biological forms—viruses, bacteria, fungus fragments and spores, algae, Protozoa, moss and fern spores, pollen, plant fragments, minute seeds, insects, and other microfauna. Table 1.1 shows the range of particles and their diameters. Most biological particles range in diameter from 0.5 to 100 μm, but the total size range of interest is from 0.0001 μm in smokes to a centimeter or so in the case of airborne insects. Biological particles may occur as single unattached organisms or as aggregates of organisms. They may also adhere to dust particles.

Generally, the numbers of living biological particles are several orders of magnitude less than the numbers of other particles. The fact that some of these biological particles (e.g., fungus spores) are capable of causing epiphytotics severe enough to destroy the greater parts of the world's wheat, corn, or rice crops, however, demonstrates their impact.

Table 1.1. Types and sizes of airborne particulate material.

Type	Diameter (μm)
Smokes	0.001 – 0.1
Condensation nuclei	0.1 – 20.0
Dusts	0.1 – cm
Viruses	0.015 – 0.45
Bacteria	0.3 – 10.0
Algae	0.5 – cm
Fungus spores	1.0 – 100
Lichen fragments	1.0 – cm
Protozoa	2.0 – cm
Moss spores	6.0 – 30.0
Fern spores	20.0 – 60.0
Pollen	10.0 – 100.0
Plant fragments, minute seeds, insects, spiders, etc.	>100

THE IMPORTANCE OF AEROBIOLOGY

Table 1.2 lists some important plant, animal, and human diseases caused by airborne microorganisms. Aeroallergens are also included. The magnitude of this list gives some idea as to the impact that airborne diseases have on our crop and fiber plants and human and animal health.

CROP AND FIBER LOSSES

Estimated average annual losses to agricultural crops due to diseases in the United States from 1951 to 1960 inclusive were $3.25 billion (LeClerg 1964). Data on dollar losses to forest ecosystems are rare, but disease losses are considerable (Davidson and Prentice 1967). The majority of these are airborne diseases and most are caused by fungi, bacteria, and viruses. Dollar losses are not the only impact—supply and demand must also be taken into account. In some areas, disease losses often have far-reaching social effects that are extremely difficult to assess. This is of particular importance in developing countries. Breeding for disease resistance and the use of pesticides have reduced some of these losses. The corn blight epidemic in the United States in 1970, however, caused by airborne spores of *Helminthosporium maydis,* resulted in 15% losses and showed the potential power of a corn blight epiphytotic.

The genetic basis of resistance of many of our crop plants is being continually narrowed, which increases the likelihood that new pathogenic strains will have a great impact. In the case of wheat rust fungi, over which we presently have little control, a new virulent strain evolves every 3–5 yr. This, in combination with the fact that many pesticides are also beginning to fail, is proving to be ecologically unsound. Murdoch (1971) points out the need for obtaining more information about the aerobiology of plant pathogens before potentially disastrous epidemics occur.

The disease loss situation ($3.25 billion annually) should be compared with direct agricultural losses to air pollutants for the same period (1959–1960), which ranged from $150 to $500 million annually (USDA 1965). Much attention has been paid to the monitoring of air pollutants and the damage they cause. That is important, but some attention must be diverted to what seems even more important—the aerobiology of plant diseases. It is anticipated that disease losses to our food and fiber production will become worse in future years and the condition will be compounded by a rise in demand for food and fiber as the world population increases.

IMPACT ON NATURAL AND MANAGED VEGETATION

Interest in plant systems is not confined to diseases; palynology has contributed largely to our understanding of how the composition and distributions of vegetation units have changed through time and has even enabled reconstruction of past climates (Webb and Bryson 1972). Studies of modern pollen dispersion and fallout in combination with data from the past will permit monitoring of vegetation production and change, and predictions of trends of changes in natural ecosystems.

HUMAN HEALTH IMPACTS

The importance of human disease transmission by biological aerosols has been in part a function of urbanization, because airborne transmission of human disease is especially common in indoor spaces (Finkelstein 1969b). Thus intramural aerobiology is important in human diseases. The human environment of crowded cities was a contributing factor in the plague epidemic of 1348. For a thousand years prior to that a low-population-density agricultural society had existed.

Bacteria, fungi, and viruses are responsible for human diseases. As modern air pollution information accumulates, it is becoming apparent that nonbiological particulates and gases in the atmosphere may aggravate the effects of disease or allergenic agents.

A conservative annual dollar resource value cost for selected diseases in the United States was calculated by Ridker (1968). The common cold costs $331 million, pneumonia $490 million, chronic bronchitis $160 million, cancer of the respiratory system $680 million, emphysema $64 million, and asthma $259 million.

The influenza epidemic of 1918–1919 resulted in 550,000 deaths in the United States alone and 20 million deaths occurred altogether. The disease is surely transmitted by airborne particles and there is accumulating evidence that winds may spread the infection several hundred kilometers under certain conditions.

Airborne allergens also cause considerable impact; 31 million Americans each year suffer from some form of allergic diseases with 86 million suffering from asthma (Davis 1972) and 10–15 million from hay fever (Finkelstein 1969a). Ragweed pollen is the cause of 90% of pollinosis in the United States, but fungus spores, house dust, algae, and the like are also important.

Hay fever, asthma, and other allergies account for one-third of all chronic conditions for children under 17 (Barkin and McGovern 1966). Many tens of millions of days are lost from work and school from childhood through adult life because of such allergies.

ANIMAL DISEASE PROBLEMS

Few diseases of commercial and domestic animals can be attributed to airborne aerosols. Most animal diseases are transmitted by contact, by insect bites, and through ingestion of contaminated food and water (Finkelstein 1969b). Several very important ones such as fowl pest and foot-and-mouth disease, however, are definitely airborne over distances of at least several tens of kilometers (Hugh-Jones and Wright 1970). Other airborne animal diseases include rinderpeste and ephemeral fever in cattle and infectious laryngotracheitis of poultry.

IMPACT ON ATMOSPHERIC PROCESSES

Biogenic materials in the atmosphere actively modify cloud and precipitation processes. Such materials, which may include bacteria, are active as (a) condensation nuclei, (b) ice nuclei, (c) coalescence centers, and (d) surface active agents. There is still, however, much uncertainty with respect to their actual contribution to atmospheric processes.

Table 1.2 Important airborne plant, animal, and human diseases, and aeroallergens.

Type of disease	Causative agent
	PLANT DISEASES
Fungal diseases	
Apple rust	*Gymnosporangium* spp.
Annosus root rot	*Fomes annosus*
Banana leaf spot (sigatoka)	*Mycosphaerella musicola*
Beet downy mildew	*Peronospora* spp.
Blossom infection	*Sclerotinia laxa*
Cedar rust	*Gymnosporangium* spp.

Table 1.2. Important . . . diseases and aeroallergens (continued).

Type of disease	Causative agent
Chestnut blight	*Endothia parasitica*
Crown rust of oats	*Puccinia coronata*
Downy mildew	*Pseudoperonospora humuli*
Fusiform rust of southern pines	*Cronartium fusiforme*
Loose smut of wheat	*Ustilago tritici*
Maize rust	*Puccinia sorghi*
Onion mildew	*Peronospora destructor*
Potato late blight	*Phytophthora infestans*
Powdery mildew of barley	*Erysiphe graminis*
Southern corn leaf blight	*Helminthosporium maydis*
Stem rust of wheat and rye	*Puccinia graminis (tritici & secalis)*
Tobacco blue mold	*Peronospora tabacina*
White pine blister rust	*Cronartium ribicola*
DISEASES OF DOMESTIC ANIMALS	
Bacterial diseases	
Tuberculosis	*Mycobacterium bovis*
Glanders	*Actinobacillus mallei*
Fungal diseases	
Aspergillosis	*Aspergillus* spp.
Cryptococcosis	*Cryptococcul neoformans*
Coccidioidomycosis	*Coccidioides* spp.
Viral diseases	
Hog cholera	
Equine influenza	
Swine influenza	
Feline distemper	
Canine distemper	
Newcastle disease	
Infectious bronchitis	
Foot-and-mouth disease	
Rinderpeste	
Ephemeral fever	
Infectious laryngotracheitis	
HUMAN DISEASES	
Bacterial diseases	
Pulmonary tuberculosis	*Mycobacterium tuberculosis*
Pulmonary anthrax	*Bacillus anthracis*
Staphylococcal respiratory infection	*Staphylococcus aureus*
Streptococcal respiratory infection	*Streptococcus pyogenes*
Meningococcal infection	*Neisseria meningitidis*
Pneumococcal pneumonia	*Diplococcus pneumoniae*
Pneumonic plague	*Pasteurella pestis*
Whooping cough	*Bordetella pertussis*
Diphtheria	*Corynebacterium diphtheriae*
Klebsiella respiratory infection	*Klebsiella pneumoniae*
Staphylococcic wound infection	*Staphylococcus aureus*

Table 1.2. Important . . . diseases and aeroallergens (continued).

Type of disease	Causative agent
Fungal diseases	
Aspergillosis	*Aspergillus fumigatus*
Blastomycosis	*Blastomyces dermatitidis*
Coccidioidomycosis	*Coccidioides immitis*
Cryptococcosis	*Cryptococcus neoformans*
Histoplasmosis	*Histoplasma capsulatum*
Nocardiosis	*Nocardia asteroides*
Sporotrichosis	*Sporotrichum schenckii*
Diseases caused by viral and related agents	
Influenza	
Febrile pharyngitis or tonsillitis	
Common cold	
Croup	
Bronchitis	
Bronchiolitis	
Pneumonia	
Febrile sore throat	
Pleurodynia	
Psittacosis	

COMMON AEROALLERGENS

Aeroallergen	Source
Pollens	Wind-pollinated plants: Grasses, weeds (including ragweed, *Ambrosia* spp.) and trees
Molds[a]	Usually saprophytic, prevalence depending upon humidity, temperature, and substrate distribution
Danders	Feathers of chickens, geese, ducks; and hair of cats, dogs, horses, sheep, cattle, laboratory animals, and humans
House dust	A composite of all dusts found about the home—probably has specific components related to mites, algae, etc.
Miscellaneous vegetable fibers and dusts	Cotton, kapok, flax, hemp, jute, straw, castor bean, coffee bean, orrisroot, rye, wheat
Cosmetics	Wave set lotions, talcs, perfumes, hair tonics
Insecticides	Insecticides containing pyrethrum as a common ingredient
Paints, varnishes, and glues	Linseed oil and organic solvents may be primary irritants; fish protein in glues still important

[a]Most common aeroallergenic fungi: *Alternaria, Aspergillus, Botrytis, Cladosporium, Curvularia, Epicoccum, Fusarium, Helminthosporium, Hormodendrum, Macrosporium, Penicillium, Phoma, Pullularia, Rhodotorula, Spondylocladium, Stemphylium, Trichoderma.*

HISTORY OF DEVELOPMENT OF AEROBIOLOGY IN THE PRE-IBP PERIOD

As a specialized field of investigation aerobiology had its first developments in the 1930s. Its origins, however, were in the pioneering experiments of Spallanzani in 1776 and the work of Pasteur, Tyndell, Miquel, and others who used aerobiological methods in the middle of the 19th century to combat the theory of spontaneous generation of life and to develop the germ theory of disease.

Much of the early work in both extramural and intramural aerobiology was presented in a symposium on aerobiology at a meeting of the American Association for the Advancement of Science held in Chicago in 1942 (Moulton 1942). A total of 37 papers were presented and the range of topics covered indicated the great diversity of fields covered under the title *aerobiology*.

The late 1950s and 1960s saw a period of active development in the field of aerobiology. P. H. Gregory published his first edition of *The Microbiology of the Atmosphere* (Gregory 1961) and the National Aeronautics and Space Administration (NASA) sponsored a conference on atmospheric biology in 1964 (Tsuchiya and Brown 1965). The Seventeenth Symposium of the Society for General Microbiology was devoted to airborne microbes (Gregory and Monteith 1967). The period, however, tended to be dominated by the efforts of the intramural aerobiologists, who formed a loose association and met at irregular intervals. Their emphasis was on defense-related research conducted at Fort Detrick, Maryland; the Naval Biological Laboratory, Oakland, California; the Defense Research Establishment, Suffield, Alberta, Canada; and the Microbiology Research Establishment, Porton Down, England. Much of the literature in intramural aerobiology is summarized in the proceedings of these conferences (Lepper and Wolfe 1966, Dimmick 1963, Silver 1970). Dimmick and Akers (1969) also produced a fine work on experimental aerobiology.

DEVELOPMENT OF AEROBIOLOGY IN THE IBP PERIOD—THE SYSTEMS APPROACH TO AEROBIOLOGY

In 1964 the International Council of Scientific Unions (ICSU) set up the International Biological Program (IBP), whose major objective was to study the biological basis for productivity of the world's ecosystems. Such a program was introduced because of the pressures facing mankind today and in the future—rapidly increasing population, food and fiber shortages, and the destruction of the environment. The US/IBP was sponsored by the National Academy of Sciences/National Research Council.

At the "Atmospheric Biology Conference" convened by NASA in 1964, the hope was raised that atmospheric dispersion of biological materials might be given attention by the IBP. Some years passed, however, before the International Aerobiology Program was established through the efforts of William S. Benninghoff (University of Michigan, Ann Arbor, Michigan) and Philip H. Gregory (Rothamsted Experimental Station, Herpenden, Herts, England) under IBP section UM (Use and Management of Biological Resources) in the summer of 1968. The US/IBP Aerobiology Program, under the directorship of Dr. William S. Benninghoff, also was established in 1968, although it was not funded until the spring of 1969.

Despite our great knowledge of many aerobiological processes and the vast amount of literature in this area, very little of the information has been synthesized in the past. It became plainly obvious at the first US/IBP Aerobiology Program conference on "Aerobiology Objectives in Atmospheric Monitoring" in June 1970 (Benninghoff and Edmonds 1971) that research in aerobiology had been done in isolated segments and in general aerobiological phenomena were not considered to be parts of ecological systems involving interactions of biological and physical components. In addition, key information (e.g., the distance that microorganisms can be carried in viable condition) for constructing an overall picture was missing.

The systems approach and mathematical modeling have been used to integrate the results of the diverse aerobiological studies reported in this volume, continuing ideas generated at a series of three workshop/conferences on "Ecological Systems Approaches to Aerobiology" (Benninghoff and Edmonds 1972, Edmonds and Benninghoff 1973a,b). Obviously, not all of the data obtained can be integrated at this time or perhaps could ever be totally integrated because of the diverse nature of aerobiology. This approach, however, serves as a tool for pointing out the blanks in our knowledge of ecological systems and thus can guide research. Prediction of the potential ecological impact of aerobiological events is also possible.

The main focus of modeling activities has been on the submodels being developed in relation to the "aerobiology pathway" as shown in Figure 1.1. Each airborne component uses this pathway. Division of aerobiological systems in this manner allows for ease of coupling of various pathways. Various aerobiological models are discussed in this volume, and meteorological dispersion models have been used extensively.

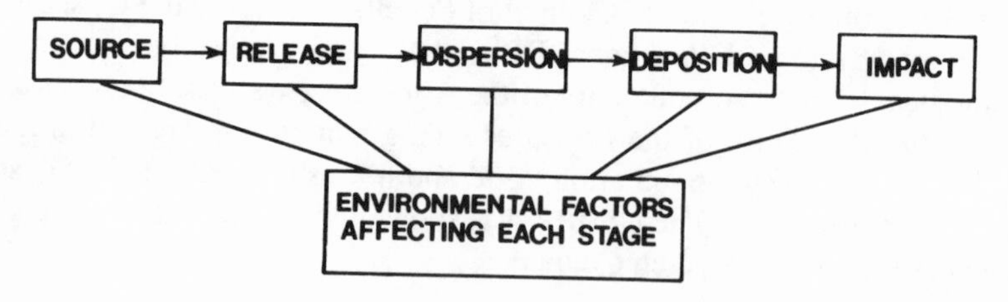

Figure 1.1. The aerobiology pathway.

The US/IBP Aerobiology Program has attempted to synthesize aerobiological information by developing unifying principles for handling aerobiological data. Although very little new research was funded under the auspices of this program, many new ideas were generated, particularly the idea of considering aerobiology in an ecological systems context. Aerobiological problems are not confined by national boundaries and we need to consider many of them on a global or regional scale. Such an approach has been taken by the European aerobiologists (Nilsson 1973, ten Houten 1972).

AEROBIOLOGY IN THE POST-IBP PERIOD

The IBP officially terminated in June 1974. With this termination it was recognized that the international aspects of aerobiology developed during the IBP should be continued. As a result, the International Association for Aerobiology (IAA) was chartered at the First International Congress of Ecology at the Hague in July 1974. The IAA was incorporated in The Netherlands with the objective of continued fostering of aerobiology through international conferences and other forms of communication.

In the United States an aerobiology panel has been set up in the National Academy of Sciences/National Research Council in Washington, D.C., to continue the development of the discipline of aerobiology.

Aerobiology is now a recognized discipline. Its import is underscored by the fact that many of the problems facing us with respect to world food and fiber supplies and health are aerobiological in nature.

LITERATURE CITED

Barkin, G. D., and J. P. McGovern. 1966. Allergy statistics. Ann. Allergy 24: 602–609.

Benninghoff, W. S., and R. L. Edmonds, eds. 1971. Aerobiology objectives in atmospheric monitoring. US/IBP Aerobiol. Program Handb. 1. Univ. Michigan, Ann Arbor. 66 p.

Benninghoff, W. S., and R. L. Edmonds. 1972. Ecological systems approaches to aerobiology. I. Identification of component elements and their functional relationships. US/IBP Aerobiol. Program Handb. 2. Univ. Michigan, Ann Arbor. 158 p.

Davidson, A. G., and R. M. Prentice, eds. 1967. Important forest insects and diseases of mutual concern to Canada, the United States, and Mexico. North Am. For. Comm., FAO. Dep. For. Rural Dev. Can. 248 p.

Davis, D. J. 1972. NIAID initiatives in allergy research. J. Allergy Clin. Immunol. 49:323.

Dimmick, R. L., ed. 1963. First Int. Symp. Aerobiol. Naval Biol. Lab., Naval Supply Cent., Oakland, Calif. 405 p.

Dimmick, R. L., and A. B. Akers, eds. 1969. An introduction to experimental aerobiology. Wiley-Interscience, New York. 494 p.

Edmonds, R. L., and W. S. Benninghoff, eds. 1973a. Ecological systems approaches to aerobiology. II. Development, demonstration, and evaluation of models. US/IBP Aerobiol. Program Handb. 3. (NTIS no. AP-USIBP-H-73-3.) Univ. Michigan, Ann Arbor. 186 p.

Edmonds, R. L., and W. S. Benninghoff, eds. 1973b. Ecological systems approaches to aerobiology. III. Further model developments. US/IBP Aerobiol. Program Handb. 4. (NTIS no. AP-USIBP-H-73-4.) Univ. Michigan, Ann Arbor. 118 p.

Finkelstein, H. 1969a. Preliminary air pollution survey of aeroallergens: A literature review. Natl. Air Pollut. Control Adm. Publ. APTD-69-23. US-DHEW/PHS, Natl. Air Pollut. Control Adm., Raleigh, N.C.

Finkelstein, H. 1969b. Preliminary air pollution survey of biological aerosols: A literature review. Natl. Air Pollut. Control Adm. Publ. APTD 69-30. US-DHEW/PHS, Natl. Air Pollut. Control Adm., Raleigh, N.C. 94 p.

Gregory, P. H. 1961. The microbiology of the atmosphere. 1st ed. Wiley-Interscience, New York. 251 p.

Gregory, P. H., and J. L. Monteith, eds. 1967. Airborne microbes. Cambridge Univ. Press, Cambridge, England. 385 p.

Hugh-Jones, M. E., and P. B. Wright. 1970. Studies on the 1967–68 foot-and-mouth disease epidemic. The relation of weather to the spread of disease. J. Hyg. 68:253–271.

LeClerg, E. L. 1964. Crop losses due to plant diseases in the United States. Phytopathology 54:1309–1313.

Lepper, M. H., and E. K. Wolfe, eds. 1966. Second International Conference on Aerobiology (airborne infection). Bacteriol. Rev. 30:485–698.

Moulton, S., ed. 1942. Aerobiology. AAAS Publ. 17. Am. Assoc. Advan. Sci., Washington, D.C. 289 p.

Murdoch, W. W. 1971. Environment. Sinauer Assoc., Inc., Sunderland, Mass. 440 p.

Nilsson, S. 1973. Scandinavian Aerobiology Bull. 18. Ecol. Res. Comm. Swed., Natl. Sci. Res. Council, Stockholm. 221 p.

Ridker, R. G. 1968. Economic costs of air pollution: Studies in measurement. Praeger Publ., New York.

Silver, I. H., ed. 1970. Aerobiology: Proc. Third Int. Symp. Academic Press, New York. 278 p.

ten Houten, J. G. 1972. Aerobiologie. Pudoc, Wageningen, The Netherlands. 119 p.

Tsuchiya, H. M., and A. H. Brown, eds. 1965. Proceedings of the Atmospheric Biology Conference. Univ. Minnesota, Minneapolis. 235 p.

USDA (U.S. Department of Agriculture). 1965. Losses in agriculture. Agric. Handb. 291. 120 p.

Webb, T., and R. A. Bryson. 1972. Late- and postglacial climatic change in the northern Midwest, U.S.A.: Quantitative estimates derived from fossil pollen spectra by multivariate statistical analysis. Quat. Res. 2:70–115.

2. SOURCES AND CHARACTERISTICS OF AIRBORNE MATERIALS

T. G. Akers, R. L. Edmonds, C. L. Kramer, B. Lighthart, M. L. McManus, H. E. Schlichting, Jr., A. M. Solomon, and J. C. Spendlove

FACTORS IN THE PRODUCTION, RELEASE, AND VIABILITY OF BIOLOGICAL PARTICLES

BACTERIA AND VIRUSES*

Bacteria and viruses are injected into the atmosphere through wind and rain action on soil, water, and plants. Man-made sources also contribute.

Characteristic virus particles are approximately spherical, polyhedral, or tadpolelike in shape, and range in size from 0.015 to 0.45 μm (Gray 1961). Viruses are little adapted to independent air dispersal.

Bacteria comprise a rather heterogeneous collection of organisms. They vary enormously in length from 0.3 μm to 10–15 μm (Gray 1961) and occur in three principal shapes: spherical cells or cocci, rod-shaped or cylindrical cells, and spiral cells.

Bacteria as well as viruses lack active discharge mechanisms for single cells. Mechanical disturbance of dust, clothing, surgical dressing, and the like can carry into the air contaminated particles of substratum acting as "rafts" that bear clumps of organisms (Lidwell 1967). Rain splash, breakers and sea spray, and processes involved with sewage treatment continuously throw minute, potentially microbe-laden droplets into the atmosphere. Droplets expelled from the respiratory tract by coughing and sneezing are important as intramural sources of pathogenic and nonpathogenic bacteria and viruses. Some viruses are spread by insects (Lidwell 1967).

Particles containing bacteria and viruses range in size from that of a single virus and its carrier milieu to comparatively large rafts containing one or more bacteria as indicated in Figure 2.1. The size of the aerosol particle is governed more by the type and percentage of solid constituents of the suspen-

---

*B. Lighthart, J. C. Spendlove, and T. G. Akers

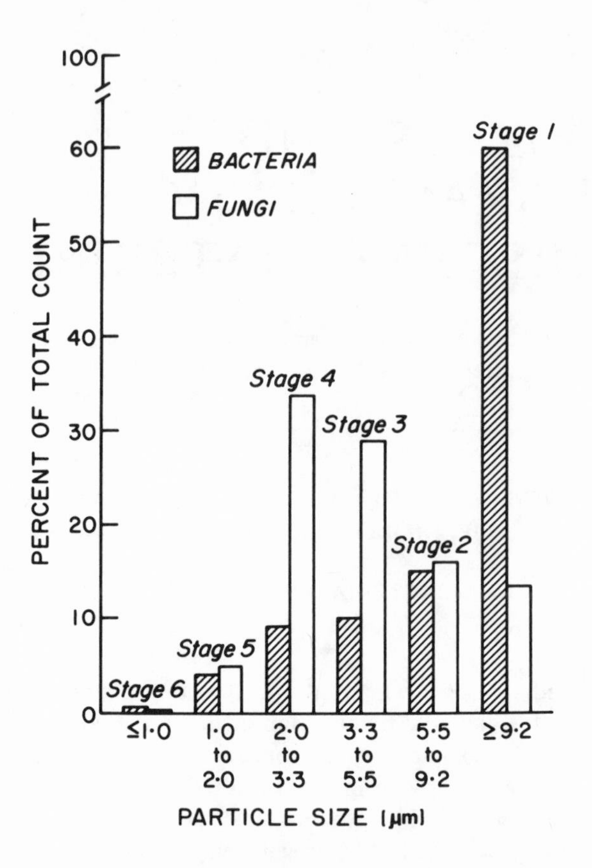

Figure 2.1. Mean airborne bacterial and fungal particle size distribution for a 6-mo period at a site in Seattle, Washington, as measured by an Andersen sampler. Sampler stage 1 is at the right and stage 6 at the left of the abscissa (B. Lighthart and M. Jadot, unpublished observations).

sion medium and the method of aerosolization than by the size of the individual microorganisms.

Constituents of the suspension medium generally impart hygroscopicity to the particle. This characteristic causes the particle size to vary with ambient temperature and relative humidity. Aerosol particles apparently come to hygroscopic equilibrium with existing relative humidity very rapidly. Milburn et al. (1957) in studies of deposition of hygroscopic salt and gelatin-phosphate particles in the human lung have shown that only 0.1 s is required for particles less than 2.0 μm in diameter to reach maximum hydration from complete dryness.

Energy imparted to the suspension medium during aerosolization causes breakup and subsequent aerosol formation, and also regulates particle size. Thus, for example, the water from the boom of a trickling filter of a sewage treatment facility probably forms comparatively larger diameter aerosols than those created in a forced-draft cooling tower where considerably more energy is brought to bear on the coolant.

Production and Release

Loading of the atmosphere with viable viruses and bacteria may range from very dense accumulations to, for all practical purposes, sterile air, depending upon proximity to the organism source, source strength, and factors affecting airborne viability. Sampled concentrations of bacteria range from none to $10^6/m^3$, but few specific data are available.

The sources of airborne bacteria and viruses may be categorized as coming from man and animals, non-man-affected loadings, and man-affected loadings involving such things as sewage treatment plants.

Release from man and animals

Talking, coughing, and sneezing result in injection of droplets containing viruses and bacteria into airstreams (Lidwell 1967). Sneezing is the most vigorous of the three and commonly as many as a million droplets below 100 μm in diameter are injected into the atmosphere (Duguid 1945). The number of droplets expelled in a cough are far fewer than in a sneeze and have a larger median diameter. Not all the droplets expelled from the mouth remain airborne for long since the larger droplets quickly fall out. Although the nose and mouth are commonly the principal sources of airborne viruses and bacteria, skin carriage release may also occur (Lidwell 1967).

Many animal virus diseases are transmitted in the atmosphere. It has been reported (Hugh-Jones and Wright 1970) that certain weather conditions, including wind and rain, played a major role in the spread of outbreaks of the 1967–1968 foot-and-mouth disease epizootics and accounted for pockets of infections more than 60 km from the main outbreak areas. Hugh-Jones et al. (1973) also presented evidence for the airborne spread of Newcastle disease of poultry.

Airborne animal transmission of nonpathogenic extramural bacteria is poorly understood although by logical inference many of the same processes producing airborne nonpathogenic bacteria from man must be operational. Thus "rafting" of epidermal bacteria on shed skin flakes and hair during movement; in saliva droplets during oral sound production, sneezing, and coughing; in fecal material during flatus ejection and defecation; and so on would be expected. Some animal pathogenic bacteria of veterinary importance transmitted by the respiratory route are tabulated and briefly discussed in Constantine (1969) and Gordon (1965). For a listing of airborne transmission and aerosol studies conducted with viruses and phages refer to the review by Akers (1969).

Non-man-affected loadings

The background loadings of airborne bacteria may be from land (Fulton 1966c), water (both fresh- and saltwater [Zobell 1946, R. L. Dimmick pers. commun.]), and possibly from atmospheric sources (see Algae and Protozoa, chap. 2, this volume). The contribution of each to the total loading depends upon the source strength, e.g., cellular numbers per unit volume, and the rate of injection from the source (see also Airborne Microbial Models, chap. 7, this volume).

Terrestrial sources

Loadings over terrestrial sources appear to be heavier compared with aquatic sources, but neither is constant in time or space. Seasonally, airborne bacteria increase in number during the warm summer and fall (Kauf 1962, Gregory 1973), probably because of combinations of convective upward transport of bacteria-laden particles from dry, hot surfaces, which become concentrated within the atmospheric inversion layer. Summer rains may also contribute large numbers of airborne cells when microdroplets are formed during raindrop impact on wetted surfaces (Blanchard and Syzdek 1970). Diurnal loading changes also occur and again convective processes may release soil-produced bacteria to the atmosphere during the day, while during quiescent nights absolute numbers of bacteria in the air column decrease, not only because they die, but because of gravitational settling. The interaction of nocturnal settling with higher relative humidities and no lethal solar radiation may also result in higher concentrations of viable coils at ground level. (The diurnal convection, injection, and gravitational settling might be simulated rather well with present state-of-the-art models [Benninghoff et al. 1973, A. S. Frisch pers. commun.]). Of course, episodes of ground level atmospheric turbulence during storms markedly contribute to soil-derived airborne bacteria loading. Frontal activity increases numbers of airborne organisms in dry regions, reduces numbers during rain episodes, and increases the number from rain splash on passage of the front (Fulton 1966b). The numbers of viable cells decrease with altitude (Fulton and Mitchell 1966) but are highly correlated with the inversion layer. The higher altitude loadings appear to have a smaller variance.

Numerous airborne bacteria may be derived from the natural epiphytic flora of plants (Preece and Dickensen 1971, Bernstein et al. 1973). Thus bacteria may be blown off the surfaces of plants by convective or frontal turbulence or by rain splash. The airborne route is conceivably an important mode of inoculation of natural epiphytic flora necessary for normal plant growth. Some phytopathogens, e.g., fire blight of apple, use the atmosphere as a transplant vector (Goodman et al. 1974). The microbial loading rate of the atmosphere is virtually unknown!

Certain bacteria, e.g., *Pseudomonas syringae* (or their volatile decomposition products), also act as ice nuclei when they are introduced into the world atmosphere (Schnell and Vali 1976, Mandrioli et al. 1973). This phenomenon could conceivably be a significant factor in regional rainfall distribution, and an understanding of the process might be used by man to modify this distribution. The effects of all large-scale terrestrial sources such as forest fires and volcanic explosions on atmospheric microbial loading is problematical.

In almost all previous studies, the numbers of viable bacteria enumerated from the atmosphere are only a minimal estimate, for cultural techniques reveal only those selected bacteria that grow in particular culture media and under certain conditions. Further, it is assumed with the culture technique that each colony is derived from one parent cell. This is not true in all cases for there are certainly particles that "raft" a number of bacteria at one time. It is anticipated that with the advent of electronic optical scanning methods the problem will be alleviated (B. Lighthart, unpublished data).

Crude quantitative estimates of airborne bacteria reveal that many species are present (Gregory 1973). The lack of comprehensive knowledge in this area might be due in part to the paucity of tools available in the past to characterize and organize such a diverse assemblage of bacterial types as are thought to occur in the atmosphere, and requirement for personnel and equipment beyond the availability of most facilities. With the advent of mass inoculation techniques (Lighthart 1968, Lighthart and Oglesby 1969) and computer clustering of large numbers of bacteria into a manageable and relatively homogeneous set of groups (Lighthart and Loew 1972), their distribution in time/space may be evaluated (Lighthart 1975).

Aquatic sources

Microbial aerosols over oceans apparently originate less frequently from aquatic sources than from terrestrial sources. It seems logical that wave action at sea, together with air movement above the sea, would be a major source of organisms in the marine atmosphere. Junge (1964) and Gregory (1964), however, argue that sea surfaces contribute very little to the microbial content of the marine atmosphere. These investigators claim that the sea acts as a sink in clearing the lower atmosphere of particles. This hypothesis would account somewhat for the decrease in particulates with distance from shore observed by several investigators. Other factors at sea that attenuate aerosols are marine fog (M. D. Hogan pers. commun., Fulton 1966a). Junge (1964) and Gregory (1964) maintain that there may be more marine organisms in the atmosphere near the shore than on the open sea because of the aerosolizing effect of the surf (Blanchard 1972) and high concentrations of microorganisms in the littoral waters because of natural and waste-water runoff. Thus surf activity is probably a major source of aerosolized marine and other aquatic microorganisms. Fulton (1966a) found that the ratio of fungi to bacteria increases with distance from the shore. This finding probably reflects the greater resistance to solar radiation and other environmental factors of fungus spores. Pady and Kelly (1953) hypothesize that numbers of organisms over the sea are probably correlated with air masses. They found that polar air masses had very low numbers of both bacteria and fungi while tropical air had higher numbers, again with fungi far outnumbering bacteria.

The movement of aerosols of marine origin over land has been observed by Zobell (1946), who found airborne bacteria capable of growing on marine agar decreased in proportion to those growing on freshwater media as open petri dishes were exposed greater distances from the sea.

In addition, Fulton and Mitchell (1966) have observed the movement of marine aerosols over land from air masses of marine origin. Invariably those air masses of marine origin had far fewer microorganisms than those originating over land. Woodcock (1948) observed that the occurrences of "red tides" were associated with local respiratory difficulties, indicating massive movement of marine organisms onshore. Blowback of microorganisms rising to the surface from submarine sewage outfalls is also possibly a significant source of potential microbial problems derived from the sea. In the light of B. Lighthart's (unpublished) observation of up to 10^5 bacteria/ml in the surface film (ca 250 μm) on Lake Whatcom, Washington, lakes and rivers may be thought of as major potential sources of airborne bacteria.

The mechanism for injection of water droplets containing bacteria into the atmosphere may occur by rain, hail, or snow splash, but possibly more significantly by subsurface bubbles that rise through the water column to burst at the surface. Breaking waves may be a significant contribution to the latter category. It has been observed (Woodcock 1955) that bubbles of 20–2000 μm in diameter burst at the surface producing about five microdroplets that are projected into the air between 0.1 and 0.15 cm from the surface. Further work by Blanchard and Syzdek (1970, 1972) have shown that bubbles concentrate cells from the water up to bubble diameters of 80 μm and that droplet ejection height may be as high as 20 cm for a bubble with a diameter of 2 mm.

Atmospheric sources

The role of the atmosphere as a source of bacteria is unknown. Whether bacteria reproduce while airborne is problematical. It is known, however, that there is a potential for reproduction; Dimmick (1960) has observed physiological activity in airborne bacteria. Specifically, he has observed the repair of bacteria damaged by ultraviolet light, suspended in a laboratory drum (R. L. Dimmick pers. commun.). If repair can occur, why not reproduction under proper circumstances?

Man-affected loadings

Both urban and rural environments have loci of human activities that significantly affect the bacterial loadings of the ambient atmosphere. The urban loadings may be integrated over time within the area described as the "heat island" (Fulton and Mitchell 1966) and quickly diluted down to background levels.

Urban sources

Major contributors to the urban loadings include vehicular activity, construction, sewage treatment plants, industrial processes especially involving animals and animal by-products, and possibly, in the future, cooling towers for nuclear and fossil fueled power plants (also see Table 2.1).

Table 2.1 Some factors found to affect airborne bacterial survival.

Factor	Level at which effects seen	Reference
Ultraviolet	1×0.75 ($\mu W/cm^2$) $\times$ sec	Riley & Kaufman (1972)
Temperature	−40°C	Ehrlich et al. (1970a)
Oxygen	0%–30% of 1 atm	Cox et al. (1974)
Helium	100% of 1 atm	Cox (1968b)
Argon	100% of 1 atm	Cox (1968b)
Carbon monoxide	85 μl/liter	Lighthart (1973)
Sulfur dioxide	2.5 mg/m^3	Lighthart et al. (1971)
Ozone		Elford & Van den Ende (1942)
Nitrogen dioxide	2.5 ppm	Chatigny et al. (1973)
Formaldehyde	1 μl/liter	Won & Ross (1969)
Open air factor	not known	Druett (1973b), Harper (1973)
Water vapor (RH)	0%–100%	Dimmick & Akers (1969)
Pressure		Druett (1973a,b)
Oxygen	0%–100%	Hess (1965)
PAN	not available	Jacumin et al. (1964)

It has been estimated that city streets may receive 62 metric tons per year of dust per square kilometer, and that along a gravel road 17 kg/ha per day may accumulate on land 46 m on either side of the center line (Roberts 1973). If the material blown up from gravel roads contains a conservative soil concentration of viable bacteria of 10^4 bacteria/g (dry wt) of soil, one would expect 6.8×10^7 bacteria to be lost daily from the air within 46 m of the road center line. Of course many of the smaller particles remain airborne and can be positively correlated with morning and afternoon peaks of atmospheric particle loadings found in the urban atmosphere, presumably because of vehicular activity (Butcher and Charlson 1972).

Urban industrial activity such as textile mills, abattoirs, rendering plants, automobile demolition, and agricultural processing must give off significant but largely unknown quantities of airborne bacteria. It is known that *Bacillus anthrasis* has been released from rendering plants and transmitted by the airborne route to infect humans downwind from the plant (Spendlove 1957). Also windborne dissemination of airborne Q-fever generated at an animal rendering plant infected individuals as far away as 16 km from the source of origin (Wellock 1960).

The spray generated as the result of sewage treatment by the activated sludge process, by bubble bursting or trickling filter, by spraying the sewage, or the rock filter is known to inject large numbers of potentially pathogenic bacteria into the atmosphere (Randall and Ledbetter 1966, Lighthart 1967). Although tremendous numbers per unit volume (10^5 bacteria/m^3) are commonly estimated to be emitted from activated sludge aeration units, the viable cells downwind are reduced dramatically as a function of wind velocity, distance from the source, time in the air, relative humidity, and species. It might also be expected that the survival of the airborne cells would be a function of sunshine (Goff et al. 1973), the gaseous composition of the atmosphere (Lighthart et al. 1971, Lighthart 1973), and the chemical composition of the sewage in the airborne particles containing the viable cells.

A new and potentially significant source of airborne bacteria are those originating from huge natural and mechanical air draft cooling towers associated with nuclear and fossil fuel fired power plants. The source of the bacteria is the water being used as the liquid coolant in the towers. It has been estimated by Lighthart and Frisch (1976) that from a 1000-MW power plant with a 150-m-high natural draft cooling tower 10^{10} viable bacteria/s could be emitted. Their work also provides a method to estimate the downwind concentration of viable bacteria emitted from a point source.

Particles containing bacteria in the urban environment and probably to a larger extent in rural atmospheres are largely in the size range ≥ 9.2 μm. Thus approximately 13% of the bacteria particles are in the size range of single pure bacteria, and approximately 77% are larger than one bacterium (Figure 2.1). It is thought that the particle sizes are large because bacteria are "rafted" on soil particles or contained within the residue from evaporated water.

Rural sources

Sources of airborne bacteria in rural locations affected by man may be categorized as agricultural and industrial. To our knowledge, few direct measurements of the bacteria injected into the atmosphere from these sources have been performed. D. T. Parker and co-workers (MS in preparation) have shown that highly concentrated aerosols result from overhead sprinkling irrigation of agricultural land with waste water from potato processing plants. Certain agricultural processes may be conjectured to be foci of airborne bacteria. For example, large numbers may be injected from the soil during disking and harrowing operations, particularly from dry soil; or during harvesting many bacteria resident on the surface of the crop may be dislodged and introduced into the atmosphere. Infectious aerosols of *Coccidioides immitis* have been recorded from air samples taken in areas of California where this organism is indigenous to the soil (M. D. Hogan pers. commun.). The movement of large herds of animals, particularly over dusty terrain, would cause an increase in the numbers of bacteria injected into the atmosphere, and many other farm practices conceivably could significantly contribute to the rural loading. A problem of direct concern to aerobiologists pertains to the use of viral pesticide aerosols for the control of insect pests. At the present time identification, efficacy of infection and host range, toxicological potentials, and field trail monitoring systems must be established (Tinsley 1975). One further note: The dirt road argument given previously for the urban environment could be even more significant in rural areas where there are large numbers of dirt roads. The effect of timber operations are also an unknown agricultural source of airborne microbes.

Rural industrial activities contributing to local airborne bacterial loads include strip mining and ore processing. To our knowledge no measurements of their contribution to the airborne microbial load have been made.

Viability

Many physical and chemical properties of the atmospheric milieu as well as the preinjection conditions affect the physiological potential and therefore the survival of airborne bacteria (Dark and Callow 1973). Table 2.1 lists some factors known to affect airborne bacteria survival. Many of these properties have been characterized in several mathematical forms that vary from detailed physiological models through regression models to theoretical models of overall cellular balance.

Effects of injection into the atmosphere

All preinjection conditions being equal, some bacterial cells metabolize and survive in the atmosphere better than others. The actual measurement of injection stress as opposed to some subsequent aerial stress has not been performed to our knowledge. With the foregoing in mind, measurements suggest that lag-phase growing cells may be the least resistant to injection and early station–phase or slow-growing cells are most resistant (e.g., Skaliy and Eagon 1972, Dark and Callow 1973). It is probable that preinjection growth conditions affect the injection and postinjection survival (Dark and Callow 1973). This might account for the rapid initial die-off of relatively rapidly growing cells after injection from sewage treatment plants (B. Lighthart, unpublished observation). Injection from the wet or dry state also affects survival (Cox 1971).

Little is known about the effects of natural suspending fluids on the survival of airborne bacteria. Laboratory experiments indicate marked enhancement or deterioration of aerial survival depending upon the suspending fluids. For example, Webb (1960, 1965) found that certain compounds with amino or secondary alcohol groups or both, and particularly substituents on six-membered rings, enhance survival, while hydroxyl groups on benzene rings were toxic although protective on pyridine rings. Inositol was found to be a particularly good protecting agent. Raffinose also protects some organisms particularly at high relative humidity (RH) (Cox 1966a). Sodium chloride seems to decrease cell survival. Finally, several additives may be used in combination to protect against multiple stress factors (Goldberg and Ford 1973). Cox (1963) presents a brief explanation of the action of protecting agents.

Postinjection survival

Once airborne, micoorganisms are thought to act as relatively naked cells and consequently are subject to deleterious physical and chemical conditions while in the air. Most of the research on these factors has been carried out in the laboratory. The results appear to hold for extramural conditions where comparisons have been possible.

Physical factors

There is evidence that survival of most organisms decreases most sharply in the range of −20°C to −40°C and above 49°C (Ehrlich et al. 1970a). Cells of certain non-spore-formers (e.g., a flavobacterium species) and bacterial spores survive well at low temperature, e.g., −30°C (Won and Ross 1968), while spores survive relatively well at high temperatures. Relative humidity as low as 0.03% to 0.2% decreases spore survival at 105°C (Brannen and Garst 1972). The survival of spores at elevated temperatures appears to be a function of water activity (Murrell and Scott 1966). Further, the effect of temperature is significantly modified by RH. The most mediating effects of RH are thought to occur at the higher levels, with one exception—a flavobacterium that was unaffected by RH (Ehrlich et al. 1970b).

The lethal and mutagenic effect of UV-IR radiation on airborne bacteria has been known for a long time, although the recovery of UV-"killed" cells has been known only a short time (Dimmick 1960, Riley and Kaufman 1972). Generally, increased exposure to simulated solar (Dorsey et al. 1970) or UV radiation results in decreased survival (Beebe and Pirsch 1958). High RH (above 60%) protects *Serratia marcescens* from most of the lethal effects of UV (Webb 1963, Riley and Kaufman 1972). To our knowledge the effects of irradiation wavelength on the survival of airborne bacteria are unknown.

Kethley et al. (1957) suggest that the composition of nonliving material associated with the airborne bacteria (a "raft") is the single most important factor affecting airborne cell survival.

Chemical factors

Many gaseous chemical factors may effect aerial survival of airborne bacteria. These gases include natural and human pollutant inorganic and organic compounds.

Natural inorganic atmospheric gases such as oxygen and water vapor have a marked effect on airborne bacteria survival. It has been found in many cases that oxygen is a major factor contributing to the death mechanism at low RH (Webb 1967); it may be a direct and linear function of concentration (Hess 1965) and follow first-order kinetics (Cox et al. 1974), while water vapor pressure (measured as RH) is a primary contributor to the death mechanism at high RH (Cox 1966a,b, Webb 1965, Wright et al. 1968). Rapid shifts in RH may also protect from or increase death depending on the shift direction (Hatch and Dimmick 1966). Druett (1973a) has investigated the effects of rapid changes in suspending gas pressure on aerial survival of bacteria. The effects of carbon dioxide on bacterial survival in the atmosphere have not been investigated, whereas nitrogen and the inert gases argon and helium may or may not affect survival (Cox 1968b).

Volatile natural organic atmospheric gases could be significant contributors to atmospheric microbial survival both as death-promoting actions or protecting mechanisms. Protection may result from the use of the volatiles as nutrient sources by airborne cells. Many natural volatile gases are widely dispersed (Rasmussen and Went 1965) and toxic or antibacterial in nature (Trust and Coombs 1973, Maruzzella 1963, Maruzzella and Sicurella 1960, Muller et al. 1964).

The effects of atmospheric pollutant gases on airborne microorganisms have been studied largely in the laboratory by suspension in the air contained within a rotating drum (Goldberg et al. 1958) or on microthreads (May and Druett 1968). These gases have been studied singly and in combination (Won and Ross 1969). Little has been published in the latter case. In general, pollutant gases such as nitrogen dioxide (Chatigny et al. 1973), sulfur dioxide (Lighthart et al. 1971), ozone (Elforde and Van den Ende 1942), and carbon monoxide (Lighthart 1973) decreased survival as indicated in laboratory measurements, although there is evidence that protection may occur under certain circumstances (Lighthart 1973). Pereira and Benjaminson (1975) found ozone to have a detectable cleansing effect on potential airborne pathogens originating from bubble bursts at a municipal waste-water treatment plant. Correlations were found by Lee et al. (1973) for airborne bacteria and carbon monoxide and hydrocarbons in the Cincinnati summer air in 1969. Partial correlations for nitric oxide, nitrogen dioxide, sulfur dioxide, and hydrocarbons also occurred when temperature and relative humidity were held constant during the analysis.

By use of the microthread technique to expose test bacteria to ambient rural and urban air, a potent, lethal, gaseous component in the urban atmosphere was found and termed the *open air factor* (Druett 1973b). It is thought to be a reaction product(s) of ozone and olefins (Druett 1973b).

The mechanism of aerial death in bacteria is many factored, but the lethal effects of RH and oxygen have received the most attention (Webb 1959). At higher RH, e.g., above ca. 40%, survival is a function of RH, probably resulting from changes in the cell wall constitutents and disruption of cellular energy metabolism (Hambleton 1970). Disruption of ribonucleic acid–protein complexes that are subsequently labile to ribonuclease disorganization (Cox 1969), disruption of an energy-requiring repair mechanism, or both may also occur (Webb 1969). At low RH, oxygen toxicity is a paramount factor in cell death, possibly through damage by free radical formation (Dimmick et al. 1961) of flavin-linked enzymes (Benbough 1967, Cox and Baldwin 1967) or because of interference with cell wall synthesis (Cox 1968a). Kinetic analysis by Cox et al. (1974) indicates the site of action of oxygen toxicity is in the interspace between the cell wall and cytoplasmic membrane.

The death mechanism(s) of airborne bacteria as a function of pollutant gases is largely unknown, although Lighthart (1973) has suggested that carbon monoxide interrupts cytochrome enzymes, while ozone may effect membrane lipids causing leakage (Scott and Lesher 1963), and sulfur dioxide and nitrogen oxide gases (Chatigny et al. 1973) are possibly associated with lethal pH changes. "Theoretical aspects of microbial survival" are discussed by Dimmick and Heckly (1969).

ALGAE AND PROTOZOA*

Nearly 100 years elapsed after the discovery of microorganisms by Leeuwenhoek in the late seventeenth century before the pioneering studies on airborne Protozoa were conducted by Spallanzani (1777), and almost two centuries passed before the controversy over spontaneous generation and the germ theory of disease between Pouchet (1860) and Pasteur (1860) led to studies on the aerial biota. Airborne Protozoa were studied by Miquel (1883) and Puschkarew (1913), and airborne algae were observed by Ehrenberg (1844) and Salisbury (1866) who speculated on airborne algae as causative agents for disease.

The first detailed scientific research on airborne algae, conducted by van Overeem and Visch in the Netherlands (Visch van Overeem 1972), laid the foundation for subsequent research on airborne algae and Protozoa as outlined by Schlichting (1961, 1964, 1969, 1974b). It has been shown that many microalgae can survive extreme environmental conditions (Schlichting 1974a) although they may undergo pleomorphic and physiological changes (Schlichting and Bruton 1970, Trainor et al. 1971). Micrometeorological conditions and geographic and topographic parameters as related to the occurrence of airborne microalgae have been discussed by Schlichting (1964), Brown et al. (1964), and Brown (1971).

---

*H. E. Schlichting, Jr.

Since the publication of reviews on airborne algae and Protozoa summarizing the history and suggesting future research in the US/IBP Aerobiology Program handbooks (Schlichting et al. 1970, 1972), four detailed studies have been conducted.

Brown (1971) studied the distribution of airborne algae over the island of Oahu, Hawaii, in 1964 and 1965. Air passing over the windward side of the island at Kailua Beach was devoid of algae. Two fallout peaks of airborne algae occurred on the island—one on the windward side and one on the leeward side. The distribution of airborne algae was related to the mountain barriers and various meteorological parameters. It was emphasized that Hawaii provides a model system for future air microbiological conditions of the islands. Sampling with petri dishes of sterile agar exposed from a moving vehicle along the Pali Highway, which transects the southeast corner of the island, demonstrated that people traveling along the highway are exposed to various concentrations of airborne algae, which may be medically significant. Airborne microalgae as causative agents in human allergies have been demonstrated by Bernstein and Safferman (1973) and this volume.

Studies in central North Carolina showed high positive correlations between wind speed, particulate matter in the air, and the numbers of microalgae found (Smith 1972, 1973). Schlichting (1974b) has reviewed research conducted through 1973.

The ejection of microalgae from the water surface via bursting bubbles, speculated by Woodcock (1948), was first demonstrated by Schlichting in 1972 (Schlichting 1974c, in press) and verified for bacterial transport by Blanchard and Syzdek (1970, 1974). Bacteria and fungi have been ejected up to 20 cm above the water surface and microalgae up to 13 cm.

The most recent study of soil algae and their appearance in the atmosphere was conducted by Carson (1975) in the vicinity of the Kilauea Iki volcanic cinder cone, showing that species of *Chlorella, Chlorococcum, Hormidium, Nannochloris,* and *Protococcus* were the principal Hawaiian soil algae. The greatest generic diversity of soil algae in the air occurred at 915 m altitude where fog-mist conditions prevailed, and it was speculated that microalgae may be trapped in the fog-mist by cohesion or electrostatic attraction to water droplets. The flux of microalgae through the air was generally greater under clear atmospheric conditions (no mist or rain) if overcast conditions prevailed.

The origins of airborne microalgae reported to date are: dust (Brown et al. 1964, Schlichting 1964), speculated adherence to moss spores (Schlichting 1961) and fern spores (J. L. Carson and R. M. Brown, Jr., pers. commun. 1975), raindrop splash on soil (J. L. Carson and R. M. Brown, Jr., pers. commun. 1975), airborne foam (Schlichting 1971), subaerial surfaces (Schlichting 1975), and water surfaces via bursting bubbles (Schlichting 1974c, in press).

Most protozoan cysts sampled from the atmosphere ranged in size from 2 to 50 μm and cyst volumes ranged from 4 to 65,000 μm^3. Weights of cysts would be proportional to volumes. When not encysted, Protozoa varied greatly in shapes from irregular plasmodial to rigid flattened or ovoid shapes.

Microalgal unicells, filaments, and colonies sampled from the atmosphere also varied in size and shape. The size range was from 1 to 150 μm for unicells but larger colonies and filaments have been sampled. Clumps of unicells sampled were 2–300 μm in diameter. Unlike the Protozoa, most microalgae in the atmosphere were in the vegetative stage, although spores and cysts were also found. The shape of microalgal unicells often were spherical, ellipsoidal and fusiform, or lunate. Filaments may be single or in packets, and colonies may be star-shaped, spherical, ellipsoidal, cuboidal, or amorphous.

Particles in urban air usually ranged in size from 1 to 90 μm and their settling velocity rates were 0.003–30 cm/s depending upon the diameter, shape, and density of the particles. Particles in this size range have been shown to be significant in medical studies of respiratory diseases and allergies because they are retained by the mucous membranes of the respiratory system.

FUNGUS, MOSS, AND FERN SPORES*[1]

The production and release of spores and reproductive propagules in the fungi, bryophytes, and pteridophytes is covered in this section. Although much is understood about factors that influence sporulation, especially in the fungi, little is known about the total number of spores that a given fruiting structure or a unit area of sporogenous development is capable of producing. In contrast, much has been published on the interesting and often unique mechanisms of spore release or liberation.

The Spore

As pointed out by Gregory (1966), it is impossible to define the word *spore* concisely. Sussman and Halvorson (1966) use the term to describe any reproductive structure found in microbes. Ingold (1971) refers to spores as microscopic reproductive units containing some food reserve and devotes considerable space to describing the various attributes of the fungal spore.

The fungi are unique in that they, along with the algae (Algae and Protozoa, chap. 2, this volume), have two distinct phases in their life history patterns: the sexual phase and a distinctly separate asexual phase. The bryophytes and pteridophytes have only a corresponding sexual cycle and lack a specialized asexual spore as found in most fungi. In many fungi, the sexual spore functions primarily as a resting or resistant structure, providing a means of getting the organism through periods when conditions are not suitable for growth and development. In others, such as most basidiomycetes, myxomy-

---

*C. L. Kramer

cetes, bryophytes, and pteridophytes, the sexual spores also function as the primary means of dispersal. Although there are some asexually produced fungal spores that are thick-walled and generally resistant to adverse conditions, many function primarily as dispersive units.

Fungal spore size ranges from <5 μm in diameter to as much as 350×115 μm for the ascospore of *Varicellaria microsticta;* most are 5–50 μm in diameter. The shape of fungal spores also varies greatly. Although most are spherical to ovoid, some are long and threadlike, coiled, branched, or with various types of appendages. Most have rigid cell walls, which vary in thickness and may be smooth or variously ornamented. Spores range in color from hyaline (commonest forms) through the light shades of yellows, tans, and reds, to various shades of green, purple, brown, and black. They are commonly multicellular with septa that may occur in either longitudinal or horizontal orientation (or both). Gregory (1973) has published several plates of colored drawings to illustrate the variation in spore morphology in the fungi, bryophytes, and pteridophytes.

Viability also varies greatly among the various groups of fungi. In a two-year study of the air spora of Kansas, Kramer and Pady (1968b) found that an average of 40% of all spores in the atmosphere were capable of germination. *Alternaria,* with multicelled, dark, thick spore walls, had the highest average germination (80%), while *Cladosporium* had 45% germination, rust urediospores 32%, basidiospores 6%, and hyphal fragments 20%. Those percentages vary somewhat with the weather conditions of the surrounding region and the resulting relative age of the spores being released into the atmosphere. When weather conditions are favorable for growth and sporulation of fungi, spores most recently produced are liberated into the air and thus probably have a higher rate of germination than older spores do.

Yarwood and Sylvester (1959) used a "half-life" rating, similar to that used for the decay of radioactivity, to evaluate the survivability of spores in the atmosphere; for the basidiospores of *Cronartium ribicola* they gave a half-life of 5 h. In contrast, Pady and Kapica (1953, 1955), during flights over northern Canada and across the northern Atlantic Ocean, collected viable spores from Arctic air masses. Spores collected obviously had originated far to the south of the Arctic, indicating a capability of extended travel in the atmosphere.

The density of fungal spores is approximately 1.0 in still air; however, they are able to remain suspended in the atmosphere and become disseminated through the action of turbulent air movement even though they are subject to gravity and fall at a rate governed by Stokes' law. Simplified, Stokes' law may be stated: The rate of fall per second of a spherical spore is proportional to the square of its radius. The terminal velocity of elongated spores is less than that of spherical spores of the same volume. For an elongated spore, with the long axis four times that of the short axis, the rate of fall is reduced by a factor of 1:1.28 (Chamberlain 1967).

In addition to spores, fungi may be dispersed by means of other structures, such as hyphal fragments. In studies of the air spora in Kansas and in England, hyphal fragments were found to compose a significant component of the population (Pady and Kramer 1960, Pady and Gregory 1963). Sinha and Kramer (1971) found that airborne hyphal fragments were primarily conidiophores of asexual forms. Isolating and cultivating trapped hyphal fragments revealed that most dematiaceous forms were *Cladosporium* and *Alternaria,* and moniliaceous forms were *Penicillium*. These organisms frequently inhabit the leaves of many plants, and fragments probably become dispersed into the atmosphere through a scrubbing action of the host plants rubbing together during wind.

Sclerotia and other similar types of fungal structures and the gemmae of mosses apparently are transported only rarely through the atmosphere. It does seem plausible, however, that such structures could be airborne in areas where high winds may create "dust storms" or in the highly turbulent air of severe weather fronts.

Lichen fungi produce sexual spores similar to those of other fungi, primarily Ascomycetes, but they apparently do not produce asexual spores. Dispersal is the primary responsibility of the soredia and isidia, reproductive structures (mostly 30–50 μm in size) composed of fungal hyphae and algal cells. Soredia are generally spherical in shape; isidia are more irregular.

Bryophyte spores are similar to fungal spores in many respects. They are generally globose, ovoid, reniform, or tetrahedral; they are mostly 6–8 μm in diameter but may reach 250 μm. The two-layered spore wall is variously textured on the surface: from smooth or wrinkled to coarsely papillose. Color ranges from pale yellowish to brown, red, or purplish. Generally rich in oil, they have a specific gravity similar to that of fungal spores. Viability ranges from 2–4 mo to 5 yr or more, depending on the spores.

Pteridophyte spores are generally somewhat larger than bryophyte spores, mostly 30–50 μm in diameter. The shape of fern spores varies, depending on the type of tetrad development. Hires (1965) has profusely illustrated the spores and discussed their development in a large number of ferns.

Spore Production

The capability of producing enormous numbers of spores is characteristic of most fungi, bryophytes, and pteridophytes; however, factual information on the numbers produced by single individuals, fruiting bodies, or unit areas of sporogenous development is scarce. Regardless of numbers produced, there is an obvious balance among spores produced, facility of dispersal, and viability of the spores, so that the survival of a given species is assured. Thus two of the primary functions of spores and vegetative propagules are dispersal and survival.

Mosses, ferns, and most fungi are stationary and must depend on reproductive propagules for dispersal, both within their geographical range to suitable ecological niches and beyond those limits, where genetic variability may be introduced into different populations of the species. In most of these organisms, the primary means of dispersal is through the atmosphere; however, water, insects, and higher animals (including man) also play a role in spore dispersal.

Spore production in the fungi

In the fungi, the process of spore production has evolved in two directions with respect to spore dispersal and survival. Some species produce extremely large numbers of very small spores that facilitate dispersal but have a comparatively smaller chance for survival; others produce fewer but larger and more resistant spores, which have a larger nutrient reserve and a greater chance for survival.

Until recently, the numbers of spores produced by these organisms, especially the fungi, have been little more than estimates based on spore size, size of the fruiting structure, and length of time that sporulation may occur. Buller (1909) counted the spores in a spore deposit of an 8-cm cap of *Agaricus campestris* (meadow mushroom) and found that approximately 1.8 billion basidiospores had been released. In a study of the same species, I have found that a basidiocarp of about the same size (cap approximately 8 cm in diameter) as that used by Buller released a total of 3667.42 million spores for the 7-day period in which sporulation occurred. The total number of spores released during each of the 24-h periods is listed in Table 2.2.

Table 2.2 Total number of basidiospores released by a single fruiting body of *Agaricus campestris*[a] for each of the seven days that sporulation occurred.[b]

Day	Millions of basidiospores
1	81.17
2	565.5
3	1092.5
4	765.0
5	648.0
6	363.75
7	151.5
Total	3667.42

[a]Diameter of the cap was approximately 8 cm. [b]The spores were collected by means of a baffling system that directed the spores to a water impinging collector somewhat similar to that described by Rockett et al. (1974).

Buller (1909) used another method to calculate the number of spores produced by the shaggy-mane mushroom, *Coprinus comatus*. Using a microscope, he made direct counts of the basidiospores on the surface of the gills; he found 34 per 0.01 mm^2. Since the spores of this species all mature at the same time, he was able to calculate that a pileus 30 cm high, with 214 gills and a surface area on each face of the gills of 1800 mm^2, produced approximately 5.24 billion basidiospores.

In an attempt to determine the number of spores produced by a giant puffball (*Lycoperdon giganteum*), measuring 40 × 28 × 20 cm and weighing 232 g, Buller (1909) suspended 0.1 g of the intact gleba (or sporogenous portion) of the puffball in 250 ml of alcohol. Using a counting chamber he then counted the spores, finding that 0.1 g contained 3245 million spores and that the entire puffball contained approximately 7 trillion spores.

White (1919) estimated that up to 30 billion spores may be released by the polypore *Ganoderma applanatum* in a single day and that this production could be maintained during the 6-mo growing season, provided sufficient moisture remained available. Ingold (1946) has reported that a perithecial stroma of *Daldinia concentrica* may discharge more than 100 million ascospores in a single day and that a colony of *Penicillium* sp. 2.5 cm in diameter may produce 400 million conidia.

Recently, techniques have been developed that have been useful in providing more nearly accurate quantitative data on the numbers of spores produced by fungi. Rockett et al. (1974) have used an automatically operated liquid impinger, which they recently developed, to study the total spore production in a number of polypores and jelly fungi (Rockett and Kramer 1974a,b). Some of the data from these studies are presented in Table 2.3.

Table 2.3. Total number of spores produced per square centimeter of exposed hymenial surface[a] during a 24-h period.

Source	Thousands of spores
Polyporus (Coricolus) pubescens	282,468
Polyporus (Coricolus) versicolor	138,900
Polyporus (Piptoporus) betulina	98,485
Lenzites (Gloeophyllum) saepiara	59,296
Fomes fomentarius	8,867
Exidia glandulosa	219
Auricularia auricula	205

[a]Exposed hymenial surface refers to the area of the underside of the fruiting body and not to the true hymenial area.

M. G. Eversmeyer and C. L. Kramer have been concerned with urediospore production in *Puccinia recondita* and *P. graminis*. Using a single-stage liquid impinger, they collected spores daily from single pustules or measured areas of infected leaf surfaces. Their studies (unpublished) were designed to help elucidate the effects of different genotypes of the rust fungi and host species, as well as the various environmental factors, on total spore production. Some data from the studies are presented in Figures 2.2 and 2.3.

In considering total spore production, one must take into account such factors as age, condition, size, and type of the fruiting structure or colony development, as well as the various environmental factors that may influence the rate of spore development and the length of time that sporulation may continue. For example, in the Gasteromycetes (puffball fungi and related forms), the number of spores produced is limited by the size of the specimen within which the spores are produced; however, the size and number of fruiting bodies developed by a given mycelial system may be influenced by environmental conditions.

In mushrooms and other fleshy fungi, spores are produced and released over a period of several days, but the numbers vary with the age and condition of the fruiting body; and production, as evidenced by release of spores, decreases as the fruiting body begins to senesce (Haard and Kramer 1970). Again, however, environmental conditions, especially moisture content of the substrate and temperature, may influence the rate of production and the length of time that the fruiting body is able to sustain production of spores.

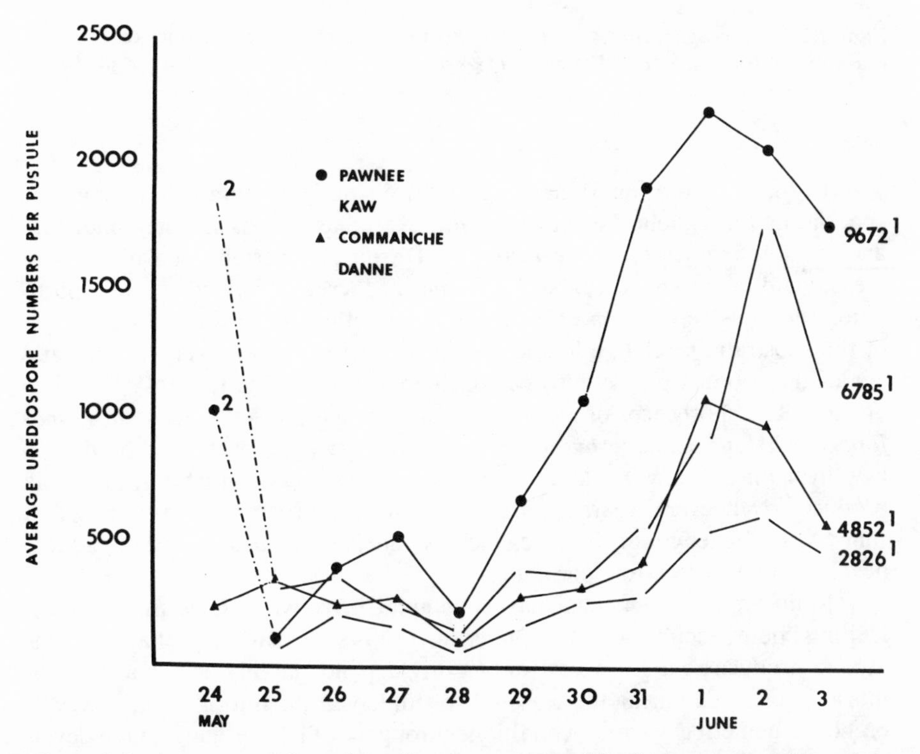

Figure 2.2. Average number of urediospores produced by four pustules of *Puccinia recondita* on four varieties of wheat. (1) Average total spores produced per pustule. (2) High numbers from first collections are from accumulated spores of first few days of sporulation.

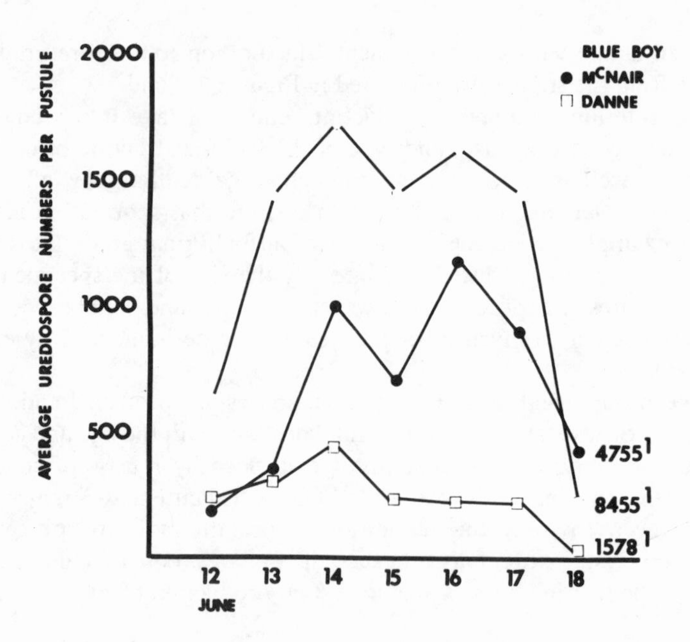

Figure 2.3. Average number of urediospores produced by four pustules of *Puccinia graminis* on four varieties of wheat. (1) Average total spores produced per pustule.

In the polypore fungi, many species are woody and are able to continue sporulating throughout the entire growing season as long as sufficient moisture and suitable temperatures are available. During dry periods, sporulation decreases and may even cease, but it resumes or increases following rain. Sporulation rate thus depends on environmental conditions such as moisture content of the substratum, relative humidity, and temperature; however, the size and age of the sporocarp must also be considered in reference to total spore production. The sporocarps of most species are annual; however, those of *Fomes fomentarius* and *Ganoderma applanatum,* for example, each year produce a new hymenial layer with the tubes of succeeding years continuous. Hymenia produced over several years apparently continue to produce spores during that time. Also, the length of the tubes increase during the growing season, adding new hymenial surface for sporulation.

Many of the large fungi have perennial mycelial systems that produce fruiting bodies each year. The number of fruit bodies that a mycelial system is able to produce in a given year apparently depends largely on the amount of moisture received during the season. We still have little information, however, on how much effect weather conditions throughout the year have on the development of the mycelial system and on how the number of fruit bodies the mycelial system produced in one season may affect fruit body development the following season.

One of the most interesting groups of fungi with regard to spore production is the Tremellales (jelly fungi). Rockett and Kramer (1974b) found that species of *Exidia* and *Auricularia,* while undergoing repeated wetting and drying, could discharge spores continuously at a relatively constant rate until the fruiting bodies themselves began to dry. A specimen of *E. glandulosa* was observed to have become dried and dehydrated, with accompanying spore production, 20 times between April and October 1972. Sporulation usually began within 2 h following wetting by rain. The length of time that sporulation continued depended on the rate of dehydration; in one instance sporulation continued for 21 days. The phenomenon of rehydration and resumed sporulation is not unique to the jelly fungi or to certain other groups, including *Schizophyllum commune,* a gill fungus; *Polystictus versicolor,* a polypore; and *Stereum hirsutum,* a thelephoraceous fungus.

In many areas where temperatures during the winter are normally below that suitable for growth and sporulation, occasionally temperatures do reach favorable levels for several days. Specimens of *Exidia* spp. and *A. auricula* brought into our laboratory during January and February in either dried or soaked conditions responded to the warmer temperatures and began to release spores when rehydrated. Individual fruiting bodies of those fungi apparently are capable of overwintering and producing spores during two growing seasons as well as occasionally during the winter months.

The release of spores in most fungi, as well as in the bryophytes and pteridophytes, does not occur at a constant level throughout a 24-h period. A greater number of spores may be released during one part of the day than at other times—in some cases entirely or partly because of a corresponding pattern of spore production. In *Erysiphe polygonum,* a single spore per conidiophore is produced in a 24-h period. The spore begins to develop during the daytime, matures through the night, and is ready for release the following morning (Pady et al. 1969, Yarwood 1936). Sporangial formation in *Pilobolus* is similar. In *Trichoderma,* certain species of *Aspergillus, Penicillium,* and others, spore development is initiated only on that mycelium that was formed during the daytime. In another species of *Erysiphe, E. cichoracearum,* however, spores are produced continuously throughout 24 h, but release normally does not occur except during the daytime, corresponding with drying and increased wind velocity.

Spore production in the bryophytes

Studies of spore production in the bryophytes are few. Müller (1954), in studying a number of liverworts, compared spore size with the numbers of spores produced within single capsules. Some of these data are presented in Table 2.4. Table 2.5 summarizes a considerable amount of information compiled by Dominick V. Basile on spore production in the liverworts.

Table 2.4 Spore production in the liverworts. Numbers of spores are those produced by single capsules (from Müller 1954).

	Spore number per capsule
Riccia gougetiana	192
Riccia crystallina	246
Riccia glauca	220
Sphaerocarpus michelii	760
Sauteria alpina	2,100
Reboulia hemisphaerica	2,500
Conocephalum conicum	5,300
Pressia quadrata	8,000
Diplophyllum albicans[a]	400,000
Scapania undulata[a]	1,000,000

[a]Spores computed.

Table 2.5. Spore production in the liverworts. Numbers of spores are the average produced by single capsules (data from Dominick V. Basile).

	No. capsules counted	Mean number spores/capsule	Range of spore production/capsule
Ptilidium pulcherinum	12	11,723	14,437 – 15,250
Calypogeia trichomanes	6	317,250	156,250 – 456,000
Lophozia barbata	2	91,750	61,000 – 122,000
Plectocolea crenulata	10	52,328	28,000 – 92,375
Scapania nemorosa	8	337,229	148,125 – 637,500
Diplophyllum apiculatum	12	92,925	51,100 – 138,750
Lophocolea cuspidata	10	67,485	40,333 – 93,375
Lophocolea heterophylla	10	127,525	93,000 – 197,375
Cephalozia bicuspidata	11	50,117	30,000 – 63,000
Cephalozia media	8	28,730	16,400 – 59,625
Nowellia curvifolia	9	97,750	75,000 – 123,750
Odontoschisma prostratum	11	64,962	24,062 – 91,875
Radula complanata	2	5,364	4,352 – 6,375
Frullania eboracensis	5	1,050	884 – 1,397
Jubula pennsylvanica	12	17,348	15,281 – 22,000
Fossombronia cristula	2	904	808 – 1,000
Fossombronia braziliensis	9	2,978	1,310 – 4,375
Blasia pusilla	10	9,822	7,341 – 13,750
Pallavicinia lyellii	8	107,344	72,500 – 173,750
Pellia epiphylla	9	2,602	1,088 – 6,803
Riccardia sp.	6	41,437	27,812 – 49,812
Sphaerocarpus taxanus	5	344	272 – 428
Conocephalum conicum	4	3,656	2,125 – 5,500
Marchantia polymorpha	6	139,680	105,750 – 167,400
Reboulia hemisphaerica	3	3,167	
Oxymitra androgyna	5	288	219 – 331
Ricciocarpus natans	3	2,787	2,530 – 2,951
Riccia fluitans	5	188	158 – 238
Notothylas orbicularis	5	790	462 – 1,448

Spore Release in the Fungi

To become disseminated in the atmosphere, the spores of fungi, as well as those of bryophytes and pteridophytes, must first overcome the problem of liberation or release from their spore-bearing structures to escape through the laminar boundary layer into the turbulent air. That they accomplish by two basic methods: (1) active liberation or discharge, with the necessary energy provided by the organism; and (2) passive liberation, with the needed energy provided from an external source such as wind.

Surrounding all solid objects in the atmosphere is a layer of still air, the laminar boundary layer (LBL). Particles, such as dust or spores, that become entrapped in this layer settle by gravity to the surface, in that no air currents move across the layer. To become airborne, spores must pass through this layer to the turbulent air layer of the atmosphere where they can be disseminated. The thickness of the LBL is not constant; it varies with the wind speed of the turbulent air layer and with the degree of roughness of the underlying surface. During very calm periods (such as may occur at night), the LBL may become extremely thick, whereas during periods of storm activity and high winds the layer may be reduced to less than a millimeter. Rough surfaces with structures projecting through the LBL may cause local eddies of turbulent air, which tend to reduce the effective thickness of the LBL.

Active discharge

Fungi contain a variety of mechanisms used in the active discharge of spores. In some members of the lower fungi, Ascomycetes and Fungi Imperfecti, cellular turgor pressure forcibly expels the spores or spore sacs (sporangia). Buller (1934) described in excellent detail the mechanism of sporangial discharge in *Pilobolus*. A large subsporangial vesicle becomes turgid and eventually ruptures at its apex, squirting its contents and carrying the sporangium of spores to distances exceeding 1 m. To demonstrate the force generated by the subsporangial vesicles, I placed a portion of horse dung with developing sporangia in one end of a 1-m length of 51-mm-ID (inside diameter) glass tubing. The tubing, wrapped in black plastic, was positioned horizontally with a light source at the opposite end. Because sporangiophores are phototrophic, sporangia were projected through the tube toward the light. Despite the extremely flat trajectory imposed by the tubing, many of the sporangia were able to travel the entire 1-m distance. Other fungi with similar means of spore discharge include *Basidiobolus* and *Entomophthora,* members of the Entomophthorales, and *Nigrospora,* a hyphomycete.

A similar mechanism involving turgor pressure is found in many of the Ascomycetes. When sufficient moisture is available, turgor pressure increases in the ascus, a saclike structure, to a point where the apex ruptures and the ascospores are forcibly discharged. In some ascomycetes, such as *Taphrina,* the apex ruptures irregularly, squirting the entire contents at once. Similarly, in the operculate discomycetes, the entire contents are expelled when an ascus cap or operculum breaks away. Ascus discharge in these fungi is usually triggered by some external factor, such as a sudden air current, which stimulates many thousands of asci to discharge their spores simultaneously, thus creating a cloud of spores and turbulent air, enabling the spores to be carried even farther.

Sufficient moisture is vital to ascospore discharge. As the substrate dries following rainy periods, spore discharge ceases in many of the smaller pyrenomycetous forms. Ascus development may continue, however, so that a comparatively large number of "mature" asci are contained in the fruiting body; when moisture again becomes available in the form of rain or dew, ascospore discharge occurs (Keitt and Jones 1926, Hirst and Stedman 1962, Pady and Kramer 1967). In many fungi, only about an hour may be required before discharge begins. In others, additional factors (such as light) may influence initiation of discharge. For example, ascospore discharge in fungi such as *Hypoxylon* (Kramer and Pady 1970), *Daldinia* (Ingold 1960), and *Bombardia* (Pady and Kramer 1969) may be light-induced or -inhibited, thus exhibiting a circadian pattern of discharge. (Further discussion of circadian rhythms of spore discharge appears later in this section.)

There are also fungi such as *Daldinia concentrica* (Ingold 1946), *Epichloe typhina* (Ingold 1948), and *Bulgaria inquinans* (Ingold 1959a) that have developed means of storing a reserve water supply or obtaining water from the vascular system of the host plant for sporulation.

In a study of *Glomerella,* Pady and Kramer (1971) demonstrated that decreasing relative humidity and sudden lowering of temperature stimulated ascospore discharge. Austin (1968) showed similar effects for *Sordaria fimicola;* the "puffing" of discocarps, when the lid is removed from moist chambers in which they may have been stored for some time, is well known.

The ascomycete fruiting body generally is designed so that the ascospores are discharged upward into the turbulent air. The ascus layer, which is not disrupted by free water, is thus exposed to rain and dew, which may stimulate spore discharge. Most basidiomycetous fruit bodies, on the other hand, are designed to protect the spore-producing (basidial) layer, which may be disrupted by free water. The basidiospores are borne externally on basidia at the tips of short, pointed sterigmata from which they are forcibly projected very short distances. For example, in mushrooms, the basidiospores are projected into the space between the gills, where they then fall downward to escape into the turbulent air beneath the cap. The possible mechanism by which the basidiospores are discharged has been discussed by Savile (1965).

Another method of spore discharge involving turgid cells is found in a number of fungi. In *Entomophthora* and *Conidiobolus* (Ingold 1971), *Arthrinium* and *Epicoccum* (Webster 1966), and *Sclerospora* (Weston 1923), the spores are propelled by the sudden rounding out of a portion of the spore wall against the conidiophore, which produces enough energy to project the spore a considerable distance. In a similar process, the aeciospores of species of *Puccinia* and *Uromyces,* first described by Zalewski (1883), are initially polyhedral in shape but, upon becoming turgid, their walls suddenly expand making them spherical. We have observed aeciospores of several species of *Puccinia* projected as much as 5 or 6 mm from the pustules.

In contrast, loss of water is involved in the violent discharge of spores in many fungi. The process involves the loss of cell water and the resulting distortion of the spores or sporophores is due to uneven thickening of the cell walls. With increased water loss, the internal pressure deficit becomes so great that the cellular water vaporizes. The sudden release of tension allows the cell walls to spring back to the normal position, projecting the spore into the atmosphere. Meredith (1965) has described this process in conidial forms.

Another mechanism that involves water loss from the sporophores is found in the downy mildew fungi, e.g., *Peronospora tabacina*. Here the sporophores are comparatively thin-walled; as water loss occurs they begin to twist jerkily, thus throwing off the spores from their points of attachment. A somewhat similar situation exists in certain of the Myxomycetes. The fruiting body contains a mass of threadlike hygroscopic structures, which may have uneven thickenings. As they gain and lose water during wet and dry periods, the sporophores twist about in sudden movements, throwing spores into the atmosphere.

Passive liberation

In contrast to those fungi that in one way or another provide their own energy for liberating their spores, many fungi rely on an external energy source, such as wind, to release their spores from the sporophores and carry them into the turbulent air of the atmosphere, where they may become disseminated. The asexual spores of most fungi are produced on hyphal stalks (conidiophores) that may rise several hundred micrometers into the air from the ground. High wind speeds may so reduce the thickness of the laminar boundary layer (LBL) that the spores are released directly into the turbulent air layer. Local eddies caused by such things as leaf hairs on the surfaces where fungi may be sporulating also reduce the thickness of the LBL and provide turbulent air currents that facilitate liberation. The whipping and scrubbing action of host plants during windy periods also facilitates spore liberation. Fungi that use these methods of release are often referred to as *day-spora*. Storm fronts passing through an area at night, however, cause the release of these spore types, both from wind action and from splash dispersal from raindrops (Hirst and Stedman 1963).

Many fungi produce spores that are sticky or are even suspended in small slime droplets and as a result are not easily liberated by wind action. One of the most important means of liberation of these spore types is splash dispersal from falling raindrops or from water dripping from an overhead structure such as a tree. In a study of this mechanism, Gregory et al. (1959) found that drops 5 mm in diameter, falling from 7.4 m onto a film of water 0.1 mm thick with suspended spores, produced over 5000 reflected droplets ranging in size from 5 to 2400 μm. Most of the droplets contained spores and were thrown distances up to 100 cm. Although those distances are relatively short, small droplets may be carried by the turbulent air, and as the water evaporates the spores are freed for dispersal in the atmosphere.

Rhythms of spore release in the fungi

A number of rhythmic patterns of spore release within the fungi have been described. Most cycle within approximately a 24-h period and are thus referred to as *circadian*. In some species with circadian patterns, maxima in spore release occur at night (nocturnal maxima); in some others, release occurs primarily during the daytime (diurnal maxima).

Circadian patterns of spore release with daytime maxima occur most frequently in those fungi that exhibit passive release, e.g., most asexual spore forms. The conidia of *Cladosporium,* among the commonest fungi in the atmosphere, are released during the daytime, with maxima occurring before noon; release is correlated with rising winds and decreasing relative humidity (Kramer et al. 1959a). Other fungi reported to have diurnal maxima include *Alternaria* (Long and Kramer 1972, Pady and Kramer 1967); *Erysiphe* (Hirst 1953, Eversmeyer and Kramer 1975, Pady et al. 1969); *Podospora* and *Sphaerotheca* (Pady 1972); *Uncinula* (Pady and Subbayya 1970); *Cercospora, Helminthosporium,* and hyphal fragments (Pady and Kramer 1967); and urediospores of *Melampsora* and *Puccinia* (Pady 1971).

There are also a number of fungi with daytime maxima that forcibly discharge their spores. In a study of the air spora of a Jamaican banana plantation, Meredith (1962) found that during dry weather fungi such as *Dieghtoniella, Nigrospora, Cordana, Corynespora, Zygosporium,* and *Zygophiala* discharge their spores between 0600 and 0900. He also observed that the typical dry-air spora were replaced to varying degrees by a variety of ascospores and splash-dispersed spores during daytime rains. Discharge of ascospores and basidiospores generally occurs at night; however, in some species, such as *Hypoxylon investiens* (Kramer and Pady 1970) and *Sordaria fimicola* (Ingold and Dring 1957), spore discharge is light induced.

Haard and Kramer (1970) studied the circadian patterns of spore discharge in a number of mushrooms and polypores in the field. In species of *Cortinarius, Lactarius, Crepidotus,* and others, maximum discharge occurred about midnight. That pattern seemed to correlate with usual fluctuations of nighttime humidity, influenced by decreasing temperature and alternating light and dark. In other species, as in the genera *Leccinum, Tylopilus,* and *Suillus,* a similar pattern was found, except that maximum spore discharge often began earlier in the day and continued for only a short time into the night. Similar patterns with nocturnal peaks are found in Ascomycetes such as *Hypoxylon rubiginosum* and *H. truncatum* (Kramer and Pady 1970); *Hypocrea gelatinosa* (Kramer and Pady 1968a); *Daldinia concentrica* (Ingold 1960); *Podospora, Xylaria, Lasiosphaeria, Nectria,* and *Melanomma* (Walkey and Harvey 1966); and *Bombardia* (Pady and Kramer 1969).

In the majority of the fungi, spore discharge is apparently directly controlled by the surrounding environmental conditions, with light perhaps the most important factor. Some fungi, however, when conditioned under alternating light and dark and then subjected to continuous light conditions, continue the pattern of spore discharge, with maxima occurring at approximately the same time in each 24-h period. Such fungi are said to have endogenous rhythms, as opposed to exogenous rhythms, which require a corresponding daily pattern in the environmental conditions to stimulate discharge. *Daldinia concentrica* (Ingold 1960) and several species of *Hypoxylon* (Kramer and Pady 1970) exhibit excellent examples of endogenous circadian rhythms. In *Daldinia* spore discharge may continue for as long as 15 days in continuous dark and 3 days in continuous light following conditioning in alternating light and dark (L/D). Similar conditions occur in *Hypoxylon*. In *H. truncatum,* however, after conditioning in alternating L/D and then removal to a continuous low light exposure, the rhythmic pattern of spore discharge continues apparently for as long as other environmental conditions remain suitable for growth and sporulation.

In addition to fungi in which circadian rhythms of spore liberation occur, a few species have spore-discharge patterns with cyclic periods other than approximately 24 h. Spore discharge in *Sphaerobolus* has a rhythmic cycle of 21 days (Alasoadura 1963). In contrast, *Ganoderma applanatum* exhibits peaks in spore discharge every 9–11 h (Kramer and Long 1970). The controlling factors for these rhythms are not known.

Patterns of spore liberation also may be reflected by the age of the sporocarp. Haard and Kramer (1970) demonstrated both circadian and "life-span" patterns of spore discharge in a number of hymenomycetes. In some cases, such as in *Inocybe* and *Mycena,* they found that the rate of spore discharge increased rapidly following initial release and then gradually declined over a period of several days as the basidiocarp began to senesce. In *Cortinarius* the reverse was true. Spore release gradually increased for several days to a maximum and then rapidly declined with senescence of the basidiocarp. These life-span patterns undoubtedly reflect the rate of spore production; temperature, available moisture, and relative humidity are very important in determing the length of time that fruiting structures may continue to sporulate.

Many studies reporting circadian patterns of spore liberation are based on collections of air spora from the atmosphere from locations that may be quite remote from the place where the spores are released from the parent fruiting structures. It is essential to distinguish between patterns of spore release and the occurrence of a spore type in the atmosphere.

Hourly occurrence of spores in the atmosphere

The population of air spora at a given location depends upon two factors: (1) the numbers and kinds of spores being released locally and (2) the air spora being transported into the area in the atmosphere from remote sources. In turn, the numbers and kinds of spores being released in a given region depend upon the type of habitat and such environmental conditions as moisture and temperature that might be influenced by the season of the year and by the time of day.

Long and Kramer (1972) studied the air spora of a prairie site and a woodland site located near each other in Kansas. They compared the periodicity in occurrence of several spore types at the two sites during periods following rain, when environmental conditions were favorable for growth and sporulation of fungi in the surrounding areas, and during prolonged dry periods.

During wet periods, the circadian pattern of occurrence of the different spore types was more pronounced at the protected woodland site, indicating that a large proportion of the spores collected were from local rather than from remote sources. In contrast, at the exposed prairie site, the hourly variation was greater and the circadian patterns were less distinct, indicating that a greater proportion of the spores collected were from remote sources. During dry periods, numbers of all spore types were low and showed no circadian pattern at the woodland site. Although the same was true for ascospores and basidiospores at the prairie site, *Cladosporium* and *Alternaria* spores, both passively released, showed distinct circadian patterns correlated with dry, windy conditions.

To further illustrate the effects of locally released spores and those transported from remote sources on the population of air spora at a given location, Long and Kramer (1972) analyzed data taken from a several-year study of air spora collected from the roof of a building on the campus of Kansas State University. When they averaged some thirteen 24-h series taken during "wet" periods and averaged twenty-five 24-h series taken during "dry" periods, they found no circadian pattern evident in the occurrence of *Cladosporium* in either. Prevailing winds at that campus location are from the west and southwest and may travel overland for several days before reaching the sampling station. As they do, *Cladosporium* spores may be picked up during the daytime and apparently can remain somewhat unevenly dispersed as "spore clouds" in the atmosphere. The passing of such spore clouds over the sampling station, expressed as peaks in the collections, may occur at any time of day. Thus it is important to distinguish between time of release of a spore form and time of its occurrence in the atmosphere. The time may or may not be the same, depending on several factors, especially the location of samplers in relation to the point of release.

Seasonal occurrence of fungal spores in the atmosphere

In addition to circadian rhythms of spore production and liberation, there is also a seasonal periodicity in the occurrence of fungal spores in the atmosphere. During a 2-yr period Kramer and others (Kramer et al. 1959b; Kramer 1974) studied quantitatively and qualitatively the population of air spora in Kansas; they made daily collections at 0900. Although the studies did not take into account the change in the population of air spora that takes place through a 24-h period, they provide a general concept of the seasonality of airborne fungi in the atmosphere. As would be expected, spore numbers were much higher during the growing season than during winter; however, some spores did occur at all seasons. In addition, a small but significant number of spores collected during winter and early spring were viable.

The magnitude of the numbers present during the growing season varied with the amount of moisture received in the region surrounding the sampling site. During periods of frequent rains, numbers of spores exceeded 100,000/m^3 of air in Kansas (Kramer 1974); during extended dry periods, numbers were often fewer than 5000/m^3 of air. As indicated earlier, the figures varied from one location to another, depending on the type of environment of the surrounding region, meteorological conditions during the past week or longer, and the time of day.

Spore Release in the Bryophytes

Unique and interesting mechanisms of active spore liberation in the bryophytes have been described. All involve drying conditions and the action of spirally thickened elaters. Two of the mechanisms, water rupture (or "gas phase") and hygroscopic movements of elaters, also are known to occur in the fungi.

In several genera of liverworts including *Cephalozia bicuspidata* (Ingold 1965), the spore-bearing capsule is situated at the apex of a stalk 1 or 2 cm high. When mature, the capsule splits longitudinally along four lines of dehiscence radiating from the apex to the base. The four sections of the capsule wall diverge, exposing the spores and elaters. The elaters, which have two thickened spiral bands, are attached at one end to the capsule wall. As they lose water, the spiral thickenings shorten into closer spirals, putting the contents of the elater into a state of tension. As water loss continues, a point is reached at which water ruptures and enters a gas phase. That releases the tension and the elaters suddenly untwist, expanding and springing into the air, throwing out spores. Similar mechanisms of active spore release in a number of fungi have been described.

In *Marchantia polymorpha* the spore-bearing capsules are also located on stalks several centimeters high. Under dry conditions the wall of the capsule splits at the apex, exposing the spores and associated elaters. The elaters have thickened spiral bands, and as they dry they begin to twist in sudden jerky movements, throwing the spores, in contact with the elaters, into the air. The hygroscopic movement of spirally thickened elaters also has been described as a means of spore release in some of the Myxomycetes.

Another mechanism, described by Kamerling (1898) and involving the drying of elaters, is found in the genus *Frullania,* which has a spore-bearing capsule borne at the apex of a short stalk. As in the two situations described above, during dry conditions the capsule splits open, exposing the spores and elaters. The elaters have spiral thickenings and are attached at both ends to the inner surface of the capsule wall. As the sections of the capsule wall continue to bend farther outward, the elaters are stretched; in fact they tend to prevent the capsule walls from opening farther. Suddenly and almost simultaneously, however, the elaters break free from their basal point of attachment and, as the free ends spring upward and outward, the associated spores are thrown into the air.

A few liverworts, such as *Riccia,* lack any specialized mechanism for active release of their spores. Mature capsules simply open during dry conditions, exposing the spores for wind, insect, and water dispersal.

Spore release in the mosses, as in the liverworts, generally occurs during drying conditions. The spore-bearing capsules are formed at the tips of stalks and each has a lid or operculum, which is circumscribed by a line of thin-walled cells that allows for the dehiscence of the operculum during dry weather. The opening of the capsule generally contains a series of hygroscopic, toothlike projections, the peristome, in which the teeth are composed of two layers of cells that differ in their ability to absorb water. In humid conditions the cells of the outer layer absorb water, causing them to lengthen and resulting in an inward curving of the teeth, partially closing the opening to the capsule. In dry conditions the teeth bend outward, but as they do they become momentarily hooked on the inner row of teeth; thus they move in a jerky fashion, which helps to dislodge the spores.

In *Funaria,* the peristome is united by a central disc at the tips of the teeth. During dry conditions, the teeth elongate forming gaps. Because the upper portion of the stalk is bent so that the tip of the capsule is pointed downward, the spores easily fall out into the atmosphere. Wind causes the stalks to vibrate and thus aids in shaking the spores free, as salt from a shaker (Ingold 1959b, 1965).

The genus *Sphagnum* has an extremely interesting mechanism of spore discharge, referred to by Ingold (1974) as the "air-gun." The spore-bearing capsule is spherical and is borne at the tip of an elongated stalk. It has a lid, delineated by a row of weak cells. When mature, the spore sac is located in the upper part of the capsule; a large air space is in the lower part. As the capsule dries, it shrinks, compressing the air within to approximately 5 atm (Nawaschin 1897). Eventually, the circle of weak cells ruptures; the air inside then expands suddenly, causing the lid to be blown off and scattering the spores to a height of 15 cm or more.

Spore Release in the Ferns

The sporangial wall of most ferns is composed of a single layer of cells with a row of specialized cells, the annulus, running from the stalk to the top of the sporangium and half the way back down the opposite side. The cells of the annulus have thick inner walls; the outer walls and the walls of other cells are thin. As drying occurs, the volume of the annular cells decreases, pulling the thin walls of the annular cells inward. That results in the annulus bending upward, pulling the thin-walled cells of the sides of the sporangium apart so that the entire structure becomes separated into two cuplike portions connected only by the annulus. As each annular cell tends to return to its original shape, extreme tension is exerted on the cellular contents. Finally, when a critical point is reached, the water of an annular cell breaks, and the resulting gas phase releases the tension suddenly, allowing the cells to return to their original shape. Apparently that causes the same process to occur in the other annular cells, so that the entire annulus springs back into its normal position, throwing spores into the air.

POLLEN*

There are two basic methods by which pollen grains are transported from anthers to stigmas in plant cross-fertilization. Although certain aquatic plants are pollinated by water, pollen grains of most seed plants are transported by either animals (zoophily) or by the atmosphere (anemophily).

The production of wind-transported (anemophilous) pollen grains is a relatively precise event in the life cycle of plants. The timing of pollen release and the quantity of pollen produced are results of genetically controlled adaptations or behavioral patterns in plants discussed in Impact of Airborne Pollen . . . (chap. 6, this volume). This allows successful cross-fertilization to occur within the biological capacities of the plants and the physical limitations of their environment.

Pollen production varies widely among anemophilous species. The maximum production of pollen grains/inflorescence is 246 times that of the minimum of the few values given in Table 2.6. Most vegetated regions of the earth that have been sampled receive (and thus discharge) 10^{12}–10^{14} pollen grains/km^2 yearly (Ritchie and Lichti-Federovich 1967, Davis and Webb 1975, Solomon and Hayes 1972). One gram of pollen contains 10^7–10^8 pollen grains (author's unpublished data), so on each square kilometer of land, between 0.01 and 10 metric tons of pollen are emitted yearly.

The seasonal pollen yield has a wide variance. If pollen deposition values are accepted as a measure of pollen production (Solomon and Hayes 1972, Solomon 1973), then the long-term pollen deposition records from Cardiff,

*A. M. Solomon

Table 2.6 Yearly pollen production of stamens, whole flowers, inflorescences, mature branches, and 100-m^2 pure stands of selected species.[a]

Species	Single stamen $\times 10^4$	Single flower $\times 10^4$	Inflorescence (group of flowers) $\times 10^5$	10-yr-old branch system (woody plants) $\times 10^7$	100-m pure stand of 10-yr-old branch systems $\times 10^9$
Herbs					
Rumex acetosa (dock)	3	18	393		
Secale cereale (rye)	1.9	5.7	43		
Trees					
Pinus sylvestris (pine)		15.8	58	34.6	12.5
Quercus sessiliflora (oak)		4.1	56	11.1	3.5
Fagus sylvatica (beech)		1.2	1.6	2.8	2.1
Betula verrucosa (birch)			56	11.8	5.6
Corylus avellana (hazelnut)			39	24.4	5.6

[a]From Erdtman (1943:175–176).

Wales, and Tucson, Arizona (Table 2.7), can be used to examine variance in pollen production. Standard deviations are normally about 80% of the mean values (range 29%–161%). No difference in this estimate of variance is apparent between the two records. Zero minimum values occur and minima average about 14% of the mean. Among the 606 values examined, a maximum value of almost 8 times the mean occurred, while average maximum values are 3.33 times the mean. The distribution of yearly pollen production values in all cases is skewed to the right, as indicated by the positive γ_1 values in the last column. Almost half the γ_1 values were significantly different from those expected due to chance alone at the $\alpha = 0.01$ level (Sokal and Rohlf 1969:171). The frequency of significant γ_1 values was about equal in both records.

Pollination Timing and Production

The problem of plant spacing for effective pollination is mitigated by precise timing of pollen production and emission. Adaptations occur that increase numbers of pollen grains produced at a given point in time and increase the instantaneous concentrations of those pollen grains when they are emitted from the flower.

The seasonality of wind pollination can be subdivided into three distinct groups: (1) temperate forest trees and shrubs emitting pollen during the spring at the beginning of the growing season, (2) annual and perennial herbaceous plants emitting pollen during late summer and early fall near the end of the

Table 2.7 Pollen yield variance measured by pollen deposition values.

Plant	$\bar{x}$	SD	SD/$\bar{x}$	min/$\bar{x}$	max/$\bar{x}$	*n*	γ^1
A. Pollen grains cm^{-2} yr^{-1} at Tucson, Arizona, 1943–1975 (from Solomon 1975)							
Trees							
Populus/Juniperus	212	200	0.94	0.12	3.47	29	1.64[a]
Fraxinus	121	53	0.44	0.12	2.01	30	0.22
Morus	883	829	0.94	0.03	2.81	32	0.48
Olea	300	301	1.00	0.00	3.25	32	0.77
Prosopis	223	170	0.76	0.03	3.34	32	1.15[a]
Pinus hualapensis	375	345	0.92	0.00	3.13	24	0.90
Pinus (native)	60	96	1.61	0.00	7.96	27	3.71[b]
Tamarix	139	135	0.97	0.16	4.92	25	2.81[b]
Eucalyptus	157	132	0.84	0.00	3.44	18	1.33[a]
Rhus lancia	155	171	1.11	0.02	4.03	14	1.95[a]
Ligustrum	226	66	0.29	0.66	1.46	14	0.23
Herbs							
Gramineae	809	452	0.56	0.24	2.04	30	0.23
Compositae	288	264	0.92	0.08	4.39	31	2.02[b]
Chenopodiaceae	587	318	0.54	0.32	2.63	30	1.08[a]
B. Pollen grains cm^{-2} yr^{-1} at Cardiff, Wales, 1942–1961 (Hyde 1963)							
Trees and shrubs							
Populus	20	10	0.50	0.21	1.82	20	0.04
Fraxinus	155	176	1.13	0.02	3.80	20	1.17[a]
Pinus	81	31	0.38	0.33	1.87	20	0.29
Alnus	82	71	0.87	0.11	3.19	20	1.37[a]
Betula	182	123	0.68	0.13	2.51	19	0.49
Carpinus	11	10	0.90	0.03	3.15	19	0.62
Castanea	30	20	0.67	0.29	2.80	20	1.56[a]
Fagus	51	80	1.56	0.02	5.08	20	1.81[b]
Quercus	317	340	1.07	0.17	4.95	20	2.67[b]
Taxus	46	25	0.55	0.21	2.35	20	0.60
Tilia	68	34	0.50	0.10	1.91	20	0.10
Ulmus	851	486	0.57	0.14	2.12	20	0.59

[a] $\alpha = 0.01$. [b] $\alpha = 0.001$.

growing season, and (3) the remaining herbaceous plants emitting pollen throughout the growing season. These three groups are particularly apparent in histograms of seasonal pollen concentrations from temperate regions (Figure 2.4) where wind pollination is most prevalent. They can also be discerned, however, in almost all areas of the world where wind pollination occurs.

The first two pollination groups result from very different adaptive approaches to reproduction. The last-named group is an artificial one dictated more by our inability to distinguish pollen to the species level than by the plant's inability to distinguish seasonality.

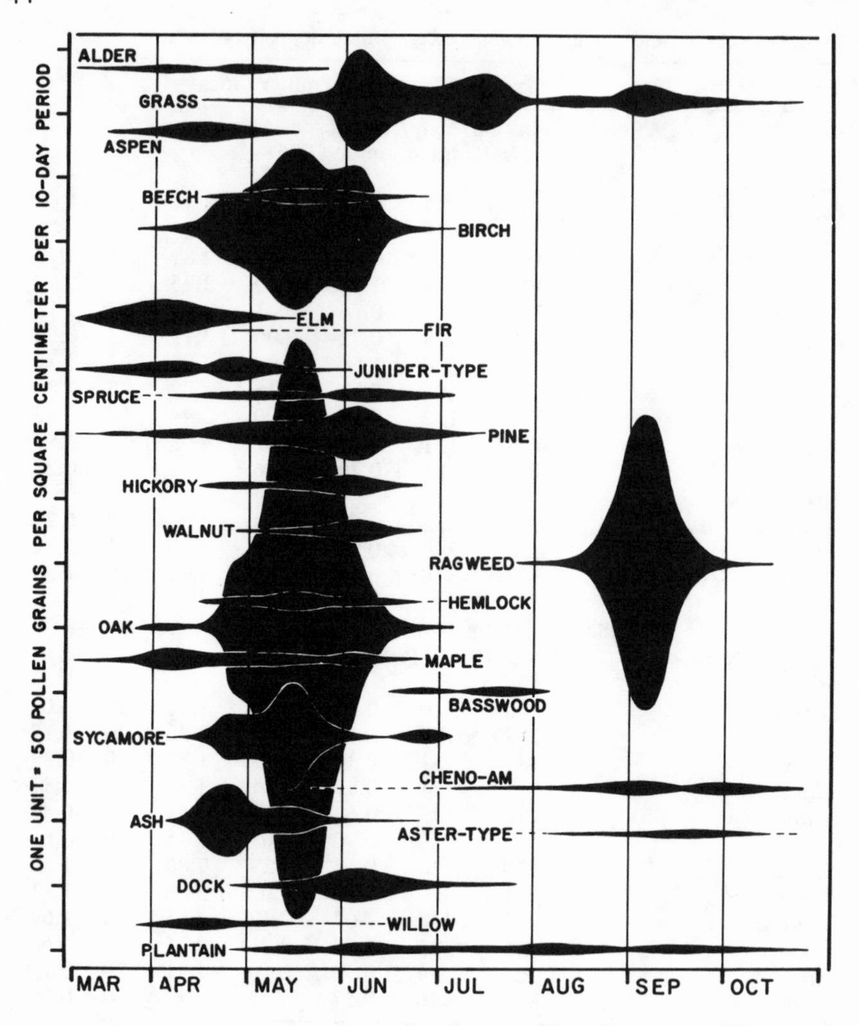

Figure 2.4. Pollen prevalence in Metropolitan New York City, from daily samples collected in 1932 (Ogden and Lewis 1960), and 1967 (author's unpublished data) at Yonkers, N.Y., and in 1968 (author's unpublished data) at Hackensack and Paramus, N.J. Each year's data were reduced to grains/cm^2 per 10-day period, then averaged for each 10-day period.

Pollen production in anemophilous trees and shrubs

Although some pines produce pollen as early as 7 yr after seed germination, most forest trees, especially hardwoods, do not flower or fruit in their first 45–50 yr. Thereafter, they flower every year until their death, 200–1000 yr later.

Life cycle

After trees flower and leaf out in the spring, they initiate leaf and flower buds for the following year. This occurs between May and July (Matthews 1963). By the end of the summer, the quantity of pollen available for spring flowering is determined because the cells designated to become pollen grains are already present.

Pollen development ceases when the first hard frost of fall induces dormancy in the tree. Pollen development is reinstituted after springtime temperatures regularly exceed a lower growth threshold temperature, usually near the freezing point. After it reaches maturity, predetermined pollen is emitted, fertilization occurs, and seed development begins.

Thereafter, the primary production of the tree is channeled into new growth, storage, and seed production. In early-blooming deciduous trees leaves do not appear until after flowering. In later blooming deciduous trees leaves appear during or before flowering. Leaf and flower buds for the following spring are again initiated several days to several weeks after seed formation begins in both deciduous and evergreen trees and shrubs.

Pollen season

See Figure 2.4. Aspen, beech, birch, elm, spruce, hickory, walnut, hemlock, oak, maple, basswood, sycamore, ash. The timing, magnitude, and length of pollination seasons are variable. Pollen emission begins at a given site as a function of accumulation of heat units above growth threshold temperatures occurring after dormancy concludes. The heat quantity needed varies from species to species. Ebell and Schmidt (1964) found that redcedar (*Thuja plicata*) required 100–200 h above 10°C, while Pacific silver fir (*Abies amabilis*) required 1300–1400 h. Boyer (1973) generated a linear regression equation that was based solely upon heat sums above 10°C occurring after 1 January in southwestern Alabama. The equation predicted peak pollen emission dates for independent data on longleaf pine (*Pinus palustris*) over a measured range of 40 days (23 February to 3 April) with an average error of 1.6 days.

Anthers extend when heat unit requirements have been met. At that point only a warm, dry day is required for pollen emission. After anther extension in white oaks, for example, pollen emission occurred in the range from 16°–24°C (Sharp and Chisman 1961).

The total pollen yield for a given species and a given season depends upon weather the previous summer and during the flowering season, and possibly upon previous seed production. Several workers (Pohl 1937, Hyde 1952, Ebell and Schmidt 1964, Kozlowski 1971) noted that maximum spring pollen yields of Temperate-Zone (winter-dormant) trees depend upon abnormally high temperatures, low precipitation, or both the previous summer. Because the cloudiness associated with high rainfall also decreases temperature, the relative importance of the two factors is not known.

During summer the bud primordia initiated in May or June differentiate into flower buds or leaf buds for the following year. If that summer is hot and dry, an abnormally high proportion of the bud primordia differentiate into flower buds with a resulting spring pollen yield greater than normal. The leaf bud:flower bud proportion shifts away from flower bud induction during cool and wet summers, providing spring pollen yields lower than normal.

Flowering season weather can either allow the predetermined pollen yield to be successfully dispersed, or reduce the pollen yield. For example, Sharp and Chisman (1961) found that 50% of unopened anthers dropped from oaks at central Pennsylvania sites after four continuous days of rain. Hyde (1952) detected no tree pollen in the air for almost a month after an unseasonably hard frost at the beginning of the 1947 tree pollen season. He did note that lighter frosts merely reduced pollination the following day without seriously disrupting the pollen season. Because the pollen yield is predetermined, unfavorable pollination-season weather can only decrease it.

The previous seed production history may also determine seasonal pollen yield. Hyde (1963) found that high spring pollen emission is frequently followed by above-normal fall season seed yields, and low spring pollen emission frequently precedes low fall season seed yields. The stored food resources of a given tree are strained when both pollen and seed are produced in large quantities during the same growing season. One, two, or even three growing seasons may be required following such a season before another high pollen yield can occur. Pohl (1937) and Hyde (1952) collected data that supported hypothesized pollen production "cycles" in oak (5-yr cycles), ash (3-yr cycles), and birch and beech (2-yr cycles).

If such cycles exist, the tree populations must be synchronized by an environmental trigger. This would most likely be a summer of either maximum or minimum warmth and dryness. If that is the case, it is not apparent in the data of either Hyde (1952) or of Pohl (1937), and indeed the existence of arboreal pollen cycles is still questionable.

The length of the pollen season for a tree population depends upon pollination season weather and topography. A few hectares of trees may require two or three days to shed their pollen (Sarvas 1962), while the whole population can take as much as a month to do so (Kozlowski 1971). A series of warm, dry days can allow all of the anthers in a population to dehisce (Sharp and Chisman 1961), while a cold, wet spring with only occasional warm, dry days produces a long pollen season (Ebell and Schmidt 1964).

Slope exposure also affects pollen emission timing. Because south-facing slopes receive more direct and longer exposure to the sun than do north-facing slopes, the former induce reactions similar to those of a more southerly climate. Jackson (1966) found that plant populations on south-facing slopes flowered an average of six days earlier than their north-exposure counterparts at two Indiana sites that were only 45.7 m apart. The time lag was equivalent to a difference of about 200 m elevation, or 177 km latitude.

Pollen production in anemophilous, late-summer, herbaceous plants

The late-summer, anemophilous, herbaceous plants are responsible for the hay-fever problem in temperate regions. Almost all are pioneer plants that invade recently disturbed substrates. Because man habitually provides that disturbance, his activities are usually responsible for the hay fever he suffers (Wodehouse 1939).

Life cycle

The life cycle of the herbaceous plants characteristically includes an initial vegetative growth stage followed by a reproductive stage that precedes senescence and death (annuals) or winter dormancy of rootstocks (biennials, perennials). The growth season begins in spring after frost becomes infrequent. Seeds of annuals germinate and overwintering roots of perennials generate new shoots. For the next several weeks, the resources of the plants are expended in production of new leaf surfaces to increase photosynthetic capacity. Vegetative maturity occurs in late June or July when daily solar energy for photosynthesis is at a maximum.

As solar energy declines, stored food resources (polymers including starch, fats, fatty acids, and the like) in roots and lower leaves are broken down and translocated to newly emerging flower buds. Development of flower bud primordia begins when the diurnal dark period reaches a critical length, which is different for each species. There are few weather events occurring during July and August that arrest plant growth; therefore the development and release of pollen, and the subsequent formation of seeds, are quite regular events.

After seeds are formed annuals die, while most perennials continue photosynthesis and food storage until the first frost of fall kills the aerial portions. Once dormancy occurs, below-freezing temperatures or alternate freezing and thawing is required by most seeds or roots to break dormancy.

Pollen season

See Figure 2.4. Ragweed. The timing of initiation of pollen emission in these plants is typically precise because of its synchronization with night length. Ragweed (*Ambrosia artemisiifolia*), for example, began local pollen production between 8 and 11 August during six consecutive years in Ann Arbor, Michigan (Sheldon and Hewson 1960). Because night length also controls the end of pollination by triggering mobilization of plant resources for seed production, the length of the pollen season is also relatively precise. By the first frost of fall, only scattered plants remain in flower.

These generalizations are valid for nonurban populations. Because night length controls flowering phenology, artificially short nights produced by urban street lighting may keep plants from emitting pollen at the beginning of the flowering season. Sheldon and Hewson 1960) demonstrated that at least the seed production response is inhibited by such conditions. When street-lighted (150-W floodlight), ragweeds continued flowering and producing pollen into late September. Adjacent unlighted ragweeds produced seeds and entered senescence on the normal solar schedule.

Year-to-year pollen yield is more variable than the seasonal timing. Late-summer-flowering weeds grow in specific synchronized stages. Inclement weather during an early stage can affect the entire population for the rest of the year. Hyde (1952) blamed seedling mortality during a late-spring frost for the subsequent 50% reduction in late-summer dock (*Rumex*) pollen concentrations. He also found that above-normal rainfall during the critical period of rapid vegetative growth (April–June) in dock directly enhanced late-summer pollen production. Ritchie and Lichti-Federovich (1963) came to the same conclusion regarding pollination by major late-summer weeds including grasses (Gramineae), ragweeds (*Ambrosia*), sage (*Artemisia*), nettles (*Urtica*), and sorrel (*Rumex*).

The most complete study of pollen production control by prepollination weather in late summer annuals was carried out by Sheldon and Hewson (1960). They found that ragweed pollen yield during late summer is a function of different weather events during each of the preceding growth stages. Abnormally wet, cold weather during seed germination in March sets back the peak pollen production date in late August. Cool or dry weather in May while shallow root systems are developing, or during June and July when ragweed competitors are maturing, reduced ragweed pollen production in late summer.

The seasonal prevalence of ragweed pollen is described by a bell-shaped profile (Figure 2.4). At the beginning, a few pollen grains may be recorded, derived from northerly areas where night length increases more rapidly and, consequently, pollen emission begins earlier. This is followed by the onset of local pollen production. Pollen concentrations rise rapidly, peak 10–20 days later, then decline as rapidly. Where more than one species makes up a recognizable pollen taxon, there is a similar number of peak pollen concentration periods. In the New York City area, for example, where giant ragweed (*Ambrosia trifida*) pollen emission peaks 8–10 days later than common ragweed (*A. artemisiifolia*), a bimodal profile of ragweed pollen concentrations can be discerned (Wodehouse 1933).

Pollen production in other anemophilous herbaceous plants

Some anemophilous plants seem to possess a wide amplitude of environmental requirements for pollen production. They exhibit a continuous capacity to germinate, grow, flower, and fruit during almost any time in the growing season. These plants are not prominent in the atmospheric pollen spectrum, consisting of the fewest species among the three groups discussed.

At a glance certain important anemophilous taxa, including the grasses (Gramineae), sedges (Cyperaceae), and the "cheno-ams" (Chenopodiaceae and genus *Amaranthus*), seem to belong here. A seasonal profile of their daily pollen concentrations from a temperate region (Figure 2.4) indicates that they emit pollen almost every day between April and October or June and October. Here the palynologists' inability to distinguish the pollen of more than one or two of the 4500 known grass species, 3000 sedge species, or 600 cheno-am species makes the classification an artificial one. In reality most grass species flower in the spring or early summer; others do so in the fall, and a few flower in midsummer. Most cheno-am species bloom in late summer and fall, with a few flowering in midsummer. The duration of the pollen season of each species is in fact relatively short, and their overlapping seasons produce the characteristic linear seasonal profile.

Little precise information is available regarding plants having a true continuous pollination season. Most species of plantain that flower from early summer through late fall (*Plantago major, P. rugelii, P. media, P. lanceolata,* and *P. aristata*) are perennials or are annuals that become perennial. The introduced English plantain (*P. lanceolata*) has the longest flowering period, from May through October. A small proportion of individuals within a local population may emit pollen on any day during that period.

The introduced perennial Bermuda grass (*Cynodon dactylon*) is another plant with a continuous pollinating season. In the Southwest, where it is used in lawns, its pollen is in the air from February through December (Figure 2.5).

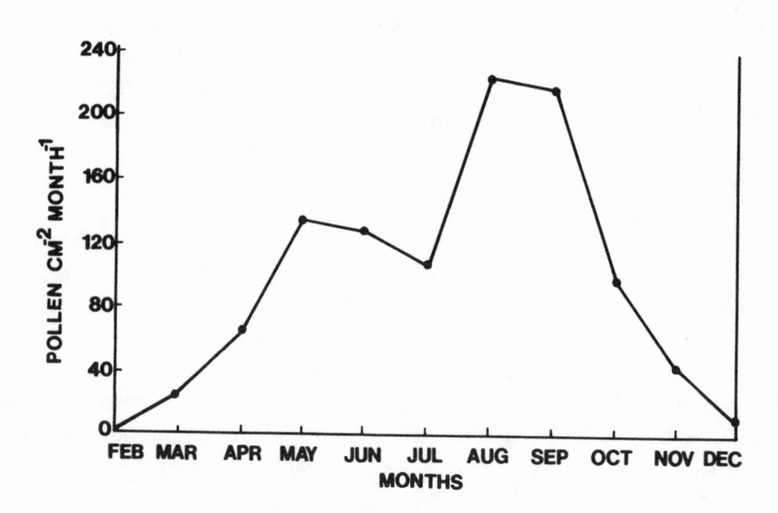

Figure 2.5. Seasonal production of grass pollen at Tucson, Ariz. Production measured by gravity slide values, 1954–1975, averaged for each month.

During this time the conditions that control variation in pollen output change with the seasons. Early-spring pollination is enhanced by warm nighttime low temperatures and high rainfall in winter. Early-summer pollen production no longer responds to winter low temperatures, but instead is increased when early-spring daytime maximum temperatures are depressed (Solomon 1973). When July through September rainfall is abnormally high (ca. 20 cm precipitation total), and when temperatures are abnormally low (20°–21°C average), increased pollen occurs from August through October. Neither the rainfall nor the temperature relationship alone induces significant variation (A. M. Solomon, MS in preparation).

Pollen Discharge

The actual transfer of pollen grains from the anthers to the atmosphere is essentially a physical process. The basic process of pollen emission consists of the following stages. First, the flower opens or anthers extend. Anther tissues dry and split. Pollen falls from the anthers and is deposited by gravity upon flower parts or leaf surfaces below the anthers. Finally, winds carry the pollen away. All of this occurs in the morning hours under fair-weather conditions. The atmospheric variables that control the process are related to the nighttime temperature inversion and to its subsequent disintegration during the morning.

Nighttime minimum temperatures exert strong control on the floral mechanisms presenting anthers to the air. In ragweed, nighttime lows of 4°C inhibit all anther extension during the following day (Sheldon and Hewson 1959). Nighttime low temperatures of 15°C must be followed by a rise of at least 14°C before anther extension occurs. A nighttime temperature of 20°C (a normal summer nighttime low in ragweed's geographical range) followed by a 24°C daytime temperature allows anther extension. Although some anemophilous plants emit pollen during other periods (e.g., Hyde and Williams 1945), the temperature-dependent mechanism adapts ragweed and probably most other similar wind-pollinators to opening just after sunrise and then only during warm weather.

Sharp and Chisman (1961) demonstrated that white oak male flowers remained closed unless temperatures during the preceding night were above 10°C. Because oaks flower in spring, periods of abnormally low temperatures early in the season (late April, early May) could arrest metabolism of the egg cells in the female flower during the few hours in which the pollen grain is viable. Similarly, the ragweed pollination season ends just before the onset of fall frosts. A mechanism that inhibits pollination except for the most propitious temperature periods has obvious adaptive advantage in the production of viable seeds.

When nighttime low temperatures remain above minimum tolerance levels, the morning temperature increase results in anther presentation. Then the splitting of extended anthers depends upon atmospheric moisture. According to Sheldon and Hewson (1959), ragweed anthers do not open when relative humidity (RH) is greater than 80%. They open gradually at 70% RH (50% open after 3 h) and more rapidly as relative humidity continues to decrease (90% open in 60% RH after 2.17 h, 90% open in 20% RH after 1.17 h).

Extended anthers did not open on white oaks when relative humidity values were greater than 45% (Sharp and Chisman 1961). The evaporative potential of air containing more than 45% saturation is not enough to dry anther walls to the breaking point. During fair-weather days, the requisite dryness appears with postsunrise temperature increases.

Once the flower anthers have extended and anther walls have split, dehiscing pollen grains may be entrained in the turbulent air that occurs with rising temperatures. If the pollen grains are intercepted by vegetative structures or other obstructions, entrainment of pollen in the atmosphere may be briefly inhibited. The grains adhere to one another to form clumps on the deposition surface, and individual pollen grains in the clump are blown away, layer by layer.

The process produces the pollen emission profiles shown in Figure 2.6. The number of flowers that open during each 15-min period (dashed line) should be a direct measure of pollen emitted. Pollen concentrations near the pollinating plant lag slightly, however. Both curves indicate a maximum at 0830, but atmospheric pollen remains concentrated long after the last flower has opened (0915–0930).

If winds are calm during anthesis, airborne pollen concentrations remain low. Once wind velocity exceeds 3–4 m/s, pollen is lifted free from the obstructions on which it rests. If wind velocities of 10 or 20 m/s occur during anthesis, there is no lag between pollen discharge and pollen entrainment in the atmosphere.

Even on windy days, few pollen grains actually become airborne. Sheldon and Hewson (1959) indicate that only 6% of the pollen emitted from a dense ragweed stand became airborne. Raynor et al. (1972) found that two-thirds of the corn pollen discharged from plants in a cultivated plot never escaped from the plot.

The proportion of pollen emission that becomes airborne is a function of wind velocity, atmospheric turbulence, anther height above the surrounding surfaces, and the density of the vegetation stand. Perhaps three-fourths of the pollen emitted from a mature tree growing in open surroundings becomes airborne on windy days. A much smaller proportion of the pollen emitted is likely to become airborne when plants grow densely. These estimates are supported by few data.

The discharge of pollen into the atmosphere is thus a series of processes, each governed by different but interrelated environmental parameters. First, high-pressure conditions must occur. Then, nighttime low temperatures control the opening of the flower and extension of the anthers. If nighttime low temperatures are not inhibiting, atmospheric moisture must be low enough to split the extended anthers. When anthers split, air movement is required to

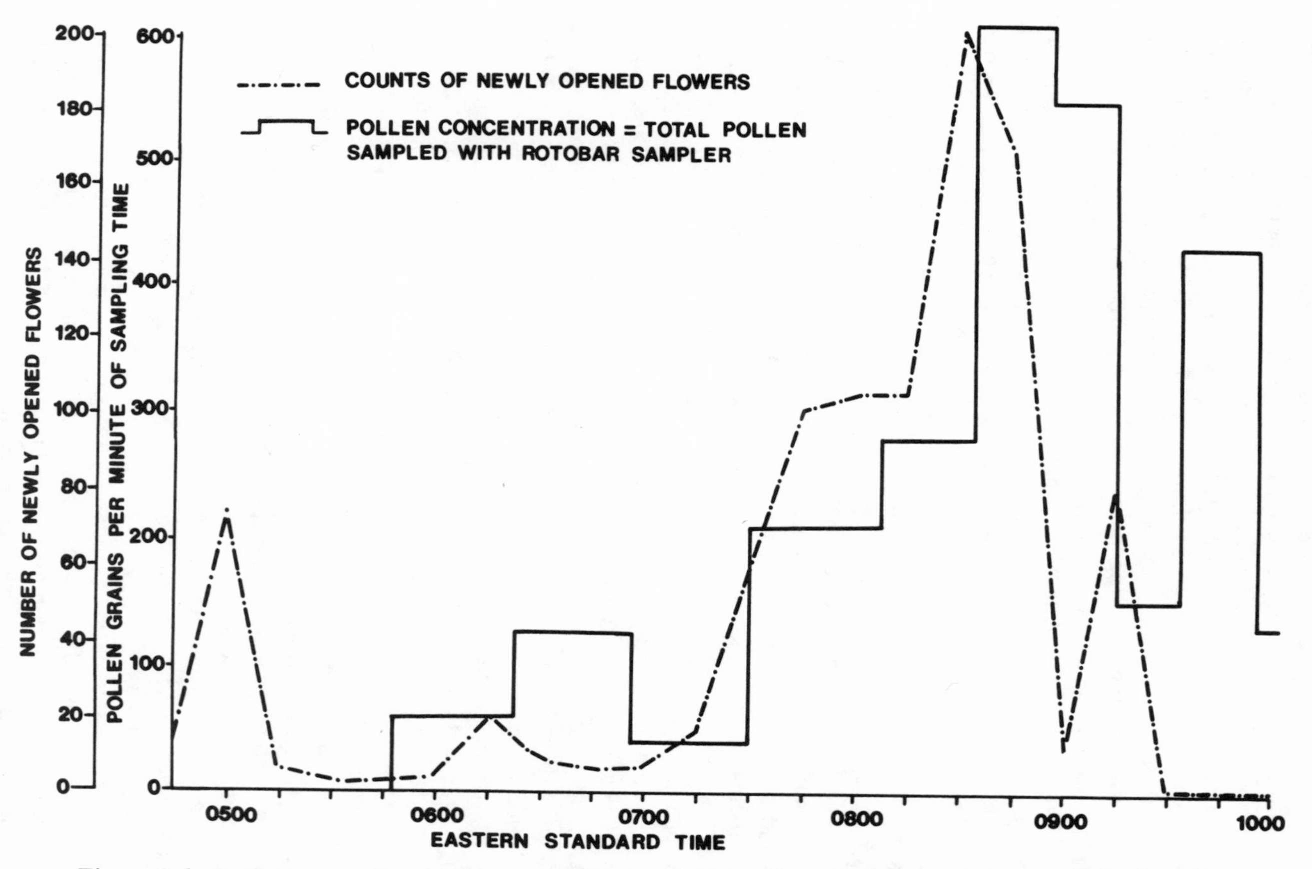

Figure 2.6. Anthers opened and pollen concentration (adapted from Sheldon and Hewson 1960, fig. 37).

entrain discharged pollen. Finally, a maximum proportion of discharged pollen can become airborne from isolated plants only on windy days.

Environmental parameters that inhibit pollen emission characteristically occur under low-pressure conditions. Temperatures are lower, the diurnal temperature range is reduced, and atmospheric moisture is high. Conversely, the environmental parameters that induce pollen emission characterize the fair-weather "microclimate" (Geiger 1966). The nighttime temperature inversion is accompanied by still air and damp, occasionally foggy conditions. After sunrise, solar energy increases air temperatures and air movement, and decreases relative humidity. Pollen emission is precisely adapted to these aspects of the fair-weather microclimate.

Pollen Viability in the Atmosphere

Data describing the length of time a pollen grain remains viable in the atmosphere are almost nonexistent. There are several studies dealing with pollen viability under special storage conditions, but these do not address the above problem.

There are indications that the atmospheric life span of anemophilous pollen may be quite short. Sheldon and Hewson (1960) studied viability of ragweed pollen stored indoors at room temperature. Pollen viability was tested at 1-h intervals by incubation in buffered sucrose agar. Maximum germination (45%) occurred 2 h after pollen dehiscence and declined to 3% after 9 h.

Werfft (1951) also used pollen germination as an index of pollen viability. She enclosed pollen grains between cellophane sheets and hung them in the sunlight. After 8 h exposure, viability was 40%–55% for maple (*Acer monospessulanum*), 10%–21% for hornbeam (*Carpinus betulus*), and 11%–35% for pine (*Pinus montana*). Viability under indoor conditions (19°–23°C and darkness) was 52% for the last species. Viability under outdoor conditions was similar in the shade and in direct sunlight and was lower in anemophilous than in zoophilous pollen. Maximum viability decline occurred between 0900 and 1200 (hours).

In separate laboratory experiments with UV radiation using wavelengths and intensities similar to those in sunlight, Werfft found that only 3 h exposure of pine pollen reduced viability to zero. Similar damage after 8 h exposure to the UV light source occurred in pollen of anemophilous alder (*Alnus glutinosa*), elm (*Ulmus glabra*), aspen (*Populus* sp.), willow (*Salix caprea*), and ash (*Fraxinus* sp.).

There are indications that the pollen life span under normal conditions in the free atmosphere may be even less, exemplified by the case of *Pinus balfouriana* and *P. longaeva*. Hybrids of these two pine species produce fertile offspring when crosses are made with plantation-grown individuals (L. C. Johnson, pers. commun.). The two species occur naturally in the Sierra and Inyo Mountains on the west and east sides, respectively, of Owens Valley

(Inyo County, California). They are separated by about 36 km. Prevailing westerly winds along the Sierra crest are strong and consistent. The trip for a *P. balfouriana* pollen grain from the Sierras to the *P. longaeva* populations in the Inyos would take no more than 2 h. Although both species flower at the same time, hybrids of the two cannot be found. At present, the most logical explanation appears to be a very short pollen life span. The adaptive strategy of wind pollination, with its requirements of dense populations, as well as the results of Werfft's study, argue in favor of this hypothesis.

Some clues on the nature of the atmospheric conditions that limit pollen viability can be gleaned. The pollen morphology of terrestrial anemophilous plants and that of emergent aquatics pollinated close to the water surface is strikingly different. The former are typically armored with a thick roof, or *tectum* that may isolate the contents of the grain from the atmosphere. The latter possess either an incomplete tectum (e.g., Cyperaceae, Sparganiaceae, Typhaceae) or no tectum at all (e.g., Ruppiaceae). The logical conclusion is that lack of atmospheric moisture limits pollen viability. The demonstration of a rapid loss of water from pollen grains as they leave the anther (Harrington and Metzger 1963), however, is at odds with this conclusion.

Ultraviolet radiation has been suggested as a major cause of pollen viability decline (Werfft 1951). Studies of the reflectance by the pollen shell or exine of impinging UV radiation may reveal some merit to the hypothesis, but the present studies do not treat the idea. Studies of pollen storage life indicate that low temperatures preserve pollen viability, so we may presume that high temperatures reduce viability. Again, there is currently no literature to confirm or deny such a hypothesis. Answers to the question are important, for they contain significant implications for studies of plant migration rates, forest tree regeneration, population genetics, and evolutionary strategies.

INSECTS AND OTHER MICROFAUNA*

In discussing the major categories of biological particles that are covered in this section, one can recognize that the insects and other microfauna represent the most difficult group to characterize. The insects are the most diverse group within the animal kingdom and their numbers are difficult to sample and quantify; furthermore, all species are somehow actively involved in the dispersal process, even if they are passively carried by the wind. Some of the behavioral adaptations that are involved in dispersal will be discussed later in this section.

Major emphasis will be devoted to the insects because more is known about their distribution in the atmosphere and, in many cases, the mass transfer of individuals to new resources may result in economic impacts on the environment. The literature base on the aerial dispersal of arthropods (insects,

---

*M. L. McManus

mites, spiders, and the like) is massive and dates back to the late 1800s; therefore I have not attempted to conduct an exhaustive review of the literature. Where applicable, I will refer to specific references that can be used to summarize the published information on a particular subject area.

Distribution and Abundance of Microfauna in the Atmosphere

Most of the early investigations on aerial arthropod populations, the so-called aeroplankton, were conducted in the United States. Studies on the passive transport of gypsy moth larvae for long distances by the wind and indications of the general direction of their spread began in the early 1900s. Burgess (1913) released newly hatched larvae that had spun silk in front of an electric fan and found that they drifted from 6 to 9 m. Later, Collins (1915) trapped on sticky-coated screens newly hatched larvae that had apparently drifted from the mainland to an island 21.7 km east of the New Hampshire coast. Collins later (1917) showed that the wind carried the larvae from 30.6 to 48.3 km inland from across Cape Cod Bay off the coast of Massachusetts.

In 1927, Felt and Chamberlain (1935) used traps on various types of kites in order to demonstrate the possibilities of capturing insects in the upper air. Later, Felt (1928a,b) made several flights using an insect trap attached to the lower wings of an airplane. This was the first documented attempt to collect microfauna from the free air at altitudes up to 4572 m.

During a five-year period from August 1926 to October 1931, the most extensive study ever made of the insect population of the air was conducted in Louisiana by the USDA Bureau of Entomology and Plant Quarantine (Glick 1939). In these investigations, some 29,000 specimens were taken from altitudes ranging from 6 to 4572 m with specially designed traps fitted to the wings of several types of aircraft. Over 1300 separate flights were made—both day and night collections are represented as are all months of the year. In Britain, Hardy and Milne (1937, 1938a,b) and Freeman (1945) sampled airborne populations with nets on the mastheads of ships at sea and from kites and tall masts over land. Berland (1935) was the first European entomologist to use an airplane to collect insects in the upper air.

Most of these exploratory studies were designed to discover the numbers and species of insects drifting in the air at various altitudes and to correlate these with meteorological factors. The results showed clearly for the first time that, on most warm days, the air over a few square kilometers of country up to 1000 m or so contains a population of millions of insects representing many species, some of which may be carried for great distances.

Zones of abundance

On the basis of its arthropod population, the atmosphere has been divided into a so-called terrestrial zone, generally extending to about 60 m above the surface, and a "plankton" zone above that (Berland 1935). Within the plankton zone, the insect population usually consists of weak fliers or smaller, flightless insects borne upon air currents (Wellington 1945b). It is generally conceded that the winds of the lower layers of the atmosphere are of most importance in the horizontal transport of insects between surface localities (Whitfield 1939). Wellington, however, emphasized that the plankton zone is populated by vertical currents, particularly at levels where the temperature is below the minimum flight temperatures of the winged transients. He proposed that a distinction should be made between the levels of the insect plankton zone, as determined by the effect of air temperature upon the relative ability of the insects to remain active, and hence airborne. He suggested that the lower levels of the zone at temperatures above 7°C might be termed the *active* plankton zone. This portion would contain the vast majority of the aerial insect population; winged individuals would be capable of maintaining a flight altitude and could be borne great distances, and wingless individuals would still be dependent upon atmospheric turbulence for any lengthy horizontal transport by the wind. The upper levels of the zone, where temperatures are below 7°C, would be called the *inert* plankton zone and the majority of insects within this region would be dependent solely upon vertical air currents and, with the exception of spiderlings or those insects suspended on silken threads, would not be carried horizontally for long distances. This kind of dispersal is referred to as *drift* by Johnson (1951).

The best single source of information on the abundance of arthropods in the upper air is still the publication by Glick (1939). The 29,000 specimens taken over Louisiana between 1926 and 1931 included representatives of 18 orders of insects (further classified into 216 families and 724 individual species) and two orders of arachnids (spiders and mites). Thirty different species of spiders representing 16 families are included in the total. The average numbers of insects taken in the systematic studies at different altitudes in 10 min of collecting in the daytime were as follows: at 61 m, 13.03 specimens; 305 m, 4.70; 610 m, 2.41; 914 m, 5.73; 610 m, 2.52; 914 m, 1.11; and 1524 m, 0.89 specimens.

In a later publication, Glick (1942) expressed the catch in terms of the average volume in cubic feet per insect. One spider was taken at 4572 m, the highest altitude at which any specimen was recovered. Flights were limited at the higher altitudes because of the limitations of the aircraft of that time. Glick (1942) concluded that the number of insects in the air at any given time depends almost entirely on combinations of one or more meteorological factors, the physiographic features of the region, and the aerostatic coefficient (relation of size, weight, and buoyancy) of the insects.

Relative abundance

Glick (1939) found that there were wide differences between the densities of different species that were recovered during the day and night flights.

There were also about one-fourth as many insects at 1524 m at night as were collected at 305 m. Although specimens were caught in every month of the year, the largest numbers were caught in May and the fewest in December and January. These results should be interpreted in light of the fact that the entire study was conducted in the southern United States where there is a continuous growing season.

The total number of individuals of any species that can be trapped at any location, altitude, or time depends on many variables: the size and species composition of the source area, the seasonal and diurnal variation in numbers, the behavior of the dispersant population, and the meteorological factors prior to and during the collection. These are only mentioned here but will be further discussed later in this section.

One major criticism of the early studies on aerial trapping is the overemphasis placed on correlating the numbers of insects trapped with the then current meteorological factors. Johnson (1951) characterized those studies as neglecting the role of terrestrial population changes in determining the numbers of insects available for aerial dispersal, while favoring the relationship between the numbers recovered and single meteorological factors such as temperature. He cited as an example a study of the bean aphid in England:

> The weather was relatively uniform from day to day; yet as the population grew very different-sized catches were obtained on successive days. When such a population change occurs with many diverse species in a differential fashion, a correlation of total numbers caught against meteorological factors alone at the time of catching is unlikely to give a very accurate interpretation.

This was, perhaps, why Glick (1939) found that "At times when the weather appeared to be ideal for insects, a few or none would be taken. Again, when weather conditions were 'bad' insects were taken in considerable numbers."

Biological Significance of Dispersal

Terminology

Because of the variation in terminology that exists in the literature, it is appropriate to define a few terms that are often used interchangeably. Elton (1927) made a distinction between *distribution,* which relates to the places that are occupied at any particular time, and *dispersal,* which refers to the movement away from a populated place resulting in the scattering of at least some of the original population. Dispersal may be on a grand scale, resulting in the redistribution of a species beyond ecological barriers (Andrewartha and Birch 1954). Wolfenbarger (1946) in summarizing a large volume of literature on dispersal found that most small organisms or disseminules tended to disperse as the logarithm of the distance from the source.

A classical ecological textbook definition of dispersal is "the movement of individuals or their disseminules into or out of the population or population area" (Odum 1959). It is then further divided into three forms: *emigration*—one-way outward movement; *immigration*—one-way inward movement; and *migration*—periodic departure and return. This use of the term *migration* is oversimplified and I prefer the definitions proposed by Southwood (1962) in his review article.

According to Southwood, locomotor movement is a characteristic of most animals and may be active (walking, flying, or swimming), or it may be passive as in the ballooning of spiders and some lepidopterous larvae. He recognizes two types of insect movements: trivial movements and migratory movements.

Migratory movements take an individual away from its habitat and frequently result in an increase in the mean distance between individuals of the source population. Therefore dispersal and migration are essentially synonymous, although Southwood prefers to say that migration leads to dispersal since there is a scattering of the source population and some individuals are left behind. He further states that, during a migratory movement, the individuals do not respond to food, a mate, or habitat. Migratory movements also have certain behavioral characteristics, e.g., insects respond to an external stimulus such as light.

Conversely, *trivial movements* are restricted to the animal's habitat and lead to only a limited increase in the mean distance between individuals. The duration of a trivial movement is completely variable and it may be terminated at almost any time on perceiving a vegetative stimulus. An individual engaging in a trivial movement may be swept up and carried away by a sudden gust of wind or transported passively in ships, planes, or by man. Southwood refers to these individuals as "vagrants"—although the ecological effect is the same as in migratory movements, the behavioral aspect is lacking.

Function of dispersal in animals

According to Odum (1959), dispersal supplements natality and mortality in shaping the growth form and density of populations. Although dispersal is constantly occurring in most animal populations and may have minimal effect, mass dispersal may completely change the structure of a stable population.

The innate tendency to disperse seems to be present, to some degree, in all species and may be accentuated by crowding, hunger, actions of predators, or adverse meteorological factors. Andrewartha and Birch (1954), however, emphasize that one must distinguish between ordinary activities of searching for food or for a mate and those movements directed toward finding a new habitat or establishing a new colony. Aside from the more obvious functions mentioned above, dispersal brings about gene exchange between populations and is therefore important in understanding population genetics and speciation.

Variability in the dispersal of arthropods

Most species of arthropods devote a large part of their collective energy to the founding of new colonies. Small species with poor power of locomotion may rely on wind or other agencies; adaptations may include behavior patterns that enable them to become airborne and an enormous fecundity that offsets the low probability of survival characteristic of this method of dispersal. Larger species may be better equipped with locomotory organs (e.g., wings) or may possess well-developed sensory organs that enable them to search more efficiently.

In addition to behavioral or structural adaptations, many species possess life-stage and sex differences in their life histories that specifically increase their probability of dispersal and survival. Examples of interest are discussed below.

Factors That Determine the Abundance of Insects and Other Arthropods

Population dynamics

The abundance of any species and therefore the number of individuals of that species available for dispersal is dependent on many factors that cumulatively determine the population dynamics of the species. *Population dynamics* is defined as the study of the changes in numbers of a population and of the factors that cause the changes. Life table studies are conducted to elucidate the lethal agents and to quantify their effect on the host population from one generation to the next. Lethal factors include biotic agents such as parasites, predators, and pathogens, or they may be abiotic in nature. Weather factors, host availability and condition, and dispersal are often included in the latter category.

In addition to the mortality factors, abundance is determined by the fecundity of the female, the number of generations per year, and the life history of the species.

In arthropods, the reproductive capacity of individual species is highly variable and may range from a few eggs per female to more than a thousand. Aphids and mites may produce 20–50 eggs per female; bark beetle species vary but usually produce fewer than 100. The gypsy moth female deposits an egg mass that may contain from 100 to 1000 eggs depending both on the quality of the population and on the host upon which it fed during the larval stage. Many species of arthropods that have a low fecundity are able to compensate by producing many generations per year. Mites, for example, though producing only 20–50 eggs per individual, have a 3-wk life cycle and are therefore capable of producing four to eight generations each season. Some species of mites are able to stimulate the production of females when the resource deteriorates; the females mate prior to dispersing and therefore each

female may establish a new colony. Other species of mites produce parthenogenetic females under similar circumstances and therefore optimize the chances of establishing new colonies. In general, most species in temperate climates produce only one generation each year although the same species may be multibrooded if it is also distributed in warmer latitudes.

The life history of a species is important in determining the number of individuals available for dispersal. For example, in those species that disperse in the adult stage, the number of offspring are reduced by the complex of mortality factors that affect the eggs and developing immature stages. Those species that disperse as immatures, however, have more progeny available for dispersal. This supposed advantage to the dispersing immature may be negated by the fact that, though fewer adults may be produced, they usually possess wings and therefore have a higher probability of reaching a new resource. An infinite number of such examples exist in the literature on arthropods.

Extent of the source area

In addition to those factors discussed above, the abundance of any species is limited by the size of its food resource. This factor is extremely important in the mass transfer of populations. For example, the potential dispersal of mites from a small isolated orchard would not be comparable to the potential number of dispersing greenbugs (aphids) from the extensive grain producing areas in Oklahoma and Texas. Both the spruce budworm and the gypsy moth have a tremendous potential for dispersal when populations are in outbreak status. The budworm in Maine and eastern Canada infests a climax spruce-fir forest that is contiguous for hundreds of kilometers—a recent gypsy moth outbreak resulted in the yearly defoliation of more than 400,000 ha of contiguous scrub oak forest from Massachusetts to Pennsylvania. In both examples, the source area of preferred food that is available for population expansion is unlimited.

Factors Affecting the Release of Insects and Related Arthropods

Active vs. passive dispersal

In general, species are categorized as being either actively or passively dispersed. As our knowledge about the behavior of individual species increases, however, it is clear that most species are in fact very actively dispersed. Southwood (1962) uses the term *passive movement* to refer to those individuals that have no control over the direction or the duration of their flight. This would certainly apply to those spiderlings, lepidopterous larvae, and mites that were recovered at various altitudes by many of the early workers. It also applies to those winged forms that are either (a) carried by convection to altitudes (plankton zone) where temperature is limiting to flight or (b) actively flying in the terrestrial zone, but the surface winds exceed the flight speed of the insect.

Therefore *passive* should refer to the state of the airborne individual, but should not apply to the dispersal process common to a species. There are many intricate behavioral mechanisms involved in the release of arthropods into the air, whereupon they are passively dispersed (discussed later in this section).

The *active* involvement of insects in dispersal refers to their ability to fly and to launch themselves into the airstream. Southwood (1962) separates flights into those that are trivial and those that are truly migratory. The trivial flights involve a search for either food or mating and are usually conducted near the individual's habitat—these are called *flits*. Many flits may develop into long-range dispersive movements, evidenced by the large numbers of small, winged species that were recovered at the higher altitudes by Glick (1939). Migratory flights usually involve the larger, winged species that are strong fliers and that can sustain flight for at least an hour.

In an effort to classify kinds of migratory flight, Taylor (1958) proposed the term *boundary layer* to describe a hypothetical layer of air near the ground where insects are able to control their movements because their flight speed exceeds wind speed. Outside the boundary layer, they move downwind. Migration is minimal in the small, winged insects that are active at night and fly within a shallow boundary layer when atmospheric turbulence is negligible. In contrast, small day-fliers such as aphids fly upward toward light at times when turbulence is high and therefore are carried through the boundary layer and oftentimes disperse over long distances (Taylor 1974).

The role of behavior

Periodicity of activity

Felt (1928a) emphasized the importance of flight periodicity when he wrote: "It may not be so generally recognized that the daily periodicity of flight, diurnal, crepuscular, and nocturnal, may have a very material bearing upon the probabilities of wind dissemination."

The importance of periodicity is best demonstrated in the aphids. A large volume of literature exists, mostly from Britain, because aphids are economically important there and are a major component of the aerial insect population over that country. According to Johnson (1951), there are three major peaks for their aerial populations. The first occurs in May–June when many aphids migrate from winter to summer hosts; the second is in July, when a redistribution of the winged forms occurs; the third is in September and October, when many aphids return to their winter hosts (Johnson 1951).

Within each population peak, there are daily peaks. Aerial populations up to 610 m build up during the day and decline at night, often leaving the air virtually clear of aphids in the early morning. This corresponds to Wellington's (1945c) explanation that, as convection reaches its maximum in the early afternoon, the maximum density of diurnal species occurs at higher altitudes in the early afternoon.

Although there are many examples of periodicity affecting dispersal, only a few will be mentioned. The potential for mass flights of female spruce budworm moths occurs in eastern North America when, at peak emergence, the moths fly upward from the canopy and may aggregate in tremendous numbers just below the inversion layer. Peak flight activity occurs between 2000 and 0100 hours over a period of 7–10 days. Under certain climatic regimes, long-range mass dispersal of egg-laden females does occur.

The gypsy moth disperses as newly hatched larvae in the spring of the year. The young larvae, which are very hairy and buoyant, spin down from the foliage on strands of silk prior to their initiating feeding, and are easily carried by the wind. McManus (1973) found that the larvae possess many behavioral traits that increase the probability of their dispersal. The newly hatched larvae usually leave the egg mass and ascend the trees in the morning so that they are at their dispersal sites in the upper tree crown when maximum turbulence occurs. Additionally, the larvae exhibit a periodicity of dropping on threads so that peak activity occurs in the early afternoon when convection and turbulence are at a maximum.

Examples in immature forms

Spiders. Although the airborne flights of spiders have been known since the time of the ancient Greeks, Blackwall (1827) was the first to document the behavioral mechanism involved. Immature spiders or "spiderlings" disperse mainly during the summer months. Richter (1970) studied dispersal in eight species of wolf spiders (genus *Pardosa*) and describes the characteristic "tiptoe" behavior as follows: "They attached their walking threads (drag-lines) to the substrate and cut the drag line with a rapid jerk. Tip-toe behavior then followed and active movement of the spinnerets occurred. The silk threads reached a length of 70 cm or more before the animals were carried up by the air current."

Richter (1970) concluded that mass dispersal in *Pardosa* populations would be restricted to a few days in the year as it is bound to a developmental stage of the life cycle of short duration. At the same time, dispersal occurs only when unseasonably warm and calm weather prevails—aeronautic behavior stopped when the wind velocity exceeded 3 m/s. Therefore, although many spiderlings are passively blown by the wind, they are quite actively involved in behavior that enhances their dispersal.

Lepidopterous larvae. Young larvae of many arboreal Lepidoptera extrude silken threads similar to that of spiders and may be carried by air currents for some distance. The threads are effective in increasing the drag of the insect. McManus (1973) found that there was a threefold decrease in the terminal velocity of a newly hatched gypsy moth larva if individuals were suspended on a 90-cm length of silk.

Both the Douglas-fir tussock moth and the spruce budworm disperse in the early instars. Greenbank (1957) felt that initial action of the larvae dropping on silken threads was caused by a mechanical stimulus. In the case of the gypsy moth and many other lepidopterans, however, the larvae are very photopositive and climb trees in response to diffuse light. Their rate of activity is controlled by ambient temperature and evaporation rate and there is considerable variation among species.

There is little doubt that most species are as actively involved in initiating dispersive movements as are the spiders. For example, if gypsy moth larvae drop from the foliage on threads when winds are insufficient to carry them off, they climb the threads and drop again thus assuring dispersal. Gypsy moth larvae also fit well within Southwood's (1962) guidelines for distinguishing a true migratory movement. The larvae do not respond to any vegetative stimulus while climbing the trees in response to light, and seem destined only to disperse.

Many other species of insects are passively blown by the wind in their immature stages. Most species of scale insects are mobile only for a brief period in their life cycle; the nymphs or "crawlers" that hatch in the spring are very active and may be blown for distances up to 4.8 km (Rabkin and Lejeune 1954, Brown 1958). Andrewartha and Birch (1954) stated that the first-stage nymphs of mites actively launch themselves into the air to enhance their probability of dispersal. Immature scale insects may also launch themselves into the air under certain meteorological conditions, although such observations have not been published.

Examples in adult forms

The best-documented example of the role of behavior in dispersal of adults is in the aphids (Kennedy 1950, Johnson 1954, Broadbent 1948). There is no doubt that dispersive flights by aphids are started by the insects' actively launching themselves into the air, especially when relatively calm conditions exist. Furthermore, Kennedy (1950) showed that aphids actively alight when they have control over their own movements. They do not settle on the first favorable food plant that they find, but rather launch themselves into the air again after a brief rest.

Many small airborne-insect groups actively launch themselves into the airstream, partly in response to light. Characteristically, the insects climb to the highest part of the vegetation, orient to the sun, and eventually take off. This is common in the orders Diptera, Thysanoptera, Homoptera, and Hemiptera.

The behavior of spruce budworm females was mentioned earlier in this section. Canadian researchers recently discovered that female moths, after mating, deposit a partial complement of eggs and then rise above the forest canopy in the evening; the moths may be concentrated in large numbers, which may then be dispersed over a large geographical area.

Although most species of aeronaut spiders disperse as immatures, representatives of one family, the Linyphiidae, disperse as adults. Dispersal is most common in the winter months and mass dispersals do occur on calm, sunny days (Duffey 1956). Apparently the "tiptoe" behavior that has been described in spiderlings is also common to the adults.

Crowding

Although crowding has been frequently mentioned as a situation that stimulates dispersal in insects, there is a large body of evidence to the contrary. Apparently the gregarious phase (locust) of many grasshoppers is induced by increased densities in their permanent breeding grounds (Kennedy 1956).

Duffey (1956), however, concluded that, in a known spider population, the amount of dispersal was not correlated with density; some species were very abundant in an area and yet there was no evidence of aerial dispersal. Furthermore, the peak dispersal of the aeronaut species did not coincide with the highest population densities. Campbell (1960) found no evidence of dispersal in two species of walkingsticks in Australia when large population buildups occurred. The same phenomenon was observed in outbreak populations of the northern walkingstick in Indiana.

In conclusion, crowding may be considered a factor in some populations of insects but is more the exception than the rule.

Condition of the host/habitat

Southwood (1962) proposed that the prime evolutionary advantage of migratory movement lies in its enabling a species to keep pace with the changes in the location of its habitats. He advanced the hypothesis that within a taxon one should find a higher level of migratory movement in those species associated with temporary habitats than in those with more permanent ones. Temporary habitats would include carrion, plant debris, or annual and perennial plants of seral communities—habitats that are in one locality for only a short time. Conversely, permanent habitats would include rivers, lakes, and perennial plants including trees of climax vegetation. Temporary habitats may also result from environmental change; for example, the vegetation varies from season to season and from year to year in arid and semiarid regions because of irregular climate.

Occasionally the impermanence is on the part of the insect whose habitat requirements change during the life cycle; for example, grasshoppers need different habitats for oviposition and for feeding. Therefore many grasshopper migrations occur when the habitat ceases to meet the environmental requirements of the species.

Deterioration of a habitat or host may result in mass migrations; this is common in many species of grasshoppers that have permanent breeding grounds or outbreak foci characterized by small areas of permanent vegetation.

Most mites live in habitats that are discontinuous and transient—populations of mites consist of separate colonies that must regularly move from an aging resource to a new one. Mite populations reproduce continuously as long as conditions are favorable, but enter a dispersal phase when conditions are adverse (Mitchell 1970). When the resource deteriorates, the reproduction of females in the family Tetranychidae is stimulated within one week. The females mate immediately after emerging and then disperse so that each female is capable of founding a new colony. This contrasts with species that disperse as unmated females, and usually disperse in small colonies in order to maximize the probability of finding both a mate and a new resource.

Richter (1970) made a comparison of the habitats of a number of aeronautic and nonaeronautic spiders and found that four species with the lowest dispersal capacity occurred in stable habitats. He further hypothesized, however, that "species living in habitats which are widespread and abundant can disperse to suitable areas by short-distance movement (i.e., locomotory activity), whereas species living in rare habitats invest in aerial dispersal to find new suitable areas."

Meteorological effects

Glick (1942) stated that "when there are great numbers of insects in the air and when insects are very active, it is quite evident that certain meteorological conditions, particularly temperature, humidity, and light, are definitely favorable to their occurrence." As stated earlier, the initial efforts to trap airborne insects probably overemphasized the correlation between trap catch and ambient meteorological conditions at the time. The role of individual factors in the release of insects and related arthropods is discussed only briefly.

Temperature

Temperature probably has the greatest direct influence on the activity of arthropods. Since these organisms lack a mechanism to regulate their body temperature, they must either respond directly to ambient conditions or seek out a location that is more favorable. Generally speaking, the activity of arthropods increases as temperature increases up to a lethal range, whereupon the individuals become comatose. The optimum range of temperature, referred to as a *preferendum,* for each species is different and even varies among stages in the life cycle of the same species.

The most important aspect of temperature related to dispersal is the threshold temperature for activity in each species. Wellington (1945a) found that the minimum temperature for flight initiation in three species of Homoptera was ca 14°C. This agrees with values recorded in England for the bean aphid (13°–15°C). The activity of nocturnal species is highly dependent on temperature as demonstrated by light-trap catches of insects.

With winged insects, temperature may be limiting for short periods but is probably not a factor in preventing the release of individuals into the airstream. In the case of immature forms that are passively blown, however, temperature may limit their predispersal behavior and therefore greatly inhibit dispersal. I observed this situation in populations of the gypsy moth—periodicity of hatch in the spring was such that the maximum number of newly hatched larvae were available for dispersal. A cyclonic disturbance accompanied by temperatures below the threshold of activity for the larvae, fog, and precipitation dominated the area for a period of three days, however, and dispersal of the larvae never occurred. As stated in an earlier section, dispersal may be limited to a brief period in the immature stages of many species; therefore, if meteorological conditions are even temporarily adverse, dispersal may be limited.

High temperature is probably the greatest single influence in stimulating aeronautic behavior in spiders (Duffey 1956, Richter 1970). The habitat of species of spiders living within the herb layer is warmed only slowly by the sun, but once the spiders begin to climb the vegetation, they encounter warmer temperatures and their rate of activity increases. Duffey suggested that the gradient of temperature may provide the initial stimulus.

Relative humidity

Insect activity is better related to evaporation rate than to absolute humidity. As with temperature, insect response to humidity varies tremendously between species and between stages of the same species. Activity of most insects is probably greater at lower relative humidities. Broadbent (1949), however, found that relative humidity affected flight activity of aphids temporarily until they became acclimated to that humidity, and that they then flew readily at all levels tested between 50% and 100% at temperatures below 27°C. One can conclude that relative humidity is not a limiting factor in the release of dispersers.

Atmospheric pressure

Wellington (1945a) conducted laboratory studies on some 15 species of insects to determine their resistance to reduced pressure over periods of 3–5 h. He measured the point at which the insects became both inactive and insensible, and related that pressure to an equivalent height in kilometers. He concluded that atmospheric pressure could be disregarded as a limiting factor in the release or aerial distribution of insects. Many species of Diptera, especially those that are bloodsucking, show increased biting activity when the barometric pressure falls rapidly; however, this does not necessarily relate to those species that are most often redistributed by wind.

Surface and upper winds

Wellington (1945b) has presented an excellent discussion and summary of the role of surface and upper winds in the distribution of insects in the atmosphere. Most of his discussion, however, relates to deposition rather than to the release of insects into the airstream, and will be covered later in this section. In reviewing the literature one can generalize that high-velocity winds are more inhibiting to the release of arthropods for dispersal; very slight breezes appear to stimulate both the flight behavior of winged forms and the dispersive behavior of those species that are passively carried by the wind. Jensen and Wallin (1965) reviewed the factors affecting the flight behavior of aphids and stated that, in controlled wind experiments, aphids were able to take off in wind speeds as high as 11.2 km/h. Winds of 4.8 km/h, however, caused a delay of 4–10 h for initial takeoff, and winds of 8 km/h delayed flight for up to 24 h. Once in the air, aphids immediately lost control of their flight direction and were blown away by velocities as low as 1.9–2.4 km/h.

Richter (1970) noted that aeronaut spiderlings exhibited most "tiptoe" behavior when wind speeds were between 0.35 and 1.70 m/s. Duffey (1956) found that maximum aerial dispersal of adult spiders occurred during a spell of mild, sunny weather with a moderate breeze (<8 km/h).

McManus (1973) found that newly hatched gypsy moth larvae more actively drop on silken threads and are carried off by wind when velocities are less than 8 km/h. As velocity increases above that point, the larvae respond by tenaciously adhering to the foliage and stems of the host plant.

In summary, surface and upper winds are important in the transport and distribution of airborne arthropds; however, relatively calm, warm conditions are most stimulating to the release of individuals into the atmosphere.

Thermal convection

Thermal convection is essentially a product of unequal surface heating. If surface wind speed is low, small masses of warm air may coalesce and result in "thermal bubbles" that move upward through the cooler, denser air. Under proper atmospheric conditions, thermal acceleration produces vertical velocities of the order of >8 m/s. Such velocities are adequate to vertically transport minute insects and weak fliers that are common to the plankton zone (Wellington 1945c). This is mainly a diurnal phenomenon and therefore those insects and other arthropods that are active during the day, especially during the late morning and early afternoon hours, may be vertically transported to high altitudes.

Convection is probably the most important process in the dispersal of weak fliers such as aphids, and lepidopterous larvae and spiderlings that are passively blown on strands of silk. Since convective currents are of relatively short duration, however, and are limited in area, they usually do not result in effective long-range transport.

The Effects of Dispersal on the Viability of Windblown Insects and Other Arthropods at Deposition

"Despite the fact that there is a sizable aerial population of insects, it is clear that a given successful distribution must overcome many hazards. The chances of a single insect alighting in a suitable habitat are not good. If this insect must be capable of carrying on the race, the outlook is even darker" (Wellington 1945d). Wellington stated further that "the long-range distribution by winds may provide many interesting or spectacular examples of species collected hundreds of miles from their points of origin, but the process cannot be of much value for successful distribution or for survival."

Dispersal is obviously a very hazardous process for many species of arthropds. The probability of survival is low for those species that are passively blown by the wind and even lower for those vagrants that are wingless. Similarly, the probability of survival is higher for those airborne insects in the terrestrial zone than for those carried into the plankton zone or higher altitudes. Some of the factors that affect the probability of survival at deposition are discussed briefly.

Lethal effects of the atmosphere

Wellington (1945a) conducted a series of laboratory experiments designed to identify the environmental conditions that would limit flight or become lethal to airborne insects. He concluded that temperature was the limiting factor in determining whether an insect could attain or survive high-altitude dispersal, since insects could not sustain flight if they were cooled below their minimum flight temperature. Any insect that is cooled below the threshold temperature for flight folds its wings in a normal resting position and then falls in a trajectory dependent upon both its terminal velocity and the vertical and horizontal components of the wind. Wellington estimated that this condition would occur at the zero isotherm, or at a height of 3–4.5 km under summer conditions in temperate North America. Soft-bodied insects such as aphids can withstand temperatures of 0°C for only a short period of time and therefore could be killed at higher altitudes. Other insects, however, can withstand repeated freezing (with ice accretion) and thawing that might be encountered in thundercloud conditions.

Convection is the only process by which an insect is likely to be exposed to lethal atmospheric conditions, although the very nature of the process normally does not expose insects to such conditions for an extended period of time. Wellington (1945c) states that convection can on occasion continue above the zero isotherm, but is normally damped out before it reaches that height. Taylor (1960) found that 99% of the insects, many of them aphids, recovered at altitudes between 305 and 1219 m were alive and capable of reproduction. One can therefore conclude that the atmosphere as a medium for dispersal of arthropods is probably not a limiting factor to the redistribution of viable individuals.

Availability of suitable habitats

According to Wellington (1945d), the successful distribution of an insect presents problems analogous to those of logistics in that the right stage of the species must arrive in the right place at the right time. For the successful establishment of most insect species, the correct stage may be a gravid female, sufficient males and females to ensure mating, or many immature forms. Those species that are vagrants, i.e., that are carried passively by the wind, have a low probability of landing on an acceptable host; this handicap may be overcome if the host is distributed over a large contiguous geographical area.

The gypsy moth disperses as newly hatched larvae in the spring of the year. All larvae disperse and losses due to dispersal are probably high; however, the current distribution of the gypsy moth in the eastern United States is superimposed on almost continuous scrub-oak forest. The species composition of this forest is made up of three to four species of oak and other species such as birch and aspen, all of which are preferred hosts of the insect. Consequently, the presence of an unlimited supply of preferred food greatly mediates the losses that one might expect from larval dispersal.

The spruce budworm in Maine and eastern Canada is also closely associated with its hosts, spruce and balsam fir. Pure spruce-fir forests dominate the landscape and the insect has an unlimited source of preferred food. Whereas major redistribution of the gypsy moth occurs in the larval stage (the females, though winged do not fly), the spruce budworm has a three-phase mechanism for dispersal. Newly emerged larvae disperse on silken threads in August and again in May of the following year. The female moths then disperse in July. According to Morris et al. (1958), losses due to larval dispersal are clearly related to the relative continuity of the forest; in typical mature balsam fir stands larval mortality through dispersal averages 80%, whereas in extremely dense balsam fir stands it may be as low as 60%. In mixed-species stands or in small stands that are isolated by clearcutting, mortality may exceed 90%. The figures clearly demonstrate that this form of dispersal is costly to the species and that the continuity of habitat greatly influences the extent of the losses. The budworm apparently compensates for these losses when the gravid female moths disperse, sometimes in massive numbers, to new areas.

Those species that can fly, even if poorly, can search out acceptable hosts after being deposited by the wind. This potential for redispersal exists in most species, even those that are wingless, and this factor distinguishes the insects and other microfauna from other airborne particulates. Behavior that leads to dispersal may be repeated, even in the case of gypsy moth larvae or aeronaut spiderlings.

Insects that are restricted to particular agricultural crops and that are windblown for distances up to a few kilometers may have a low probability of survival upon deposition. Therefore the availability of suitable habitats is probably the most limiting factor to the successful establishment of dispersing populations of arthropods.

SOURCE AND RELEASE OF AIR POLLUTANTS*

Although this volume deals primarily with biological material in the atmosphere, it is also important to consider in the context of aerobiology the nonbiological material in the atmosphere, much of which we consider as the typical air pollutants. Air pollutants include a wide array of toxic substances, including gases, particulates, and radioactive materials, which affect our food and fiber supply, health, and economy. They also interact with many of the biological particulates in the atmosphere, many of which can themselves be considered as pollutants (e.g., ragweed pollen).

Many articles cover the topic of sources of air pollution. Stern (1968) probably has the widest coverage and no attempt has been made here to do more than summarize this topic.

Air pollutants can be divided into two main categories: primary pollutants—emitted directly from identifiable sources, and secondary pollutants—those produced by interaction among two or more normal constituents of the atmosphere, with or without photochemical reactions (Bryson and Kutzbach 1968). Figure 2.7 gives examples of the sources of primary pollutants.

---

*R. L. Edmonds

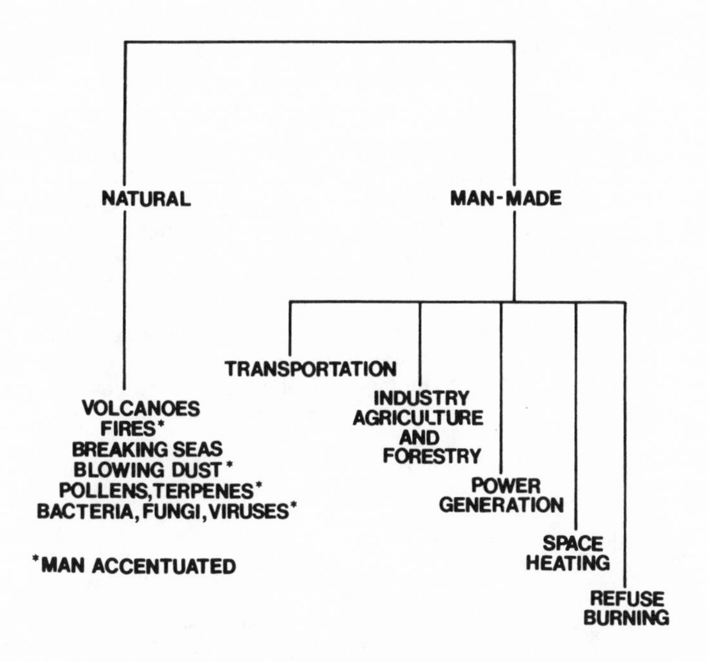

Figure 2.7. Sources of primary pollutants.

Generally, environmental pollution is a man-made or man-accentuated problem. The motor vehicle is the most important source of atmospheric pollution (100 million tons in 1970) with industrial sources second (26 million tons in 1970; Mudd and Kozlowski 1975). Steam and electricity generation produce slightly less than industry (22 million tons in 1970) and space heating produced only 9 million tons. Man-accentuated sources are also important. For example, overgrazing of deserts in northwest India and Pakistan has caused tremendous dust problems resulting in far more particulates in the air in that region than that in the air over polluted cities in the United States (Bryson and Kutzbach 1968).

The most common air pollutants are hydrocarbons, chlorine, hydrogen chloride, fluorine, hydrogen fluoride, hydrogen sulfide, oxides of nitrogen, ozone, peroxyacetyl nitrate (PAN), sulfur dioxide, carbon monoxide, and particulates. The composition of pollutants from the various sources varies, with industry emitting the most diversified pollutants. Table 2.8 shows a list of pollutants and sources. Carbon monoxide is the major primary component of pollution by automobiles; sulfur oxides are primary pollutants of industry, power generation, and space heating. Various oxides of nitrogen and hydrocarbons are also produced by combustion.

There are two major secondary types of pollutants produced by the interactions of the primary ones. The first of these is produced in a reducing type of atmosphere involving sulfur dioxide and sulfuric acid and the other is produced in an oxidizing atmosphere and involves nitrogen oxides and hydrocar-

Table 2.8. Sources of major air pollutants known to affect plants, animals, and humans.[a]

Major pollutant	Source
Sulfur oxides and sulfur dioxide	Combustion of coal, production, refining, utilization of natural gas, manufacturing and industrial utilization of sulfuric and sulfur-pulping smelting and refining of ores - Cu, Zn, Ni
Nitrogen oxides	Coal, fuel oil, natural gas, gasoline, combustion, petroleum refining, incineration of organic wastes, manufacture of HNO_3, H_2SO_4, paint, rubber, soap, nylon intermediates
Carbon monoxide	Combustion of gasoline, coal, petroleum refining, hydrocarbons
Fluorine and hydrogen fluoride	Al reduction phosphate fertilizer manufacture brick plants pottery and ferro enamel-works steel manufacture refineries rocket fuel combustion
Chlorine and HCI chlorine	Refineries glass making waste incineration scrap buring disinfection of pools HCI-incineration of polyvinyl chlorides
Hydrocarbons	Motor vehicles, plants (terpenes)
Ozone and oxidants	Upper atmosphere electrical storms photochemical reactions
Particulates	Combustion of coal, gasoline, fuel oil, cement production, lime kiln operation, incineration, agriculture
Fe, Mn, Ti[b]	Heavy industry, steel making
Pb, NO_3 and benzene soluble particulates	Internal combustion engines
SO_4, suspended particulates	Fuel burning - e.g., coal
Ni, Pb	Petroleum refining
Zn-Sn, Cr, Cu	Plating processes

[a]After Mudd and Kozlowski (1975), Stern (1968), and Wood (1968). [b]Manning (1976).

bons. London smog is a good example of the first type. Los Angeles smog is an example of the second (and newer) type. The nitrogen oxides and hydrocarbons in automobile exhaust react in the presence of ultraviolet radiation to produce ozone, PAN, and other oxidation products.

There are many other types of urban air pollution (e.g.,the particulates) not included in the above two categories, and they can be classified on the basis of their chemical composition as indicated in Table 2.8.

Although air pollutants are most commonly thought of as emanating from ground-based sources, there are several ways in which pollutants can be injected into the upper atmosphere. Aircraft and rocket exhausts may contribute in this area and we currently know little about dispersion and removal of pollutants at this level.

Airborne contamination from nuclear weapons tests and vented radioactive wastes from reactors also present additional problems. As we increasingly use nuclear energy the latter problem could increase in scope.

FOOTNOTE

[1]Contribution No. 1273-B, Division of Biology, Kansas Agricultural Experiment Station, Manhattan, Kansas 66506.

LITERATURE CITED

Akers, T. G. 1969. Survival of airborne virus, phage and other minute microbes. Pages 296–339 *in* R. L. Dimmick and A. B. Akers, eds. An introduction to experimental aerobiology. Wiley-Interscience, New York.

Alasoadura, S. O. 1963. Fruiting in *Sphaerobolus* with special reference to light. Ann. Bot. 27:125–145.

Anderson, R., B. Bergström, and B. Bucht. 1973. Outdoor sampling of airborne bacteria: Results and experiences. Pages 58–62 *in* J. F. P. Hers and K. C. Winkler, eds. Airborne transmission and airborne infection. Oosthoek Publ. Co., Utrecht, The Netherlands.

Andrewartha, H. G., and L. C. Birch. 1954. The distribution and abundance of animals. Univ. Chicago Press, Chicago. 782 p.

Austin, B. 1968. Effects of air speed and humidity changes on spore discharge in *Sordaria fimicola*. Ann. Bot. 32:251–260.

Beebe, J. M., and G. W. Pirsch. 1958. Response of air-borne species of *Pasteurella* to artificial radiation simulating sunlight under different conditions of relative humidity. Appl. Microbiol. 6.

Benbough, J. E. 1967. Death mechanisms in airborne *Escherichia coli*. J. Gen. Microbiol. 47:325–333.

Benninghoff, W. S., G. F. Estabrook, J. B. Harrington, H. Nichols, and A. M. Solomon. 1973. Report of palynology group. Page 84 *in* R. L. Edmonds and W. S. Benninghoff, eds. Ecological systems approaches to aerobiology. III. Further model developments. US/IBP Aerobiol. Program Handb. 4. (NTIS no. AP-USIBP-H-73-4.) Univ. Michigan, Ann Arbor.

Berland, L. 1935. Premiers résultats de mes recherches en avion sur la faune et la flore atmosphériques. Ann. Entomol. Soc. France 104:73–96.

Bernstein, I. L., and R. S. Safferman. 1973. A model system for studying the aerobiology and biomedical impact of green algae. Page 156 *in* R. L. Edmonds and W. S. Benninghoff, eds. Ecological systems approaches to aerobiology. II. Development, demonstration, and evaluation of models. US/IBP Aerobiol. Program Handb. 3. (NTIS no. AP-USIBP-H-73-3.) Univ. Michigan, Ann Arbor.

Bernstein, M. E., H. M. Howard, and G. C. Carroll. 1973. Fluorescence microscopy of Douglas fir foliage epiflora. Can. J. Microbiol. 19:1129–1130.

Blackwall, J. 1827. Observations and experiments made with a view to ascertain the means by which the spiders that produce gossamer effect their aerial excursions. Trans. Linn. Soc. (Lond.) 15:449–459.

Blanchard, D. C. 1972. The borderland of burning bubbles. Saturday Rev. 55: 60–63.

Blanchard, D. C., and L. Syzdek. 1970. Mechanics for the water-to-air transfer and concentration of bacteria. Science 170:626–628.

Blanchard, D. C., and L. D. Syzdek. 1972. Concentration of bacteria in jet drops from bursting bubbles. J. Geophys. Res. 77:501–513.

Blanchard, D. C., and L. D. Syzdek. 1974. Bubble tube: Apparatus for determining rate of collection of bacteria by an air bubble rising in water. Limnol. Oceanogr. 19:133–138.

Boyer, W. D. 1973. Air temperature, heat sums, and pollen shedding phenology of long leaf pine. Ecology 54:420–426.

Brannen, J. P., and D. M. Garst. 1972. Dry heat inactivation of *Bacillus subtilis* var. *niger* spores as a function of relative humidity. Appl. Microbiol. 23:1125–1130.

Broadbent, L. 1948. Aphis migration and the efficiency of the trapping method. Ann. Appl. Biol. 35:379–394.

Broadbent, L. 1949. Factors affecting the activity of alatae of the aphids *Myzus persicae* (Sulzer) and *Brevicoryne brassicae* (L.). Ann. Appl. Biol. 36:40–62.

Brown, C. E. 1958. Dispersal of the pine needle scale, *Phenacaspis pinifoliae* (Fitch) (Diaspididae:Homoptera). Can. Entomol. 90:685–690.

Brown, R. M., Jr. 1971. Studies of Hawaiian freshwater and soil algae. I. The atmospheric dispersal of algae and fern spores across the island of Oahu, Hawaii. Pages 175–188 *in*, Contributions in phycology. Allen Press, Lawrence, Kans.

Brown, R. M., Jr., D. A. Larson, and H. C. Bold. 1964. Airborne algae: Their abundance and heterogeneity. Science 143:583–585.

Bryson, R. A., and J. E. Kutzbach. 1968. Air pollution. Assoc. Am. Geogr., Comm. Coll. Geogr. Washington, D.C. 42 p.

Buller, A. H. R. 1909. Researches on fungi. Longmans, Green, & Co., New York. 287 p.

Buller, A. H. R. 1934. Researches on fungi, vol. 6. Longmans, Green, & Co., New York. 513 p.

Burgess, A. F. 1913. The dispersion of the gypsy moth. USDA Bur. Entomol. Bull. 119. 62 p.

Butcher, S. S., and R. J. Charlson. 1972. An introduction to air chemistry. Academic Press, New York. 241 p.

Campbell, K. G. 1960. Preliminary studies in population estimation of two species of stick insects (Phasmatidae, Phasmatodea) occurring in plague numbers in highland forest areas of southeastern Australia. Proc. Linn. Soc. (NSW) 85:121–137.

Carson, J. L. 1975. Investigations of the soil and airborne algae of Hawaii. Ph.D. dissertation, Univ. North Carolina, Chapel Hill.

Chamberlain, A. C. 1967. Deposition of particles to natural surfaces. Pages 138–164 *in* P. H. Gregory and J. L. Monteith, eds. Airborne microbes. Cambridge Univ. Press, Cambridge, England.

Chatigny, M. A., A. Wolochow, W. R. Leif, and J. Hebert. 1973. The toxicity of nitrogen oxides for an airborne microbe: Effects of relative humidity test procedures and contaminant and composition of spray suspension. Pages 94–97 *in* J. F. P. Hers and K. C. Winkler, eds. Airborne transmission and airborne infection. Oosthoek Publ. Co., Utrecht, The Netherlands.

Collins, C. W. 1915. Dispersion of gypsy moth larvae by the wind. USDA Bull. 273. 23 p.

Collins, C. W. 1917. Methods used in determining wind dispersion of the gypsy moth and some other insects. J. Econ. Entomol. 10:170–177.

Constantine, D. G. 1969. Airborne microorganisms: Their relevance to veterinary medicine. Chapter 19 *in* R. L. Dimmick and A. B. Akers, eds. An introduction to experimental aerobiology. Wiley-Interscience, New York.

Cox, C. S. 1963. Protecting agents and their mode of action. Pages 345–368 *in* R. L. Dimmick, ed. First symposium on aerobiology. Naval Biol. Lab., U.S. Naval Supply Cent., Oakland, Calif.

Cox, C. S. 1966a. The survival of *Escherichia coli* atomized into air and nitrogen from distilled water and from solutions of protecting agents, as a function of relative humidity. J. Gen. Microbiol. 43:383–399.

Cox, C. S. 1966b. The survival of *Escherichia coli* in nitrogen atmospheres under changing conditions of relative humidity. J. Gen. Microbiol. 45:283–288.

Cox, C. S. 1968a. The aerosol survival and cause of death of *Escherichia coli* K12. J. Gen. Microbiol. 54:169–175.

Cox, C. S. 1968b. The aerosol survival of *Escherichia coli* B in nitrogen, argon, and helium atmospheres and the influence of relative humidity. J. Gen. Microbiol. 50:139–147.

Cox, C. S. 1969. The cause of loss of viability of airborne *Escherichia coli* K12. J. Gen. Microbiol. 57:77–80.

Cox, C. S. 1971. Aerosol survival of *Pasteurella tularensis* disseminated from wet and dry states. Appl. Microbiol. 21:483–486.

Cox, C. S., and P. Baldwin. 1967. The toxic effect of oxygen upon the aerosol survival of *Escherichia coli* B. J. Gen. Microbiol. 49:115–117.

Cox, C. S., S. J. Gagen, and J. Baxter. 1974. Aerosol survival of *Serratia marcescens* as a function of oxygen concentration, relative humidity, and time. Can. J. Microbiol. 20:1529–1534.

Dark, F. A., and D. S. Callow. 1973. The effect of growth conditions on the survival of airborne *E. coli*. Pages 97–99 *in* J. F. P. Hers and K. C. Winkler, eds. Airborne transmission and airborne infection. Oosthoek Publ. Co., Utrecht, The Netherlands.

Davis, R. B., and T. Webb, III. 1975. The contemporary distribution of pollen in eastern North America: A comparison with the vegetation. Quat. Res. 5:395–434.

Dimmick, R. L. 1960. Delayed recovery of airborne *Serratia marcescens* after short-time exposure to ultraviolet irradiation. Nature (Lond.) 187:25–252.

Dimmick, R. L., and A. B. Akers, eds. 1969. An introduction to experimental aerobiology. Wiley-Interscience, New York. 494 p.

Dimmick, R. L., and R. J. Heckly. 1969. Theoretical aspects of microbial survival. Pages 347–374 *in* R. L. Dimmick and A. B. Akers, eds. Introduction to experimental aerobiology. Wiley-Interscience, New York.

Dimmick, R. L., R. J. Heckly, and D. P. Hollis. 1961. Free-radical formation during storage of freeze-dried *Serratia marcescens*. Nature (Lond.) 192:766–777.

Dorsey, E. L., R. F. Berenat, and E. L. Neff, Jr. 1970. Effect of sodium fluorescein and plating medium on recovery of irradiated *Escherichia coli* and *Serratia marcescens* from aerosols. Appl. Microbiol. 20:834–838.

Druett, H. A. 1973a. Effect on the viability of microorganisms in aerosols of the rapid rarefaction of the surrounding air. Pages 90–94 *in* J. F. P. Hers and K. C. Winkler, eds. Airborne transmission and airborne infection. Oosthoek Publ. Co., Utrecht, The Netherlands.

Druett, H. A. 1973b. The open air factor. Pages 141–149 *in* J. F. P. Hers and K. C. Winkler, eds. Airborne transmission and airborne infection. Oosthoek Publ. Co., Utrecht, The Netherlands.

Duffey, E. 1956. Aerial dispersion in a known spider population. J. Anim. Ecol. 25:85–111.

Duguid, J. P. 1945. The numbers and sites of origin of the droplets expelled during expiratory activities. Edinb. Med. J. 52:385.

Ebell, L. F., and R. L. Schmidt. 1964. Meteorological factors affecting conifer pollen dispersal on Vancouver Island. Dep. For. Publ. 1036. For. Res. Branch, Can. Dep. For. 30 p.

Ehrenberg, C. G. 1844. Bericht über die zur Bekanntmachung geeigneten Verhandlungen der Königl. Preuss. Akad. Wiss. (Berlin) 9:194–197.

Ehrlich, R., S. Miller, and R. L. Walker. 1970a. Relationship between atmospheric temperature and survival of airborne bacteria. Appl. Microbiol. 19:245–249.

Ehrlich, R., S. Miller, and R. L. Walker. 1970b. Effects of atmospheric humidity and temperature on the survival of airborne *Flavobacterium*. Appl. Microbiol. 20:884–887.

Elforde, N. J., and J. Van den Ende. 1942. An investigation of the merits of ozone as an aerial disinfectant. J. Hyg. 42:240.

Elton, C. S. 1927. Animal ecology. 3rd ed. (1947). Macmillan Co., New York.

Erdtman, G. 1943. An introduction to pollen analysis. Chronica Botanica, Waltham, Mass. (Reprinted 1954 by Ronald Press, New York.) 239 p.

Eversmeyer, M. G., and C. L. Kramer. 1975. Air-spora above a Kansas wheat field. Phytopathology 65:490–492.

Felt, E. P. 1928a. Dispersal of insects by air currents. N.Y. State Mus. Bull. 274:59–129.

Felt, E. P. 1928b. Insect inhabitants of the upper air. Trans. 4th Int. Congr. Entomol. V. 2:869–872.

Felt, E. P., and K. F. Chamberlain. 1935. The occurrence of insects at some height in the air, especially on the roofs of high buildings. N.Y. State Mus. Circ. 17. 70 p.

Freeman, J. A. 1945. Studies in the distribution of insects by aerial currents. The insect population of the air from ground level to 300 feet. J. Anim. Ecol. 14: 128–154.

Fulton, J. D. 1966a. Microorganisms of the upper atmosphere. IV. Microorganisms of a land air mass as it traverses an ocean. Appl. Microbiol. 14:241–244.

Fulton, J. D. 1966b. Microorganisms of the upper atmosphere. V. Relationship between frontal activity and the micropopulation at altitude. Appl. Microbiol. 14: 245–250.

Fulton, J. D. 1966c. Microorganisms of the upper atmosphere. III. Relationship between altitude and micropopulation. Appl. Microbiol. 14:237–240.

Fulton, J. D., and R. B. Mitchell. 1966. Microorganisms of the upper atmosphere. II. Microorganisms in two types of air masses at 690 meters over a city. Appl. Microbiol. 14:232–236.

Geiger, R. 1966. The climate near the ground. 2nd ed. Harvard Univ. Press, Cambridge, Mass. 611 p.

Glick, P. A. 1939. The distribution of insects, spiders and mites in the air. USDA Tech. Bull. 673. 150 p.

Glick, P. A. 1942. Insect population and migration in the air. Pages 88–98 *in* S. Moulton, ed. Aerobiology. AAAS Publ. 17. Am. Assoc. Advan. Sci., Washington, D.C.

Goff, G. D., J. C. Spendlove, A. P. Adams, and P. S. Nicholes. 1973. Emission of microbial aerosols from sewage treatment plants that use trickling filters. Health Serv. Rep. 88:640–652.

Goldberg, L. J., and I. Ford. 1973. The function of chemical additives in enhancing microbial survival in aerosols. Pages 86–89 *in* J. F. P. Hers and K. C. Winkler, eds. Airborne transmission and airborne infection. Oosthoek Publ. Co., Utrecht, The Netherlands.

Goldberg, L. J., H. M. S. Watkins, E. E. Boerke, and M. A. Chatigny. 1958. The use of a rotating drum for the study of aerosols over extended periods of time. Am. J. Hyg. 68:85–93.

Goodman, R. N., J. S. Huang, and P. Huang. 1974. Host-specific phytotoxic polysaccharide from apple tissue infected by *Erwinia amylovora*. Science 183: 1081–1082.

Gordon, J. E., ed. 1965. Control of communicable diseases in man. American Public Health Assoc., Washington, D.C. 282 p.

Gray, P., ed. 1961. The encyclopedia of the biological sciences. Reinhold & Co., New York. 1119 p.

Greenbank, D. O. 1957. The role of climate and dispersal in the initiation of outbreaks of the spruce budworm in New Brunswick. II. The role of dispersal. Can. J. Zool. 35:385–403.

Gregory, P. H. 1964. Problems of sampling for atmospheric microbes. Proc. Conf. Atmos. Biol. p. 165–169. Univ. Minnesota, Minneapolis.

Gregory, P. H. 1966. The fungus spore: What it is and what it does. Pages 1–13 *in* M. F. Madelin, ed. The fungus spore. Butterworth, London.

Gregory, P. H. 1973. The microbiology of the atmosphere, 2nd ed. Halsted Press Div., John Wiley, New York. 377 p.

Gregory, P. H., E. J. Guthrie, and E. Bunce. 1959. Experiments on splash dispersal of fungus spores. J. Gen. Microbiol. 20:328–354.

Haard, R. T., and C. L. Kramer. 1970. Periodicity of spore discharge in the Hymenomycetes. Mycologia 62:1145–1169.

Hambleton, P. 1970. The sensitivity of Gram-negative bacteria, recovered from aerosols, to lysozyme and other hydrolytic enzymes. J. Gen. Microbiol. 61: 197–204.

Hardy, A. C., and P. S. Milne. 1937. Insect drift over the North Sea. Nature 139:510–511.

Hardy, A. C., and P. S. Milne. 1938a. Studies in the distribution of insects by aerial currents: Experiments in aerial townetting from kites. J. Anim. Ecol. 7:199–229.

Hardy, A. C., and P. S. Milne. 1938b. Aerial drift of insects. Nature (Lond.) 141:602–603.

Harper, G. J. 1973. The influence of urban and rural air on the survival of microorganisms exposed on microthreads. *In* J. F. P. Hers and K. C. Winkler, eds. Airborne transmission and airborne infection. Oosthoek Publ. Co., Utrecht, The Netherlands.

Harrington, J. B., and K. Metzger. 1963. Ragweed pollen density. Am. J. Bot. 50:532–539.

Hatch, M. T., and R. L. Dimmick. 1966. Physiological responses of airborne bacteria to shifts in relative humidity. Bacteriol. Rev. 30:597–602.

Hess, G. E. 1965. Effects of oxygen on aerosolized *Serratia marcescens*. Appl. Microbiol. 13:781–787.

Hires, C. S. 1965. Spores, ferns, microscopic illusions analyzed, vol. 1. Mistaire Labs., Milburn, N.J. 548 p.

Hirst, J. M. 1953. Changes in atmospheric spore content: Diurnal periodicity and the effects of weather. Trans. Br. Mycol. Soc. 36:375–393.

Hirst, J. M., and O. J. Stedman. 1962. The epidemiology of apple scab (*Venturia inaequalis* [Cke.] Wint.). II. Observations on the liberation of ascospores. Ann. Appl. Biol. 50:525–550.

Hirst, J. M., and O. J. Stedman. 1963. Dry liberation of fungus spores by raindrops. J. Gen. Microbiol. 33:335–344.

Hugh-Jones, M. E., W. H. Allan, F. A. Dark, and G. J. Harper. 1973. The evidence for the airborne spread of Newcastle disease. J. Hyg. 71:325–329.

Hugh-Jones, M. E., and P. B. Wright. 1970. Studies on the 1967–68 foot-and-mouth disease epidemic: The relation of weather to the spread of disease. J. Hyg. 68:253–271.

Hyde, H. A. 1952. Studies in atmospheric pollen. V. A daily census of pollens at Cardiff for 1943–48. New Phytol. 51:281–293.

Hyde, H. A. 1963. Pollen-fall as a means of seed prediction in certain trees. Grana Palynol. 4:217–230.

Hyde, H. A., and D. A. Williams. 1945. Studies in atmospheric pollen. II. Diurnal variation in the incidence of grass pollen. New Phytol. 44:83–94.

Ingold, C. T. 1946. Spore discharge in *Daldinia concentrica*. Trans. Br. Mycol. Soc. 29:43–51.

Ingold, C. T. 1948. The water-relations of spore discharge in *Epichloe*. Trans. Br. Mycol. Soc. 31:277–280.

Ingold, C. T. 1959a. Jelly as a water-reserve in fungi. Trans. Br. Mycol. Soc. 42:475–478.

Ingold, C. T. 1959b. Peristome teeth and spore discharge in mosses. Trans. Bot. Soc. Edinb. 38:76–88.

Ingold, C. T. 1960. Spore discharge in Pyrenomycetes. Friesia 6:148–163.

Ingold, C. T. 1965. Spore liberation. Clarendon Press, Oxford, England. 210 p.

Ingold, C. T. 1971. Fungal spores: Their liberation and dispersal. Clarendon Press, Oxford, England. 302 p.

Ingold, C. T. 1974. Spore liberation in cryptogams. Oxford Biology Readers no. 49. Oxford Univ. Press, Oxford, England. 16 p.

Ingold, C. T., and V. J. Dring. 1957. An analysis of spore discharge in *Sordaria*. Ann. Bot. 21:465–477.

Jackson, M. T. 1966. Effects of microclimate on spring flowering phenology. Ecology 47:407–416.

Jacumin, W. J., D. R. Johnston, and L. A. Ripperton. 1964. Exposure of microorganisms to low concentrations of various pollutants. J. Am. Ind. Hyg. Assoc. 25:595–600.

Jensen, R. E., and J. R. Wallin. 1965. Weather and aphids: A review. USDC Weather Bur. Tech. Note 5. Agmet. 1:1–19.

Johnson, C. G. 1951. The study of wind-borne insect populations in relation to terrestrial ecology, flight periodicity and the estimation of aerial populations. Sci. Prog. 39:41–62.

Johnson, C. G. 1954. Aphid migration in relation to weather. Biol. Rev. 29: 87–118.

Junge, C. E. 1964. Large-scale distribution of microorganisms in atmosphere. Pages 117–125 *in* Proc. Conf. Atmos. Biol. Univ. Minnesota, Minneapolis.

Kamerling, Z. 1898. Der Bewegungsmechanismus der Lebermooselateren. Flora 85:157–169.

Kauf, H. 1962. Organic condensation nuclei and weather conditions. Pages 585–588 *in* Proc. 1st Natl. Conf. Aerosols, Phys., Chem. Sci., Liblice, Czechoslovakia.

Keitt, G. W., and L. K. Jones. 1926. Studies of the epidemiology and control of apple scab. Wisc. Agric. Exp. Stn. Res. Bull. 73.

Kennedy, J. S. 1950. Host-finding and host-alternation in aphids. Pages 423–426 *in* Proc. 8th Int. Congr. Entomol.

Kennedy, J. S. 1956. Phase transformation in locust biology. Biol. Rev. 31: 349–370.

Kethley, T. W., E. L. Fincher, and W. B. Cown. 1957. The effects of low temperatures on the survival of airborne bacteria. Rep. Proj. No. 8-7958. Arct. Aeromed. Lab., Ladd AFB, Alaska. 62 p.

Kozlowski, T. T. 1971. Growth and development of trees. II. Cambial growth, root growth, and reproductive growth. Academic Press, New York. 514 p.

Kramer, C. L. 1974. Seasonality of airborne fungi. Pages 415–424 *in* H. Lieth, ed. Phenology and seasonality modeling. Springer-Verlag, New York.

Kramer, C. L., and D. L. Long. 1970. An endogenous rhythm of spore discharge in *Ganoderma applanatum*. Mycologia 62:1138–1144.

Kramer, C. L., and S. M. Pady. 1968a. Spore discharge in *Hypocrea gelatinosa*. Mycologia 60:208–210.

Kramer, C. L., and S. M. Pady. 1968b. Viability of airborne spores. Mycologia 60:448–449.

Kramer, C. L., and S. M. Pady. 1970. Ascospore discharge in *Hypoxylon*. Mycologia 62:1170–1186.

Kramer, C. L., S. M. Pady, and C. T. Rogerson. 1959a. Kansas aeromycology. III. *Cladosporium*. Trans. Kans. Acad. Sci. 62:200–207.

Kramer, C. L., S. M. Pady, C. T. Rogerson, and L. G. Ouye. 1959b. Kansas aeromycology. II. Materials, methods and general results. Trans. Kans. Acad. Sci. 62:184–189.

Lee, R. E., Jr., K. Harris, and G. Akland. 1973. Relationship between viable bacteria and air pollutants in an urban atmosphere. J. Am. Ind. Hyg. Assoc. 34: 164–170.

Lewis, B. G. 1974. On the question of airborne transmission of pathogenic organisms in cooling tower drift. Proc. Ann. Meet. Cooling Tower Inst., 28–30 January 1974, New Orleans, La.

Lidwell, O. M. 1967. Take-off of bacteria and viruses. Pages 116–137 *in* P. H. Gregory and J. L. Monteith, eds. Airborne microbes. Cambridge Univ. Press, Cambridge, England.

Lighthart, B. 1967. A bacteriological monitoring of sewage on passage through an activated sludge sewage treatment plant. Ph.D. dissertation, Univ. Washington, Seattle. 119 p.

Lighthart, B. 1968. Multipoint inoculate system. Appl. Microbiol. 16:1797–1798.

Lighthart, B. 1973. Survival of airborne bacteria in a high urban concentration of carbon monoxide. Appl. Microbiol. 25:86–91.

Lighthart, B. 1975. A cluster analysis of some bacteria in the water column of Green Lake, Washington. Can. J. Microbiol. 21:392–394.

Lighthart, B., and A. S. Frisch. 1976. Estimation of viable airborne microbes downwind from a point source. Appl. Environ. Microbiol. 31:700–704.

Lighthart, B., V. E. Hiatt, and A. T. Rossano, Jr. 1971. The survival of airborne *Serratia marcescens* in urban concentrations of sulfur dioxide. J. Air Pollut. Control Assoc. 21:639–642.

Lighthart, B., and G. A. Loew. 1972. Identification key for bacterial clusters from an activated sludge plant. J. Water Pollut. Control Fed. 44:2078–2085.

Lighthart, B., and R. T. Oglesby. 1969. Bacteriology of an activated sludge wastewater treatment plant—A guide to methodology. J. Water Pollut. Control Fed. 41, Part I:R267–R281.

Long, D. L., and C. L. Kramer. 1972. Air spora of two contrasting ecological sites in Kansas. J. Allergy Clin. Immunol. 49:255–266.

McManus, M. L. 1973. The role of behavior in the dispersal of newly hatched gypsy moth larvae. USDA For. Serv. Res. Pap. NE-267. USDA For. Serv., Northeast. For. and Range Exp. Stn., Upper Darby, Pa. 10 p.

Mandrioli, P., G. L. Puppi, N. Bagni, and F. Prodi. 1973. Distribution of microorganisms in hailstones. Nature (Lond.) 246:416–417.

Manning, W. J. 1976. The influence of ozone on plant surface microfloras. Pages 159–172 *in* C. H. Dickinson and T. F. Preece, eds. Microbiology of aerial plant surfaces. Academic Press, New York. 669 p.

Maruzzella, J. C. 1963. Antifungal properties of perfume oils. J. Pharm. Sci. 52:601–602.

Maruzzella, J. C., and N. A. Sicurella. 1960. Antibacterial activity of essential oil vapors. J. Am. Pharm. Assoc. 49:692–694.

Matthews, J. D. 1963. Factors affecting the production of seed by forest trees. For. Abstr. 24:i–xiii.

May, K. R., and H. A. Druett. 1968. A microthread technique for studying the viability of microbes in a simulated airborne state. J. Gen. Microbiol. 51:353–366.

Meredith, D. S. 1962. Some components of the air-spora in Jamaican banana plantations. Ann. Appl. Biol. 50:577–594.

Meredith, D. S. 1965. Violent spore release in *Helminthosporium turcicum*. Phytopathology 55:1009–1012.

Milburn, R. H., W. L. Crider, and S. D. Morton. 1957. The retention of hygroscopic dusts in human beings. AMA Arch. Ind. Health 16:59–62.

Miquel, M. P. 1883. Les organismes vivants de l'atmosphere. Gauthier-Villars, Paris.

Mitchell, R. 1970. An analysis of dispersal in mites. Am. Nat. 104:425–431.

Morris, R. F., C. A. Miller, D. O. Greenbank, and D. G. Mott. 1958. The population dynamics of the spruce budworm in eastern Canada. Proc. 10th Int. Congr. Entomol. 4:137–149.

Mudd, J. B., and T. T. Kozlowski, eds. 1975. Responses of plants to air pollution. Academic Press, New York. 381 p.

Muller, C. H., N. H. Muller, and B. L. Haines. 1964. Volatile growth inhibitors produced by aromatic shrubs. Science 143:471–473.

Müller, K. 1954. Die Lebermoose Europas. Rabenhorst's Kryptogamen-Flora, Bd. 6, 1. Leipzig. 756 p.

Murrell, W. G., and W. J. Scott. 1966. The heat resistance of bacterial spores at various water activities. J. Gen. Microbiol. 43:411–425.

Nawaschin, S. 1897. Über der sporenausschleuderung bei der torfmoosen. Flora 83:151–159.

Odum, E. P. 1959. Fundamentals of ecology, 2nd ed. Saunders, Philadelphia. 546 p.

Ogden, E. C., and D. M. Lewis. 1960. Airborne pollen and fungus spores of New York State. N.Y. State Mus. & Sci. Serv. Bull. 378. 104 p.

Pady, S. M. 1971. Urediospore release in *Melampsora euphorbiae, M. lini* and *Puccinia pelargonii-zonalis*. Mycologia 63:1019–1023.

Pady, S. M. 1972. Spore release in powdery mildews. Phytopathology 62: 1099–1100.

Pady, S. M., and P. H. Gregory. 1963. Numbers and viability of airborne hyphal fragments in England. Trans. Br. Mycol. Soc. 46:609–613.

Pady, S. M., and L. Kapica. 1953. Airborne fungi in the Arctic and other parts of Canada. Can. J. Bot. 31:309–323.

Pady, S. M., and L. Kapica. 1955. Fungi in the air over the Atlantic Ocean. Mycologia 47:34–50.

Pady, S. M., and C. D. Kelly. 1953. Numbers of fungi and bacteria in transatlantic air. Science 117:607–609.

Pady, S. M., and C. L. Kramer. 1960. Kansas aeromycology. VI. Hyphal fragments. Mycologia 52:681–687.

Pady, S. M., and C. L. Kramer. 1967. Sampling airborne fungi in Kansas for diurnal periodicity. Rev. Palaeobot. Palynol. 4:227–232.

Pady, S. M., and C. L. Kramer. 1969. Periodicity in ascospore discharge in *Bombardia*. Trans. Br. Mycol. Soc. 53:449–454.

Pady, S. M., and C. L. Kramer. 1971. Spore discharge in *Glomerella*. Trans. Br. Mycol. Soc. 56:81–87.

Pady, S. M., C. L. Kramer, and R. Clary. 1969. Sporulation in some species of *Erysiphe*. Phytopathology 59:844–848.

Pady, S. M., and J. Subbayya. 1970. Spore release in *Uncinula necator*. Phytopathology 60:1702–1703.

Pasteur, L. 1860. Expériences relatives aux generations dites spontanees. Compt. Rend. 50:303–307.

Pereira, M. R., and M. A. Benjaminson. 1975. Broadcast of microbial aerosols by stacks of sewage treatment plants and effects of ozonation on bacteria in the gaseous effluent. Public Health Rep. 90:208–212.

Pohl, F. 1937. Die pollenerzeugung der windblüter. Beih. Bot. Centralbl. 56: 366–470.

Pouchet, F. 1860. Atmosphérique-Moyen de rassembler dans un espace infinitement petit tous les corpuscules normalement envisibles contenus dans un volume d'air déterminé. Compt. Rend. 50:748–750.

Preece, T. P., and C. H. Dickinson. 1971. Ecology of leaf-surface microorganisms. Academic Press, New York.

Puschkarew, B. 1913. Über die Verbreitung der Süsswasser-protozoen durch die Luft. Arch. Protistenkd. 23:323–362.

Rabkin, F. B., and R. R. Lejeune. 1954. Some aspects of the biology and dispersal of the pink tortoise scale, *Toumeyella numismaticum* (Pettit and McDaniel) (Homoptera:Coccidae). Can. Entomol. 86:570–575.

Randall, C. W., and J. O. Ledbetter. 1966. Bacterial air pollution from activated sludge units. J. Am. Ind. Hyg. Assoc. 27:506–519.

Rasmussen, R. A., and F. W. Went. 1965. Volatile organic material of plant origin in the atmosphere. Proc. Natl. Acad. Sci. USA 53:215–220.

Raynor, G. S., J. V. Hayes, and E. C. Ogden. 1972. Dispersion and deposition of corn pollen from experimental sources. Agron. J. 64:420–427.

Richter, C. J. J. 1970. Aerial dispersal in relation to habitat in eight wolf spider species (*Pardosa, Araneae*, and *Lycosidae*). Oecologia 5:200–214.

Riley, R. L., and J. E. Kaufman. 1972. Effect of relative humidity on the inactivation of airborne *Serratia marcescens* by ultraviolet radiation. Appl. Microbiol. 23:1113–1120.

Ritchie, J. C., and S. Lichti-Federovich. 1963. Contemporary pollen spectra in central Canada. I. Atmospheric samples at Winnepeg, Manitoba. Pollen Spores 5: 95–114.

Ritchie, J. C., and S. Lichti-Federovich. 1967. Pollen dispersal phenomena in arctic-subarctic Canada. Rev. Palaeobot. Palynol. 3:255–266.

Roberts, J. W. 1973. The measurement, cost and control of air pollution from unpaved roads and parking lots in Seattle's Duwamish valley. M.S. thesis, Univ. Washington, Seattle. 75 p.

Rockett, T. R., and C. L. Kramer. 1974a. Periodicity and total spore production by lignicolous basidiomycetes. Mycologia 66:817–829.

Rockett, T. R., and C. L. Kramer. 1974b. The biology of sporulation of selected Tremellales. Mycologia 66:926–941.

Rockett, T. R., C. L. Kramer, and T. I. Collins. 1974. A new collector to measure total spore production. Mycologia 66:526–530.

Salisbury, J. H. 1866. On the cause of intermittent and remittent fevers, with investigations which tend to prove that these affections are caused by certain species of *Palmellae*. Am. J. Med. Sci. 51:51–75.

Sarvas, R. 1962. Investigations on the flowering and seed crop of *Pinus sylvestris*. Commun. Inst. Forstl. Fenn. 53:1–198.

Savile, D. B. O. 1965. Spore discharge in Basidiomycetes: A unified theory. Science 147:165–166.

Schlichting, H. E., Jr. 1961. Viable species of algae and Protozoa in the atmosphere. Lloydia 24:81–88.

Schlichting, H. E., Jr. 1964. Meteorological conditions affecting the dispersal of airborne algae and Protozoa. Lloydia 27:64–78.

Schlichting, H. E., Jr. 1969. The importance of airborne algae and Protozoa. J. Air Pollut. Control Assoc. 19:946–951.

Schlichting, H. E., Jr. 1971. A preliminary study of algae and Protozoa in seafoam. Bot. Mar. 14:24–29.

Schlichting, H. E., Jr. 1974a. Survival of some freshwater algae under extreme environmental conditions. Trans. Am. Microsc. Soc. 93:610–613.

Schlichting, H. E., Jr. 1974b. Periodicity and seasonality of airborne algae and Protozoa. Pages 620–634 *in* Phenology and seasonality modeling. Springer-Verlag, Heidelberg.

Schlichting, H. E., Jr. 1974c. The ejection of microalgae into the air via bursting bubbles. J. Allergy Clin. Immunol. 53:185–188.

Schlichting, H. E., Jr. 1975. Some subaerial algae from Ireland. Br. Phycol. J. 10:110–115.

Schlichting, H. E., Jr. (In press.) The release of microalgae from water surfaces. Proc. Int. Seaweed Symp. 8th. Bangor, Wales.

Schlichting, H. E., Jr., R. M. Brown, Jr., and P. E. Smith. 1972. Airborne algae of Hawaii: A model for coordinated aerobiological research. Pages 63–68 *in* R. L. Edmonds and W. S. Benninghoff, eds. Ecological systems approaches to aerobiology. II. Development, demonstration, and evaluation of models. US/IBP Aerobiol. Program Handb. 3. (NTIS no. AP-USIBP-H-73-3.) Univ. Michigan, Ann Arbor.

Schlichting, H. E., Jr., and B. A. Bruton. 1970. Some problems of pleomorphism and algal taxonomy. Lloydia 33:356–359.

Schlichting, H. E., Jr., G. S. Raynor, and W. R. Solomon. 1971. Recommendations for aerobiology sampling in a coherent monitoring system: Algae and Protozoa in the atmosphere. Pages 60–66 *in* W. S. Benninghoff and R. L. Edmonds, eds. Aerobiology objectives in atmospheric monitoring. US/IBP Aerobiol. Program Handb. 1. Univ. Michigan, Ann Arbor.

Schnell, R. C., and G. Vali. 1976. Biogenic ice nuclei. I. Terrestrial and marine sources. J. Atmos. Sci. 33:1554–1564.

Scott, D. B. M., and E. C. Lesher. 1963. Effect of some ozone on survival and permeability of *Escherichia coli*. J. Bacteriol. 85:567.

Sharp, W. M., and H. H. Chisman. 1961. Flowering and fruiting in white oaks. I. Staminate flowering through pollen dispersal. Ecology 42:365–372.

Sheldon, J. M., and E. W. Hewson. 1959. Atmospheric pollution by aeroallergens. Univ. Mich. Res. Inst. Prog. Rep. 3. Univ. Michigan, Ann Arbor. 103 p.

Sheldon, J. M., and E. W. Hewson. 1960. Atmospheric pollution by aeroallergens. Univ. Mich. Res. Inst. Prog. Rep. 4. Univ. Michigan, Ann Arbor. 191 p.

Sinha, R. J., and C. L. Kramer. 1971. Identifying hyphal fragments in the atmosphere. Trans. Kans. Acad. Sci. 74:48–51.

Skaliy, P., and R. G. Eagon. 1972. Effect of physiological age and state on survival of desiccated *Pseudomonas aeruginosa*. Appl. Microbiol. 24:763–767.

Smith, P. E. 1972. The effects of some air pollutants and meteorological conditions on airborne algae and Protozoa. Ph.D. dissertation, North Carolina State Univ., Raleigh.

Smith, P. E. 1973. The effects of some air pollutants and meteorological conditions on airborne algae and Protozoa. J. Air Pollut. Control Assoc. 35:66–71.

Sokal, R. R., and F. J. Rohlf. 1969. Biometry: The principles and practice of statistics in biological research. W. H. Freeman, San Francisco. 776 p.

Solomon, A. M. 1973. Predictive models of airborne pollen concentrations: Uncertainties in pollen production estimates. Pages 99–116 *in* R. L. Edmonds and W. S. Benninghoff, eds. Ecological systems approaches to aerobiology. II. Development, demonstration, and evaluation of models. US/IBP Aerobiol. Program Handb. 3. (NTIS no. AP-USIBP-H-73-3.) Univ. Michigan, Ann Arbor.

Solomon, A. M. 1975. A rational model for interpretation of pollen samples from arid regions. Paper presented at 8th Ann. Meet., Am. Assoc. Stratigr. Palynol. October 1975, Houston, Tex. (Abstract.)

Solomon, A. M., and H. D. Hayes. 1972. Desert pollen production. I. Qualitative influence of moisture. J. Ariz. Acad. Sci. 7:52–74.

Southwood, T. R. E. 1962. Migration of terrestrial arthropods in relation to habitat. Biol. Rev. 37:171–214.

Spallanzani, L. 1777. Opuscules de Physique animale et vegetale. Trad. Ital. Jean Sennebeir.

Spendlove, J. C. 1957. Production of bacterial aerosols in a rendering plant process. Public Health Rep. 72:176–180.

Spendlove, J. C. 1974. Industrial, agricultural, and municipal microbial aerosol problems. Dev. Ind. Microbiol. 15:20–27.

Stern, A. C., ed. 1968. Air pollution, vol. 1. Air pollution and its effects. Academic Press, New York. 694 p.

Sussman, A. S., and H. O. Halvorson. 1966. Spores: Their dormancy and germination. Harper & Row, New York. 354 p.

Taylor, L. R. 1958. Aphid dispersal and diurnal periodicity. Proc. Linn. Soc. Lond. 169:67–73.

Taylor, L. R. 1960. The distribution of insects at low levels in the air. J. Anim. Ecol. 29:45–63.

Taylor, L. R. 1974. Insect migration, flight periodicity and the boundary layer. J. Anim. Ecol. 43:225–238.

Tinsley, T. W. 1975. A new way to slay old pests. Nature 257:2–3.

Trainor, F. R., R. C. Rowland, J. C. Lylis, P. A. Winter, and P. L. Bonanomi. 1971. Some examples of polymorphism in algae. Phycologia 10:113–119.

Trust, T. J., and R. W. Coombs. 1973. Antibacterial activity of B-thujaplicin. Can. J. Microbiol. 19:1341–1346.

Visch van Overeem, M. A. 1972. Onderzoekingen naar groene organismen en protozoën in de atmosfeer. Pages 54–57 *in* Aerobiologie. Centrum voor Landbouwpublikaties en Landbouwdocumentatie, The Netherlands.

Walkey, D. G. A., and R. Harvey. 1966. Studies of the ballistics of ascospores. New Phytol. 65:59–74.

Webb, S. J. 1959. Factors affecting the viability of airborne bacteria. I. Bacteria aerosolized from distilled water. Can. J. Microbiol. 5:649–669.

Webb, S. J. 1960. Factors affecting the viability of airborne bacteria. II. The effect of chemical additives on the behavior of airborne cells. Can. J. Microbiol. 6:71–87.

Webb, S. J. 1963. The effect of relative humidity and light on air-dried organisms. J. Appl. Bacteriol. 26:307–313.

Webb, S. J. 1965. Bound water in biological integrity. C. C. Thomas, Springfield, Ill. 187 p.

Webb, S. J. 1967. The influence of oxygen and inositol on the survival of semi-dried microorganisms. Can. J. Microbiol. 13:733–742.

Webb, S. J. 1969. The effects of oxygen on the possible repair of dehydration damage of *Escherichia coli*. J. Gen. Microbiol. 58:317–326.

Webster, J. 1966. Spore projection in *Epicoccum* and *Arthrinium*. Trans. Br. Mycol. Soc. 49:339–343.

Wellington, W. G. 1945a. Conditions governing the distribution of insects in the free atmosphere. I. Atmospheric pressure, temperature and humidity. Can. Entomol. 77:7–15.

Wellington, W. G. 1945b. Conditions governing the distribution of insects in the free atmosphere. II. Surface and upper winds. Can. Entomol. 77:21–28.

Wellington, W. G. 1945c. Conditions governing the distribution of insects in the free atmosphere. III. Thermal convection. Can. Entomol. 77:44–49.

Wellington, W. G. 1945d. Conditions governing the distribution of insects in the free atmosphere. IV. Distributive processes of economic significance. Can. Entomol. 77:69–74.

Wellock, C. E. 1960. Epidemiology of Q fever in the urban East Bay area. Calif. Health 18:72–76.

Werfft, R. 1951. Über die Lebensdauer der Pollenkörner in der freien Atmosphäre. Biol. Zentralbl. 70:354–367.

Weston, W. H. 1923. Production and dispersal of conidia in the Philippine sclerosporas of maize. J. Agric. Res. 23:239–278.

White, J. H. 1919. On the biology of *Fomes applanatus*. Trans. R. Can. Inst. 12:133–174.

Whitfield, F. G. 1939. Air transport, insects and disease. Bull. Entomol. Res. 30:365–442.

Wodehouse, R. P. 1933. Atmospheric pollen. J. Allergy 3:220–226.

Wodehouse, R. P. 1939. Weeds, waste and hayfever. Nat. Hist. 63:150–163, 178.

Wolfenbarger, D. O. 1946. Dispersion of small organisms. Distance dispersion rates of bacteria, spores, seed, pollen, and insects: Incidence rates of diseases and injuries. Am. Midl. Nat. 35:1–152.

Won, W. D., and H. Ross. 1968. Behavior of microbial aerosols in a 30°C environment. Cryobiology 4:337–340.

Won, W. D., and H. Ross. 1969. Reaction of airborne *Rhizobium meliloti* to some environmental factors. Appl. Microbiol. 18:555–557.

Wood, F. A. 1968. Sources of plant-pathogenic pollutants. Phytopathology 58: 1075–1084.

Woodcock, A. H. 1948. Note concerning human respiratory irritation associated with high concentrations of plankton and mass mortality of marine organisms. J. Mar. Res. 7:56–62.

Woodcock, A. H. 1955. Bursting bubbles and air pollution. Sewage Ind. Wastes 27:1189–1192.

Wright, D. N., G. D. Bailey, and M. J. Hatch. 1968. Survival of airborne mycoplasma as affected by relative humidity. J. Bacteriol. 95:251–252.

Yarwood, C. E. 1936. The diurnal cycle of the powdery mildew *Erysiphe polygoni*. J. Agric. Res. 52:645–657.

Yarwood, C. E., and E. S. Sylvester. 1959. The half-life concept of longevity of plant pathogens. Plant Dis. Rep. 42:125–128.

Zalewski, A. 1883. Über sporenabschnurung und Sporenabfallen bei den Pilzen. Flora (Jena) 66:268–270.

Zobell, C. E. 1946. Marine microbiology. Chronica Botanica, Waltham, Mass. 240 p.

3. ATMOSPHERIC TRANSPORT

M. A. Chatigny, R. L. Dimmick, and C. J. Mason

PRINCIPLES OF ATMOSPHERIC TRANSPORT*

The transport of material by the atmosphere is brought about by the exchange of momentum between the dispersing elements and the moving medium in which they are entrained. As a result of this exchange, energy is abstracted from the atmosphere and given to the dispersing elements as kinetic energy, i.e., energy of motion, which is utilized to move an element from one position to another. Sufficient energy can be gained in this manner to cause a significant extension in range as compared with dispersion in a stationary medium. The transport process is passive in that no coherent pattern is observed in the individual movements of any given element; i.e., each "step" is random in duration and direction. As we shall see later, this randomness is a result of the fluctuations in flow that persist everywhere in the lower atmosphere. The dispersing elements particpate in these chaotic motions as if they were indistinguishable from the atmospheric elements themselves. In the mean, or on the average, coherent dispersal patterns emerge from these random motions since the atmosphere itself possesses physically meaningful averages over time and space such as mean wind direction and mean wind speed.

TURBULENCE IN THE ATMOSPHERE

The most important characteristic of the atmosphere with regard to dispersion is its turbulent structure. The interaction of weather systems on all scales results in a fluctuating wind, both in speed and direction, at any given point. For example, if we stand in one place for any length of time, we note that the wind does not blow steadily from the same direction, even on windy days. Instead, it first comes from one direction, then momentarily shifts to another in a completely unpredictable fashion. Simultaneously, the wind speed also changes abruptly and, again, unpredictably. This continuous fluctuation is called *turbulence*.

---

*Conrad J. Mason

How turbulence serves to disperse material is illustrated in Figure 3.1, which depicts what happens to a puff of material as it moves through the atmosphere and is acted upon by horizontal, turbulent eddies of various sizes. In Figure 3.1A the eddies are much smaller than the dimensions of the puff; in this case, the puff disperses only laterally as it moves in a straight line. In Figure 3.1B the eddies are much larger than the puff. Here the puff disperses very little, however it is acted upon as a whole by the eddy structure and moves as a unit in the local wind field. Finally, in Figure 3.1C, the eddies are approximately the same size as the puff; in this case, the puff undergoes severe distortion and does not necessarily move in a straight line. Since eddies of different sizes exist in the atmosphere at any one time, any given puff experiences all of the above dispersive modes simultaneously. Since the size of the puff is changing continuously, the size of eddies most effective in dispersing it also changes continuously. Eventually, the puff is completely dissipated.

A similar dispersive capability involving vertical eddies also exists so that any given quantity of material is simultaneously dispersing both laterally and vertically.

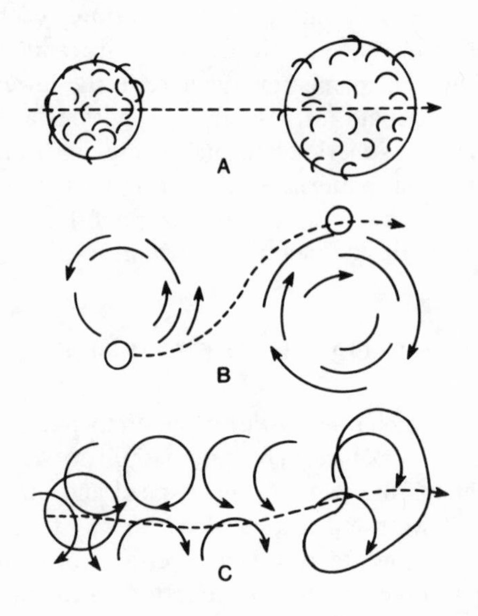

Figure 3.1. Turbulent dispersion: (A) large puff in field of small eddies; (B) small puff in field of large eddies; (C) puff in field of eddies of same size (after Slade 1968).

DISPERSION IN THE ATMOSPHERE IN RELATION TO ROUGHNESS, VERTICAL WIND SHEAR, AND VERTICAL TEMPERATURE PROFILES

The dispersive power of the atmosphere is related to *turbulent intensity,* a measure of the energy in the eddies that exist at any given time, which in general depends upon three factors. The first is the roughness of the terrain over which the wind is blowing. Roughness and obstructions to flow introduce mechanical turbulence. This type of turbulence increases as the wind speed increases, but decreases with altitude since it is generated by flow over surface features. The second factor is vertical wind shear, i.e., the change of wind speed with height. Turbulence is generated by wind shear as follows: If a parcel of air is displaced vertically to a new level, it still retains its initial speed; as a result of the wind shear, it is surrounded by neighboring parcels having either higher or lower speeds and perturbs their flow. If these perturbations are summed, the result is a chaotic field of horizontal and vertical turbulence.

The third factor, and perhaps the most important, is the vertical temperature profile of the atmosphere. Well above the surface, air mass structure determines this profile. Near the surface the temperature profile is determined principally by the *insolation* (i.e., the amount of radiant energy reaching the surface) and the wind speed. With regard to insolation, the incident radiant energy is more effective in heating the ground than the atmosphere; but, as the ground warms, it heats the layer of air immediately adjacent to it. This air becomes less dense and starts to rise; it does so in a series of eddies and chaotic motions, i.e., thermal turbulence. The more the ground is heated, the more the air above it is heated, and the higher the turbulent intensity. With regard to wind, the lower the wind speed, the higher the turbulent intensity for a given amount of insolation. In the absence of wind, the air at the surface can become heated very strongly and can rise easily; however, if the air is in motion the heating effect is not so pronounced, and less turbulence is generated. Also, wind flow tends to "even out" temperature differences of adjacent air layers because of the mechanical and wind-shear turbulence induced.

The atmosphere can be classified as *unstable* or *stable* to characterize its dispersive ability. An unstable atmosphere is one in which a small vertical displacement of an air parcel results in forces acting on the parcel that enhance or reinforce its vertical motion; as a result, turbulent motions are easily induced. On the other hand, a stable atmosphere is one in which small vertical displacements result in forces acting on the parcel that suppress its motion and restore it to its initial position; turbulence is not favored in this situation. The turbulent intensity, and consequently the dispersive ability, is greater in an unstable atmosphere as compared with a stable atmosphere.

To categorize the degree of stability of the atmosphere, its temperature profile or, more precisely, the lapse rate (i.e., the rate of change of temperature with height) is used. The relationship between lapse rate and stability can be demonstrated as follows: First, consider what would happen if a parcel of air were to be lifted from the surface without being allowed to absorb any heat from the surrounding air. Since atmospheric pressure decreases with altitude, the parcel begins to expand; and, since no heat can flow into it, the expansion causes the parcel to cool. In fact, its temperature changes at a definite rate called the *adiabatic lapse rate*, approximately equal to 1 °C/100 m. Now, if the parcel is immersed in an atmosphere whose lapse rate is equal to the adiabatic rate, then no matter how the parcel is displaced from one level to another, it always maintains the same temperature and density as the surrounding air. The cooling or warming of the parcel as it is moved up or down exactly matches the temperature gradient of the ambient environment. Consequently, there are no buoyant forces acting on the parcel either to elevate or depress it; it is stable. An atmosphere whose lapse rate is equal to the adiabatic lapse rate is said to have *neutral stability*.

It is entirely possible, however, that the ambient lapse rate may not equal the adiabatic rate. Suppose the ambient air temperature decreases with height faster than the adiabatic rate; this is called a *superadiabatic* lapse rate. Now, if a parcel of air is displaced upward, again without being allowed to exchange heat with the surrounding air, it still cools at the adiabatic rate. At the end of the displacement, however, its temperature is higher than that of the surrounding air, it is less dense and is, in fact, acted upon by a buoyant force that continues to make it rise. Once the parcel of air is put into motion, it tends to remain in motion—an unstable situation favoring the formation of eddies. Thus superadiabatic lapse rates are associated with unstable atmospheres.

On the other hand, suppose the ambient lapse rate is less than adiabatic; here, the air temperature decreases more slowly with height. In this case, at the end of an upward displacement, the parcel is colder than the surrounding air, more dense, and sinks back to its original level. This is stable behavior and suppresses turbulence; i.e., a stable atmosphere has a subadiabatic lapse rate. Finally, we can consider an extremely stable case, characterized by a temperature *inversion*. Here, instead of cooling, the air temperature actually increases with height. This is a condition of extreme stability because small upward displacements of air parcels that cool adiabatically produce temperature differentials even more intense than in the preceding example, with the parcels always cooler. Here, turbulence is suppressed even more strongly and little dispersion would occur.

Figure 3.2 depicts how stability affects dispersion, in which various plume patterns are correlated with atmospheric stability as characterized by different lapse rates. The dashed lines at the left show the adiabatic lapse rate; the solid curves are the observed profiles. For a neutral atmosphere, i.e., one with an adiabatic lapse rate, the plume disperses as shown in Figure 3.2D, developing a "cone" shape. If the lapse rate is superadiabatic, turbulent intensity is high and the plume disperses much more rapidly; Figure 3.2C illustrates dispersion under such conditions. In this case, the turbulent eddies are sometimes so large and pronounced that the plume can loop and touch the ground for short intervals of time. Figure 3.2A illustrates a subadiabatic or

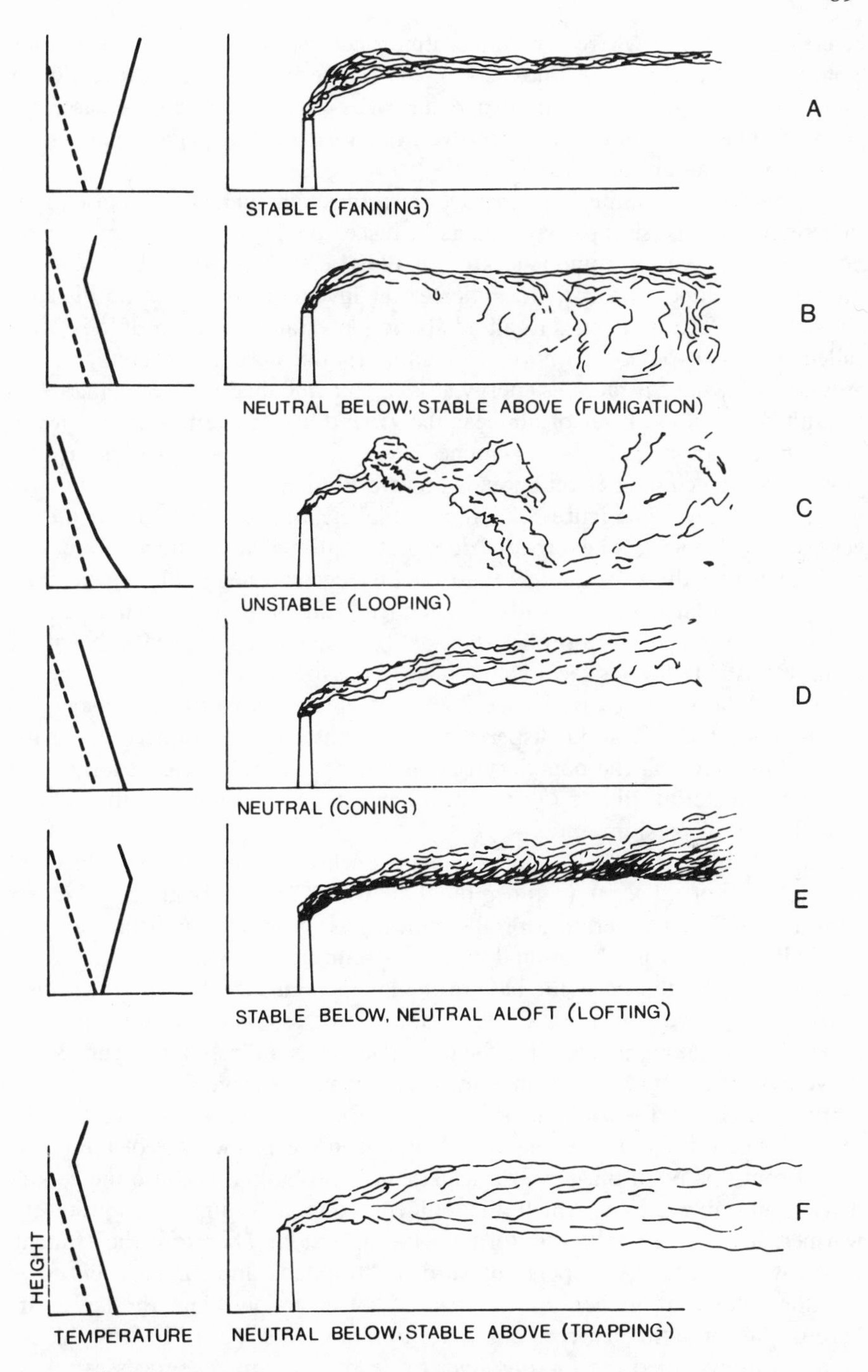

Figure 3.2. Temperature profiles, associated atmospheric stability, and plume dispersal (after Slade 1968).

inversion condition where the stable atmosphere suppresses vertical mixing. Note that, as a result, the thickness of the plume in the vertical direction is small. Since some horizontal dispersion takes place even in this case, the plume assumes a fanlike shape; however, the horizontal dispersion here is not as great as for an unstable atmosphere.

Thus far all examples are characterized by single unchanging lapse rates. In actuality, the lapse rate may change in successive layers of the atmosphere. For instance, a very common situation is illustrated in Figure 3.2E. Near the surface the temperature profile indicates an inversion, but at some distance above the surface it changes to an adiabatic lapse rate. Such a profile is often called a *surface-based radiation inversion*. In the absence of sunlight, the ground cools, since it radiates energy at a greater rate than it is receiving it. As a result, it cools the layer of air near the ground and, since this cold layer of air is more dense than the air immediately above it, it stays close to the ground. As the cooling effect spreads upward in the atmosphere, the temperature profile shown in Figure 3.2E forms; that is, a temperature inversion develops and deepens. The process does not continue indefinitely; there is a point at which the effects of ground cooling become negligible and a more nearly normal lapse rate prevails. Consequently the lapse rate changes at this point, as the temperature profile shows. The result is a stratified atmosphere with an unstable layer over a stable one. If the top of the inversion is below the top of the stack, as in Figure 3.2E, a favorable situation for dispersion occurs since there is good dispersion in the unstable atmosphere while the warm stable layer at the boundary acts as a barrier to downward transport of the plume material. Plume *lofting* occurs and there is no limit on the volume available for dispersion.

Now, it is possible to have an inversion layer aloft, that is, a layer of atmosphere not adjacent to the ground in which the temperature increases with height. The temperature profile shown in Figure 3.2F indicates an adiabatic lapse rate from the ground up to a certain height at which point an inversion starts. In this case, the base of the inversion is elevated and it acts as a barrier to upward transport. Close to the source, emission into an unstable atmosphere is taking place. Thus the dispersion is as indicated in Figure 3.2D, as we have seen, but as the plume moves downwind it begins to spread in the vertical direction (as well as the horizontal). Eventually a point will be reached where the plume boundaries begin to intercept the base of the inversion. From this point on, upward dispersion is prohibited because the cooler, denser, unstable air in which the effluent is mixed cannot penetrate the warmer, less dense, stable air forming the inversion. Therefore the effluent, which would normally disperse upward, is "trapped" and forced back down into the volume of air beneath the base of the inversion. Since the lapse rate here is adiabatic, this material continues to mix in this region. In fact, this region is simply called the *mixing layer* to denote the mixing processes taking place. Obviously the height of the mixing layer is equal to the altitude of the base of the inversion. Further downwind, the effluent continues to be mixed within the mixing layer until it becomes impossible to discern any plume outline; in other words, the effluent is now uniformly mixed throughout the mixing layer and, for example, the concentration at the surface is the same as that at the top of the mixing layer. As one moves still farther downwind of the source, the concentration decreases because horizontal dispersion continues; but if the inversion aloft persists for any period of time, then pollutant concentrations can build up beneath it.

Inversions aloft are quite common and can be formed by three different mechanisms. The first is by *burning-off* a surface-based radiation inversion such as is depicted in Figure 3.3. Suppose a surface-based inversion has formed at night (Figure 3.3A); then, in the morning at sunrise, the sun heats the soil and the layer of air adjacent to it is warmed (Figure 3.3B). As the sun continues to heat the surface, the temperature of the air adjacent to the surface and for some distance above it begins to increase (Figure 3.3C). As it does so, however, the air immediately above the warming layer is still characterized by a temperature inversion, that is, what is left of the surface-based inversion that has not as yet burned off. Thereby an inversion aloft is formed. A second mechanism is *subsidence,* in which a descending cold air mass produces an inversion aloft. The details of this process are not discussed here. Normally subsidence is found in high-pressure systems and, as such, can persist for a period of a few days. Such is the situation in Los Angeles, for example, whose summer weather patterns are dominated by the semipermanent Pacific high-pressure system and the lack of ventilation in the valley basin. Any stagnating high-pressure system, however, can cause similar episodes in other areas. The third mechanism for producing an inversion aloft is the passage of a front; this phenomenon is usually short-lived and transient.

One other situation that can influence dispersion is the "fumigation" process, in which the height of an inversion base changes with time. Initially the base of the inversion is below the height of the stack, with emission into a stable atmosphere. Then, for some reason, perhaps radiation inversion burnoff or lifting of a subsidence inversion, the base of the inversion begins to rise. At some instant of time it begins to intersect the plume. Below the inversion an unstable atmosphere exists, characterized by turbulence that mixes this part of the plume downward. As the inversion continues to rise, it reaches a height where its base is just above the top of the stack. At this time the plume is being emitted into an unstable atmosphere, is vigorously mixed downward, and is brought to the ground as depicted in Figure 3.2B. Very high concentrations can persist for short periods of time. Usually the base of the inversion continues to rise, increasing the dispersion volume and decreasing the ground-level concentration. The fumigation process is generally of short duration, perhaps lasting 10–15 min, but it can persist for an hour or more. In some localities, however, it can be an almost daily occurrence associated with the lifting of a surface-based inversion.

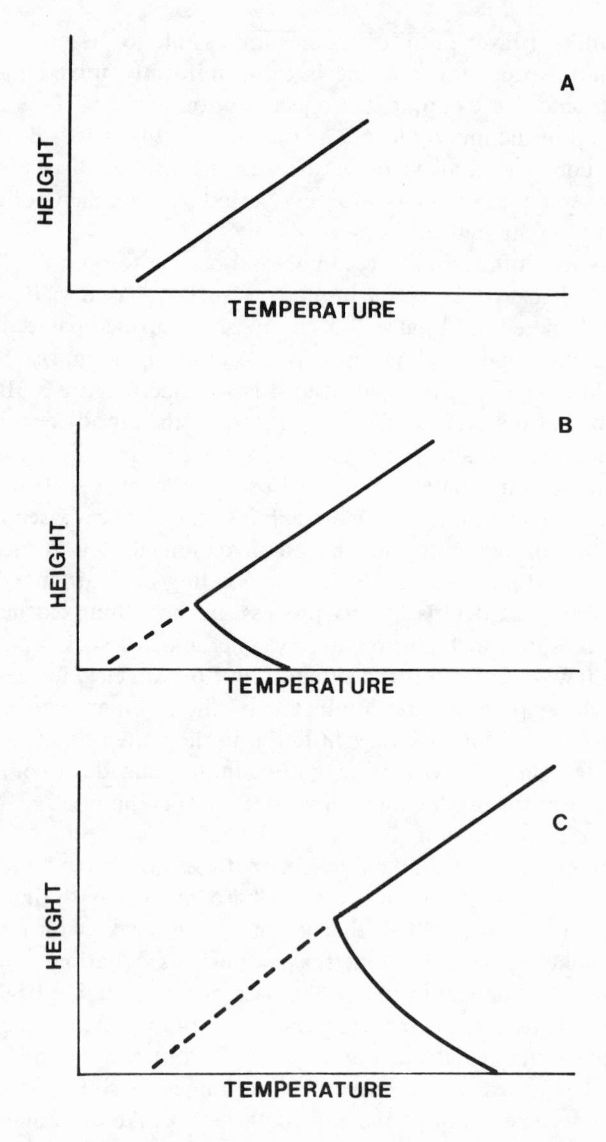

Figure 3.3 "Burning off" of a surface-based inversion.

SCALES OF ATMOSPHERIC TRANSPORT

Atmospheric transport can be conveniently categorized as microscale, mesoscale, or large-scale transport phenomena. Each of these is characterized by flow patterns possessing definite time and space scales.

Microscale Transport

Microscale transport is limited to very small time and space scales, i.e., time scales on the order of an hour or less in duration and space scales of a few hundred meters in extent. The characteristics of microscale transport are determined simply by the average wind and its turbulent fluctuations. The average wind direction defines the direction in which the transport takes place, the average wind speed determines the concentration of dispersing material (plume stretching), and the intensity of the turbulent fluctuations establishes the rate at which lateral and vertical dispersion occurs. Generally, the magnitude of the microscale time and space bounds are set by the requirement that the transport process be both time and space independent. Thus the time scale is set as the longest interval during which the meteorological conditions are unchanging while the space scale is limited to the distance over which homogeneous turbulence persists. The spatial extent of the microscale transport at a given location is determined by topographic features in the vicinity of the source, since homogeneous turbulence can be maintained only over terrain with similar surface characteristics. Obviously, the effects of diurnal variations cannot be considered on this scale. In a sense, all atmospheric transport regardless of scale is initially a microscale process; close to the source and for a suitably short time interval, any transport process is time and space independent.

In particular instances, the space and time scales for microscale transport can be extended to very large values. These exceptional cases occur only where the surface roughness (i.e., the terrain) remains homogeneous for the distance involved and when meteorological conditions are unvarying for the length of time it takes the effluent to be transported this distance. Under these circumstances, even though the time and space scales are large, the same dispersive mechanisms operate and the process is describable in microscale terms. An example of long-range transport under these conditions is given by Peterson (1968), who was able to trace the radioactive plume emitted by the Brookhaven nuclear reactor on Long Island 160 nautical miles (296 km) out over the Atlantic Ocean.

Mesoscale Transport

For mesoscale transport, the appropriate time and space scales are days and a few hundred kilometers or less. Often, within these bounds, dispersion is constrained by the existence of a well-mixed surface layer extending upward several hundred meters so that ultimately a uniform distribution (at least vertically) of effluent prevails downwind of the source. Diurnal variations in the height of the mixing layer take place, however, since, as discussed previously, daytime heating promotes vertical mixing while nighttime cooling suppresses it. Also, on this scale, the direction and speed of the wind vary with both height and location, and usually the surface roughness over which the transport occurs is not uniform. Finally, various loss mechanisms operate on the mesoscale that remove the effluent from the atmosphere. These include, for example, deposition as a result of gravitational settling or direct impact on surfaces, precipitation scavenging, and chemical reactions. A detailed discussion of these processes is presented in subsequent chapters. For the present, it is sufficient to know that loss mechanisms exist that gradually remove dispersion material from the atmosphere. In microscale transport, these sinks are usually ignored.

An excellent example of mesoscale transport, which demonstrates terrain effects as well as diurnal variation, is provided by the plume from the Trail smelter in British Columbia. Early in this century, extensive crop damage was noted in the Columbia River valley at distances up to 55 km from the smelter. Subsequent analysis of the plume showed that it was transported down the river valley at night by cold drainage winds associated with a nocturnal radiation inversion (Dean et al. 1944). Because of the extremely stable atmosphere, dispersion was at a minimum. In the morning, as the sun rose, the surface-based inversion was burned off from below causing a fumigation condition to occur almost simultaneously along the length of the valley. More recent examples of mesoscale transport can be found in the generation of "urban plumes" downwind from large urban complexes, in which the pollution from many sources becomes well mixed to form a single plume after traveling a few hundred kilometers.

Large-Scale Transport

Large-scale transport includes time and space scales sufficiently great to include global circulation patterns. In addition to the complications that must be considered for the mesoscale, here we find that the transport is essentially controlled by large-scale, three-dimensional atmospheric motions. Generally, lateral dispersion on a global scale is unpredictable and the uncertainties in estimating vertical diffusion are large because of the temporal and spatial variability in mixing height itself. The most successful approach toward understanding the long-range transport process is trajectory analysis, in which after-the-fact air trajectories are determined from prevailing circulation patterns as derived from worldwide synoptic observations. Many examples of long-range transport are in evidence. Rahn (1971) has traced the path of trace elements in aerosols from industrial areas of the United States to remote regions of northern Canada. Rodhe (1972) concludes that most of the sulfur deposited in southern Sweden has its origins in the United Kingdom and the continent of Europe. Munn and Bolin (1971) discuss many other cases in their survey paper. Perhaps some of the more spectacular instances are related by Humphreys (1940) and are associated with massive volcanic eruptions. In these events, volcanic debris was so widespread and persistent that mean surface temperatures were depressed by several tenths of a degree causing marked changes in the weather, and led to, for instance in 1816, the famous "year without a summer."

TRANSPORT OF AEROSOLS IN THE INTRAMURAL ENVIRONMENT*

The transport of aerosols in intramural environments housing human and animal activities, the sources of such aerosols, and some of the measures for their control are considered here. The aerosol particles of concern are primarily those in the respirable size range (1–8 μm) that travel with air masses and produce infections of man or animal.

Since the environmental conditions of temperature, humidity, wind velocity, and so on, prevailing in the outdoors may have a considerably wider range than those in intramural environments, translation of outdoor transport phenomena to indoor conditions is not simply one of considering a reduction in scale. For example, it is generally conceded that the modal size range of particles indoors is less than that of the projected range of 10^5 for the outdoor environment, as expressed by the formula of Junge (1963). Davies (1974), however, believes that a size distribution of atmospheric particles cannot be represented by the Junge power law as a coherent function, because this "law" is independent of the larger particles. Almost by coincidence, this is substantially true of the intramural environment. Since the source of large particles in the intramural environment is not as strong as that outdoors, we can safely confine ourselves to the respirable particle size range provided we do not ignore "very large" particles or "very small" particles, both of which have unique sources and distributional problems, and provided we recognize that neither are in major numbers nor do they usually create major problems. This assumption is further amplified in the discussion of in-hospital infections by Leedom and Loosli (chap. 6, this volume).

This rather arbitrary selection is borne out by the work of Lum and Graedel (1973), who showed that significant differences were observed in the characteristics exhibited by indoor and outdoor or natural aerosols. Their measurements indicated that indoor aerosols had a modal size distribution spanning a much narrower range than that of naturally occurring aerosols. They developed a theoretical model for indoor aerosol particle size distribution that takes into account air recirculation cycles and filtration systems used in controlled areas. The objective of their work was to develop engineering techniques by which the quality of indoor aerosol environments might be tailored to requirements for particular applications. Their tests indicated that major sources of indoor aerosols are people and their activities. They found a

*M. A. Chatigny and R. L. Dimmick

variation in particle count as much as fivefold from the average of viable particle counts. The variation is important but the correlation of numbers of particles with the number of airborne viable particles is weak at best. As shown by Oxborrow et al. (1975), such a correlation should be limited to areas where a high degree of environmental control is exercised and, even so, its relationship to airborne pathogenic organisms, which are of primary interest, is nebulous.

SOURCES OF INTRAMURAL AEROSOLS

The outdoor environment is an obvious source of intramural contamination. Gregory (1973) has shown that a wide variety of plant pathogens are readily transported in a viable and infective state throughout the environment, and their penetration into occupied spaces should be assumed. Studies of foot-and-mouth disease outbreaks in Scandanavian countries (Fogedby et al. 1960) and in England in 1967–1968 (Hugh-Jones and Wright 1970) have demonstrated the transport of that disease virus more than 60 km from the main epidemic areas.

Wellock (1960) demonstrated that Q fever was disseminated from a rendering and hide tanning plant in the San Francisco–East Bay area and some residents of housing and apartments in the downwind area became infected. Airborne etiology was clearly indicated when cases were diagnosed in people in hospitals or confined to their habitats. The rickettsia of Q fever is extremely hardy, and Delay et al. (1950) demonstrated the recovery of *Coxiella burnetii* from naturally infected airborne dust in 1950.

Adams and Spendlove (1970) have demonstrated the presence of viable coliform aerosols downwind of sewage treatment plants. Enteric pathogens have been found in such aerosols by Ledbetter and Randall (1965). There is little question that some viruses, e.g., poliomyelitis and infectious hepatitis, both of which have been demonstrated to be present in ordinary sewage, are probably also aerosolized from such plants.

Perhaps the most prolific and important sources of aerosols within usual intramural spaces are the personnel or occupants (e.g., animals, plants), both directly and indirectly. They must be considered as both carriers and disseminators of material. These factors have been studied most extensively and quantitatively in hospital air contamination with respect to the spread of nosocomial (hospital) infections. Leedom and Loosli (chap. 6, this volume) discuss many of the infectious agents, their sources, and effects of some control measures. Hambraeus (1973) showed that patients with severe burns produced staphylococcus outputs of more than 1000 colony-forming units (cfu) $m^{-3} \cdot min^{-1}$. Sellers et al. (1970) showed recovery of foot-and-mouth disease virus from the nostrils of workers handling infected animals and from sneezes by those workers.

The aerosols created by humans can be augmented by the medical treatment process. Indeed, the term *ergasteric* infection has been coined to define this phenomenon (Sutnick et al. 1971). It has been shown that substantial contamination of air can be created during the process of hemodialysis and, as might be expected, that such aerial contamination appears to influence the spread of hepatitis, an infection that rarely has been attributed to airborne virus.

It is interesting to note that, with some exceptions, the particle sizes of most such aerosols are in the respirable size range. Particles disseminated from the skin of a person tend to be in the 18- to 20-μm size range and carry multiple skin bacteria with them (Leedom and Loosli, chap. 6, this volume). Kethley (1967) has shown that even these large particles, which tend to be flake or nonspherical, are rapidly dispersed through a typical space and can remain airborne over relatively long periods of time.

Man's activities in the clinical and microbiology laboratories also produce aerosols, usually in the respirable size range. Dimmick et al. (1973) have shown that virtually every process in the microbiology laboratory produces such aerosols. In a review, Chatigny (1961) summarized the work of others that showed that almost insignificant actions involving microorganisms produced aerosols within the laboratory. In a limited epidemiological study made in 1949 and extended over some years, Sulkin and Pike (1949, 1951a,b) pointed out that from 26% to 80% of laboratory-incurred infections were not attributable to any specific action or source. They conclude that the infection must come from "incidental" aerosols and intramural transmission of the agents in question. The experiences in the early days of the study of Q fever bear this out. The etiological agent of Q fever was identified (Huebner 1947), as have been many other such agents, by infections in laboratory workers. Almost without exception, infections have occurred in laboratory workers from every pathogenic agent studied.

The frequency of ornithosis diseases in workers in turkey processing plants is well known, as is the incidence of brucellosis in workers in meat packing plants, Q fever in the abattoire, and innumerable cases of serological conversion in blood processing plants in the pharmaceutical industry. These examples relate only to overt infectious diseases. The incidence of antibody development with subsequent allergic responses in these activities and in such activities as fermentation plants producing milk products (Heldman et al. 1964), enzymes, and other products are too numerous to enumerate.

Several conclusions concerning the sources of intramural contamination can be drawn. (1) Virtually every action of animal or man that involves a microbial agent produces aerosols. (2) These aerosols, with some exceptions, are predominantly in the respirable size range, with a relatively long settling time. Accordingly, they are transported much in the manner of gases. Transport and mixing laws describing gaseous diffusion and transport apply reasonably well to such aerosols. The exception, of course, is notably the skin scale output from personnel, of consequence in hospitals, where rapid settling of these large, 18- to 20-μm particles onto wound sites has caused considerable concern and is the rationale for the imposition of severe control measures. (3) Considering the diversity of sources, method of generation, and the variety of suspending menstrua, all of which can affect the viability of airborne microbial particulates, one must conclude that it is not possible to develop a simple model for the biological degradation of the aerosols produced. We do not believe that it is an oversimplification to conclude that a substantial fraction of the particles aerosolized remains viable and infective for the periods of minutes or hours necessary for transport and deposition in the intramural environment. This is probably the only safe assumption to make if one is to evaluate the aerosol hazard in the laboratory or hospital. (4) The purity of outdoor air entering occupied spaces should not be taken for granted. While it is not expected that a substantial dosage of human pathogens is frequently encountered, it is reasonable to assume that any process sensitive to contamination (e.g., surgery, cell culture, fermentation processes) must take this source of contaminants into consideration. (5) Contamination arising from industrial processes must also be considered. Documented case histories probably reflect a small fraction of the actual numbers of incidents; however, they serve to emphasize that the problem of microbial airborne contamination is not unique to laboratories or hospitals, but is ubiquitous to the intramural environment.

TRANSPORT MECHANISMS

For the moment let us consider only particles of respirable size and assume that their travel and physical state are described adequately by the physical laws governing diffusion of gases without errors exceeding those to be expected when assessing aerosols of viable materials. This assumption has been further confirmed by the work of Allander and Strindehag (1973), who examined several different types of room air inlets including perforated ceilings providing near laminar flow. The conventional sets of parameters describing mixed reactor performance used by engineers will suffice to describe such transport phenomena. We do not attempt to repeat all the descriptive parameters, but describe a few and their applications here.

Penetration of contaminants from the outdoors is by infiltration from the wind pressure against the structure and via the intakes of the ventilating systems. Infiltration parameters are adequately described in the engineering guides (ASHRAE 1970) and some effects of wind pressure and penetration into spaces are further described by Daws (1967) and Spendlove (1975).

With respect to intramural contamination, all of these outside sources must be considered diffuse sources as differentiated from a point source within a building, the major difference being that outdoor contamination can penetrate an entire building within a very short time. It is probably not well recognized that the aerosol concentration $\times$ time relationship inside a building exposed to an aerosol cloud traveling outside is very nearly the same as the outside total. With the exception of materials removed by settling and by impaction in ductwork, against window sills, and so on (primarily those particles of the larger size), the total dose received by an occupant of a building is the same as that received by a person outside a building; i.e., dosage = (concentration × time × respiration rate × retention). The dosage outside is likely to be a high concentration for a short period of time, whereas inside a space there is usually an exponential buildup followed by an exponential reduction. The integrated area under the buildup-decay curve as f(time) is very nearly the same as that for outdoors. This has been discussed by Spendlove (1975), who described a model by Calder (1957) of the ventilation transfer process inherent in structure penetration.

Within a structure, even though it may be mechanically ventilated, some nonmechanical aspects of distribution are also effective. In a tall structure, thermal rise or chimney effect can distribute aerosols throughout a large facility in a short time. An example of this is the staphylococcus infective material that was shown by Hurst et al. (1958) to have been distributed throughout a hospital facility by means of the laundry chute. Daws (1967) has described the effects of convective currents in creation of vortices and air mixing within rooms and buildings. He considered the heat outputs from people as heating systems and demonstrated the movement of aerosols by heat sources and sinks. A high degree of mixing of aerosols occurs by convection in rooms, even though the thermal gradients from air to walls, heaters, windows, or people are relatively small. In our own work we have shown that within a room, even with no air movers used, there is a rapid diffusion of small particles from one end of the room to the other (Dimmick et al. 1973). Using point sources we were able to show that aerosols were distributed throughout rooms within a period of 10 min after an aerosol burst was created. Although there was a higher concentration of aerosol near the aerosol source for at least the first 5 min after burst generation, the cloud dissipated very rapidly; within 10 min aerosol concentration throughout a room was reasonably uniform to the degree consistent with the biological sampling accuracy available. In large structures being built currently, the trend has been to use mechanical ventilation systems. The term *mechanical ventilation* is used here as a simple description of any system that can include scrubbing, filtering, air conditioning, humidification, dehumidification, and other processing of both fresh air from outdoors and recirculated air. The ventilating system can be looked at as a mixed blessing. In control or transport of microbial particles there is no question that mechanical ventilation accelerates the distribution of particles from any point source to diverse areas of the structure, often in ways quite mystifying. Phillips (1965) described several incidents wherein the mechanical ventilation systems rapidly transported aerosols from a point source in a single room throughout entire buildings, thereby infecting not just the occupants of the room, but also visitors in reception rooms and in corridors. Miller et al. (1963) demonstrated that an aerosol generated when drilling a patient's tooth was rapidly transported throughout the entire dental suite and that people in the waiting room received a substantial dosage of the material used for tracing purposes. On the other hand, mechanical ventilation systems have a capability for the reduction of the microbial burden. The effects of treatment of incoming air are to "clean" it and to reduce the viability of airborne microbes. In recirculation modes at high flow rates, such systems can quickly remove particles from the environment (Chatigny et al. 1974). If one considers the extremes of mechanical ventilation, e.g., the clean room in which the principles of recirculation, air filtration, and so on are carried to great lengths, one can produce virtually a sterile environment, even in the presence of heavy contaminating sources. This is perhaps an exception to the rule and it may be better to consider some of the basic premises on which conventional mechanical ventilating system designs are based and their effects on intramural transport.

One premise used by most design engineers is that it is possible to achieve near-perfect mixing in each space served by the mechanical ventilation system. This is important for proper design, since one of the engineer's objectives is to provide cool or warm air throughout the space without drafts or hot spots that would be obvious to the occupants. To do this, he most frequently has the air enter high in the room, usually on or near the ceiling, and for economy and design convenience exhausts air from the same general area. The design of grills and diffusers to provide thorough air mixing without drafts has been given extensive consideration, and, suffice to say, we will concede that the engineer can do an excellent job in reaching his objectives. That the objective may be at odds with the maintenance of minimization of aerosol transport is something else again. Acceptance of these mixing criteria for ventilation systems allows use of some of the classical mixed reactor equations, e.g., $C_t = C_O(1 - \exp kt)$, where C_t = aerosol concentration at time t, C_O = initial aerosol concentration, k = removal factor, where k comprises $k_1 + k_2 + k_3 + k_4$ and k_1 = air change rate room volumes/unit time, k_2 = settling rate removal/unit time, k_3 = biological "decay" or loss of infectivity rate/unit time, $k_4 \simeq$ electrostatic action, thermal deposition, etc. (These factors are usually negligible except for small (<1.0-μm) particles.

Variations and expansions of this basic removal expression can include the assumption that C_O is a function of time and therefore reflects continuing inputs from outdoors or indoors, or the exceptional condition in which the ventilation system itself can be a source of contamination. For example, *Pseudomonas aeruginosa* has been known to proliferate in humidification systems of large ventilating devices and in the cooling coils if the space is air conditioned (Solberg et al. 1956). Recognizing the relative infrequency of these events, such analysis of mixing in ventilated spaces using these simple models is used by Kethley (1967) and by Heldman (1968), who modeled the convective movement of particle-laden air through doorways and corridors.

There are many types of mechanical ventilation systems used. These can range from a once-through 100% air supply for a hospital surgery (as has been required in the past to clear the surgery of explosive fumes; Leedom and

Loosli, chap. 6, this volume), to a single system for an entire building or complex, which can include various sized ventilating zones, each of which may be isolated by air locks or pressure gradients. Sometimes this isolation can be most helpful in providing aerosol control. The simplest and most frequently used control measure is the maintenance of the space or volume at a positive pressure to the outdoors to reduce infiltration of uncontrolled air. Given the economics of ventilation and air conditioning, this almost automatically leads to a good grade of construction to minimize leakage and losses of costly treated air.

To summarize the transporting mechanisms in intramural spaces (excepting laminar flow rooms) the following conclusions may be drawn:

1. In a space or volume without mechanical ventilation systems, there is a relatively rapid convection and thermal diffusion of respirable-size-particle aerosols throughout the volume regardless of the fact that no mechanical movement of air mass is created.

2. The outdoor air entering intramural spaces must be considered a diffuse source that will contaminate the entire interior volume, and the human or animal dose of contaminants indoors must be considered approximately equivalent to any dose received outdoors.

3. Mechanical ventilation systems increase both the rapidity of transport and removal of small-particle aerosols throughout the intramural environment. They further aid the transport of larger particles and are, for the most part, directed toward providing perfect mixing "within the spaces treated."

CONTROL AND TRANSPORT MECHANISMS

A discussion of control factors is essential because the concept is inherent in any engineered mechanical ventilating system. Design manuals treat both general principles and specific requirements for design of facilities for laboratories and hospitals. It is in the latter areas where the concept of control of the sanitary aspects of the air has been most thoroughly treated.

Although one objective of control may be the provision of highly purified air within an environment, this is usually the exception rather than the rule. In modern practice, it is customary to filter incoming air and an average value of ca 0.42 cfu/m^3 of filtered supply air is not uncommon. Infiltration from unknown sources may exist and can change this number by as much as 10^2 on a short-term basis. The practical limitation on ventilation rates can be a controlling factor in transport of infectious particles. As Kethley (1963) has pointed out,

> When sources of airborne contamination exist within enclosed spaces, the principal method of removal of such contamination is by dilution ventilation.

> Particles, even very large ones, do not go in straight predictable lines, but may be distributed randomly and rapidly throughout the room; the concentration of particles contained is dependent on the rate of production of particles and the volume flow of clean air employed for dilution; the rate of ventilation, in changes per hour, determines the rapidity with which a room may be cleared of contamination, but it is the volume flow in cubic feet per minute which determines the final concentration of airborne particles.

This is a valid generalization. Practical limitations of ventilation rates, however, limit the usual air exchange rates in rooms to something less than 20 changes per hour. Aerosol "dilution" removal rates are relatively insensitive to air change rates up to six per hour. For increased air exchange rates, i.e., (k) greater than approximately six changes per hour (with the assumption that all air is exchanged for "clean" fresh air), the classic mixing equation, i.e., [$c_t = c_0(1 - \exp -kt)$], predicts a leveling of removal rates with increasing ventilation rates. Concurrently the costs and disadvantages of the increased ventilation (e.g., energy costs, noise, space requirements for ductwork, draft control, and so on) are cumulative in effect. Thus treating air by dilution has practical limits to its effectiveness. There have been attempts to increase the effective "sanitary air exchange" rate by the use of ultraviolet radiation (Leedom and Loosli, chap. 6, this volume). The efficacy of ultraviolet (UV) is dependent on many factors. Suffice to say that UV can be effective when used under well-controlled conditions. Those conditions are probably that of air which is already clean and for which the ultraviolet is used as a "polishing" agent. In any event, discussion of UV as "equivalent air changes" (Leedom and Loosli, chap. 6, this volume) is only incidental to the discussion of transport of particles.

Foord and Lidwell (1972, 1975b) have developed a reasonable mathematical analysis of the transport by ventilation of airborne bacteria between hospital patients in partitioned wards, divided wards, and isolation units with various combinations of natural ventilation, mechanical ventilation, and air conditioning. Their data show a significant benefit in control by the use of air locks between isolation rooms and a general corridor. The assumption of perfect mixing in the individual spaces holds, however, and the quantity of air removed from air locks is the most effective factor in determining transit of particular material through the locks. Since the establishment of ventilated air locks from rooms to corridors is the exception rather than the rule in most buildings, leakage through doorways can be significant. It has been shown that a temperature difference as small as 0.1 °C across a 2-m by 1-m doorway can generate as much as 350 m^3/h of air transfer through the doorway. There appears to be a considerable difference of opinion on the effects of temperature differences across doorways. Recently Whyte and Shaw (1973a) have developed models describing such convective flow through doorways. The introduction of a substantial forced directional flow to the convective exchange through doorways or hallways can improve containment of aerosols by as much as 100-fold.

Most of the tests used gas tracers and particulate tracers, e.g., sodium iodide, fluorescent particles (Foord and Lidwell 1972). Concurrently, other studies have been done comparing these data to those gained by sampling for infectious particles in practical situations. Ayliffe et al. (1971) describe studies in a hospital ward, with sections treated in different ways. They tested separate suites of rooms, self-contained isolation suites, with plenum ventilation at 20 changes per hour, UV barriers at doorways, and some with airlocks. They traced bacteria and showed that few of these bacteria entered the plenum recirculation ventilated rooms. Bacteria released inside conventionally ventilated cubicles or rooms escaped into the corridor, and this transfer was reduced by the presence of an air lock. They concluded that a single bedroom without mechanical ventilation could provide adequate structural isolation for patients requiring special protection if they were located in a ward where the staphylococcal infection rates were already low. Although the aerial transfer of infectious organisms to patients in such rooms is apparently not prevented, the numbers of bacteria are smaller than those found in the open ward even with the doors open. Under these conditions, the importance of the transfer of infectious organisms by the nurse-patient contact mechanism as described by Lowbury et al. (1971) and Lidwell et al. (1974, 1975) becomes apparent. The generally accepted standard of approximately 176 viable infectious organisms per cubic meter for operating theaters and for rooms housing immunologically suppressed patients is probably adequate if one considers that the nurse-patient contact mechanism can effectively produce a higher infection rate than would be found at this level of airborne contamination. Further, the correlation of airborne viable organism count with airborne pathogen count is poor and, as Reid et al. (1956) have concluded, the number of pathogens present is the fact of concern. In a continuation of these early studies Foord and Lidwell (1972, 1975a,b, Lidwell et al. 1974, Lidwell 1975) modeled and analyzed airborne infection in a fully air-conditioned hospital discussing air transfer between rooms and transfer of airborne particulates along passages. Their tests using tracer gases showed that when rooms were ventilated at about six changes per hour with an excess airflow out the door of about 0.1 m/s and a temperature difference between rooms and corridors of $<0.5°C$, concentrations of the tracer in rooms close to where the tracer was being liberated were 10^3 or more lower than those in the source rooms. This ratio fell to about 200-fold in the absence of excess airflow out of each of the target rooms and doorways, and dilution along the corridors reduced concentration by approximately 10-fold for every 10 m of corridor. When similar experiments were repeated using particulate matter, the protection effect or reduction in aerosol concentration was between 4 and 25 times greater than that for gaseous contamination. This was attributed to the existence of inertial deposition of particulate matter and the settling rate of the larger particles in the hospital environment. A settling rate of about 0.3 m/min ensued, however. It is interesting to note that Lidwell (1975) developed "diffusion constants" in air of approximately 0.06 m^2/s at a drift velocity of 0.04 m/s. Although the format of his mathematical models is somewhat different from that previously described, the models are consistent with the expectation that for conventional mechanical ventilation systems directional control, mixing, and particulate settling all occur simultaneously.

In cases where the "mixed" transport of infective particles, even in diluted form, may be of severe consequence, a radically different approach may be necessary. The objectives of such systems would be to provide removal of any internally or externally generated contaminants with little or no general mixing. Such systems have been developed and extensively evaluated. Whitfield et al. (1962) developed the so-called laminar-flow concept wherein a process was maintained in a stream of clean air emanating directly from a very-high-efficiency (99.9% collection HEPA) filter. He used velocities of approximately 30.5 linear m/min to bathe the process to be protected. This "piston" of air was smooth and offered little lateral mixing. As a result, there was little contamination within this airstream and little or no flow upstream against the air current. The concept was rapidly extended to entire rooms, and standards (USPHS 1966) for so-called clean rooms of class 100, 1000, and up to 100,000 have been developed, specifying the allowed particulate burden and particle size for each class of room. Although originally used in the space industry to provide particulate-free air, the inference that particulate-free air would also be bacteriologically clean rapidly projected the use of such techniques to surgical suites and holding spaces for immunologically incompetent patients where danger of infection was extremely high. In Britain, Blowers and Crew (1960) reviewed the utility of various turbulent-type ventilation systems for surgical suites and concluded that, whatever method was used for removing airborne bacteria, it should always be secondary to efforts at preventing their dissemination. Charnley (1970) also recognized the need for control of the dissemination sources in operating rooms (OR). He further recognized that changing the air in an enclosure as many as 100 times an hour and directing removal of air from the bottom of the enclosure would not avoid mixing or provide a clear downward laminar flow displacement. The presence of equipment and occupants in these spaces destroyed any concept of "clean" laminar flow. An essential part of his concept of control of infection was to introduce vacuum extraction masks to carry away the expired air from the personnel in the OR and, later, the addition of clothing that did not permit a rapid dispersal of skin fragment source microorganisms. A number of studies on the effects of obstructions and thermals and laminar-flow systems (Whyte and Shaw 1974), partial-wall laminar-flow rooms (Whyte et al. 1974), and the "direct transfer" mechanisms (Lidwell et al. 1974) show that considerable care must be taken to ensure that such systems function properly. The stringency of air cleanliness expected in the surgical theater or OR is not applied as specifically in other situations, but laminar flow has been applied to the animal care laboratory (McGarrity et al. 1969) to pharmaceutical processing plants, and to cell culture laboratories (Barile 1973). It is interesting to note that McGarrity et al. (1969) demonstrated effective reduction of airborne microbes to near-zero levels in animal holding rooms by use of 4.6-m/min downflow laminar air supply. This is in sharp contrast to the 30.5 m/min deemed desirable by Charnley for the OR and the thermal at 18.3 m/min of particles over a human (Daws 1967). It probably reflects the effect of a low level of human activity in the animal room compared with the OR. Comments offered by Laufman (1973) and Whyte and Shaw (1973b) regarding the design and advantages of laminar-flow systems are pertinent. They looked at the laminar-flow room technique in surgery and concluded that advantages of laminar-flow ventilation in operating theaters do exist, but they are often overestimated. The 100-fold reduction in contamination levels has not been achieved, as supposedly has occurred in industrial applications. They list some deficiencies in the laminar-flow OR, and describe other techniques such as the sterile jet airflow system espoused by Allander and Able (1973) to alleviate certain deficiencies. Fox (1969), after concluding an exhaustive study of test setup and actual neurosurgery in laminar-flow operating rooms, observed that although the general level of air contamination was reduced from 3.5–353 organisms/m^3 to 3.5–353 organisms/100 m^3, there were still areas of heavy air contamination near the surgeons. He also concludes, however, that one can approach clean air concepts very closely and economically by conventional means and by such special techniques as laminar flow or clean air jets, but in the intramural environment the presence of human sources cannot be discounted.

If we are to maintain a clean environment, we must maintain an environment with clean air at positive pressure. Such a positive pressure can easily be gained by blowing up a balloon and, indeed, if one had sterile breath, the balloon would have sterile air inside. Living or working inside a balloon, or a reasonable facsimile thereof, however, would be difficult to say the least, not to mention the fact that anyone living in such a space is himself a generator of contaminants and, unless he were to lead a monastic life, he would rapidly interchange flora with any other occupants. This has been demonstrated through the most complicated of "sealed" spaces; literally, a fleet ballistic missile submarine, as reported by Watkins et al. (1970) and by Morris and Fallon (1973). In such "balloons" subdivision into smaller compartments is essential, and if we wish to subdivide into smaller "balloons" we must establish airflow velocities, directions, and concomitant pressures. *Pressure,* it must be noted, has not been described herein as a transport control mechanism. It is airflow and quantity of air moved (or treated) per unit time that is of major effect. Therefore the dilution principle must be applied. Years ago, when the concepts of mechanical ventilation systems were remote, our public buildings were cavernous, high-ceilinged affairs, wherein the volume of space per person permitted excellent dilution, particularly when this cavernous space was ventilated both by poor construction and a series of windows, usually placed both high and low, to stimulate outdoor air supply and convective currents. This was usually aided by placing the heating source below the windows with results as described by Daws (1967). At present, the cost of construction and the availability of mechanical systems have changed the basic design of many interior spaces, and we find that architects attempt to minimize the volume of spaces we inhabit and approach our "balloon" model. It is not always clear that these systems have improved air sanitation in the intramural environment. It is likely that a clear case for such a requirement will not be made. On the general principle that clean is better, we shall probably continue to press for improved air quality indoors, particularly in hospitals, pharmaceutical plants, and other similar areas.

In summary, there is adequate information to describe mechanisms, sources, and equipment affecting airborne transport of infectious microorganisms in the intramural environment. With a few exceptions, usually attributed to crowding or special conditions, the demonstration of viable infectious microbes in the air around infected patients has remained difficult, even with large-volume samplers or sentinel animals. The case for true airborne transmission of infectious disease, particularly acute respiratory disease, remains weak. Mixing and transport of airborne infectious particles occur in the intramural environment but are not the sole or major cause of the spread of infectious disease within occupied spaces. The origins of intramural aerial contamination are diverse and include diffuse sources from the outdoors, point and traveling sources on the inside, and vectors from personnel, animals, and other inhabitants in the spaces. Literally no activity of man or animal is free of the production of viable aerosols. Those actions involving the use of pathogens usually result in the aerosolization of those pathogens unless stringent precautions are taken. The rapidity of spread within a space depends in great part on the degree and rate of ventilation in a space and removal, for the most part, is dependent on the quantity of air removed and replaced with clean air (discharged or cleaned). Engineering information is available for both qualitative and quantitative description of transport and for development of aerosol control measures. Correct use of laminar-flow techniques can be an effective control measure. Such techniques require a somewhat different approach for description of aerosol transport but are subject to many of the same limitations as conventional ventilation. Problems in prediction of transport and the economical and practical application of control measures remain.

LITERATURE CITED

Adams, P. A., and J. C. Spendlove. 1970. Coliform aerosols emitted by sewage treatment plants. Science 169:1218–1220.

Allander, C. G., and E. Able. 1973. Some aspects on ventilating clean rooms. Pages 587–591 *in* J. F. P. Hers and K. C. Winkler, eds. Airborne transmission and airborne infection. Oosthoek Publ. Co., Utrecht, The Netherlands.

Allander, C. G., and O. Strindehag. 1973. Particle concentrations in patient rooms with various types of ventilation. J. Hyg. 71:633–640.

ASHRAE. 1970. Air conditioning in the prevention and treatment of disease. Pages 127–136 *in* Handbook of fundamentals. Am. Soc. Heat. Refrig. Air-Cond. Eng., New York.

Ayliffe, G. A. J., B. J. Collins, E. J. L. Lowbury, and M. Wall. 1971. Protective isolation in a single-bed room: Studies in a modified hospital ward. J. Hyg. 69:511–527.

Barile, M. F. 1973. Mycoplasmal contamination of cell cultures: Mycoplasma–virus–cell culture interactions. Pages 131–172 *in* J. Fogh, ed. Contamination in tissue culture. Academic Press, New York.

Blowers, R., and B. Crew. 1960. Ventilation of operating theatres. J. Hyg. 58: 427–448.

Calder, K. L. 1957. A numerical analysis of the protection afforded by buildings against BW aerosol attack. BWL Tech. Study no. 2, U.S. Army Biol. Lab., Fort Detrick, Md. (As cited by Spendlove 1975.)

Charnley, J. 1970. Experience with germ-free environments in surgery in relation to design. Pages 191–198 *in* I. H. Silver, ed. Aerobiology: Proceedings of the third international symposium. Academic Press, New York.

Chatigny, M. A. 1961. Protection against infection in the microbiological laboratory: Devices and procedures. Pages 131–192 *in* W. W. Umbreit, ed. Advances in applied microbiology, vol. 3. Academic Press, New York.

Chatigny, M. A., W. E. Barkley, and W. A. Vogl. 1974. Aerosol biohazard in microbiological laboratories and how it is affected by air conditioning systems. Pages 463–469 *in* Symposium on air conditioning for particulate control in industrial processes. Am. Soc. Heat. Refrig. Air-Cond. Eng., New York.

Davies, C. N. 1974. Size distribution of atmospheric particles. J. Aerosol Sci. 5:293–300.

Daws, L. F. 1967. Movement of air streams indoors. Pages 31–59 *in* P. H. Gregory and J. L. Monteith, eds. Airborne microbes. Cambridge Univ. Press, Cambridge, England.

Dean, R. S., R. E. Swain, E. W. Hewson, and G. C. Gill. 1944. Report submitted to the Trail smelter arbitral tribunal. U.S. Bur. Mines Bull. 453.

Delay, P. D., E. H. Lennette, and K. B. DeOme. 1950. Q fever in California. II. Recovery of *Coxiella burnetii* from naturally-infected airborne dust. J. Immunol. 65:211–220.

Dimmick, R. L., W. A. Vogl, and M. A. Chatigny. 1973. Potential for accidental microbial aerosol transmission in the biological laboratory. Pages 246–266 *in* A. Hellman, M. N. Oxman, and R. Pollack, eds. Biohazards in biological research. Cold Spring Harbor Lab., Cold Spring Harbor, N.Y.

Fogedby, E. G., W. A. Malmquist, O. L. Osteen, and M. L. Johnson. 1960. Air-borne transmission of foot-and-mouth disease virus. Nord. Veterinaermed. 12: 490–498.

Foord, N., and O. M. Lidwell. 1972. The control by ventilation of airborne bacterial transfer between hospital patients, and its assessment by means of a particle tracer. I. An airborne-particle tracer for cross-infection studies. J. Hyg. 70:279–286.

Foord, N., and O. M. Lidwell. 1975a. Airborne infection in a fully air-conditioned hospital. I. Air transfer between rooms. J. Hyg. 75:15–30.

Foord, N., and O. M. Lidwell. 1975b. Airborne infection in a fully air-conditioned hospital. II. Transfer of airborne particles between rooms resulting from the movement of air from one room to another. J. Hyg. 75:31–44.

Fox, D. G. 1969. A study of the application of laminar flow ventilation to operating rooms. USDHEW/PHS Monogr. 78. U.S. Gov. Print. Off., Washington, D.C. 50 p.

Gregory, P. H. 1973. The microbiology of the atmosphere, 2nd ed. Leonard Hill Books, Aylesbury, England. 377 p.

Hambraeus. A. 1973. Studies on transmission of *Staphylococcus aureus* in an isolation ward for burned patients. J. Hyg. 71:171–183.

Heldman, D. R. 1968. A stochastic model describing bacterial aerosol concentrations in enclosed spaces. J. Am. Ind. Hyg. Assoc. 29:285–292.

Heldman, D. R., T. I. Hedrick, and C. W. Hall. 1964. Air-borne microorganism populations in food packaging areas. J. Milk Food Technol. 27:245–251.

Huebner, R. J. 1947. Report of an outbreak of Q fever at the National Institutes of Health. II. Epidemiological features. Am. J. Public Health 37:431–440.

Hugh-Jones, M. E., and P. B. Wright. 1970. Studies on the 1967–68 foot-and-mouth disease epidemic. The relation of weather to the spread of disease. J. Hyg. 68:253–271.

Humphreys, C. E. 1940. Physics of the air. McGraw-Hill, New York. 614 p.

Hurst, V., M. Grossman, and F. R. Ingram. 1958. Hospital laundry and refuse chutes as sources of staphylococci cross-infection. J. Am. Med. Assoc. 167:1223–1229.

Junge, C. E. 1963. Air chemistry and radioactivity. Academic Press, New York. 382 p.

Kethley, T. W. 1963. Air: Its importance and control. Pages 35–46 *in* Proc. Natl. Conf. Inst. Acquired Inf., USDA/PHS Bull. 1188.

Kethley, T. W. 1967. Effect of ventilation on distribution of airborne microbial contamination—Laboratory studies. Pages 271–278 *in* B. R. Fish, ed. Surface contamination. Pergamon Press, New York.

Laufman, H. 1973. Confusion in application of clean air systems to operating rooms. Pages 575–580 *in* J. F. P. Hers and K. C. Winkler, eds. Airborne transmission and airborne infection. Oosthoek Publ. Co., Utrecht, The Netherlands.

Ledbetter, J. O., and C. W. Randall. 1965. Bacterial emissions from activated sludge units. Ind. Med. Surg. 34:130–133.

Lidwell, O. M. 1972. The control of ventilation of airborne bacterial transfer between hospital patients, and its assessment by means of a particle tracer. II. Ventilation in subdivided isolation units. J. Hyg. 70:287–297.

Lidwell, O. M. 1975. Airborne infection in a fully air-conditioned hospital. III. Transport of gaseous and airborne particulate material along ventilated passageways. J. Hyg. 75:45–56.

Lidwell, O. M., B. Brock, R. A. Shooter, E. M. Cooke, and G. E. Thomas. 1975. Airborne infection in a fully air-conditioned hospital. IV. Airborne dispersal of *Staphylococcus aureus* and its nasal acquisition by patients. J. Hyg. 75:445–474.

Lidwell, O. M., A. G. Towers, J. Ballard, and B. Gladstone. 1974. Transfer of micro-organisms between nurses and patients in a clean air environment. J. Appl. Bacteriol. 37:649–656.

Lowbury, E. J. L., J. R. Babb, and P. M. Ford. 1971. Protective isolation in a burns unit: The use of plastic isolators and air curtains. J. Hyg. 69:529–546.

Lum, R. M., and T. E. Graedel. 1973. Measurements and models of indoor aerosol size spectra. Atmos. Environ. 7:827–842.

McGarrity, G. J., L. L. Coriell, R. W. Schaedler, R. J. Mandle, and A. E. Greene. 1969. Medical applications of dust-free rooms. III. Use in an animal care laboratory. Appl. Microbiol. 18:142–146.

Miller, R. L., W. E. Burton, and R. W. Spore. 1963. Aerosols produced by dental instrumentation. Pages 97–120 *in* R. L. Dimmick, ed. First symposium on aerobiology. Naval Biol. Lab., U.S. Naval Supply Cent., Oakland, Calif.

Morris, J. E. W., and R. J. Fallon. 1973. Studies on the microbial flora in the air of submarines and the nasopharyngeal flora of the crew. J. Hyg. 71:761–770.

Munn, R. E., and B. Bolin. 1971. Global air pollution—Meteorological aspects. Atmos. Environ. 5:363.

Oxborrow, G. S., N. D. Fields, J. R. Puled, and C. M. Herring. 1975. Quantitative relationship between airborne viable and total particles. Health Lab. Sci. 12: 47–51.

Peterson, K. R. 1968. Continuous point source plume behavior out to 160 miles. J. Appl. Meteorol. 7:217.

Phillips, G. B. 1965. Safety in the chemical laboratory. XIII. Microbiological hazards in the laboratory. Part one—Control. J. Chem. Educ. 42:A43–A44, A46–A48.

Rahn, K. A. 1971. Sources of trace elements in aerosols—An approach to clean air. Ph.D. thesis, Univ. Michigan, Ann Arbor.

Reid, D. D., O. M. Lidwell, and R. E. O. Williams. 1956. Counts of airborne bacteria as indices of air hygiene. J. Hyg. 54:524–532.

Rodhe, H. 1972. A study of the sulfur budget for the atmosphere over northern Europe. Tellus 24:128.

Sellers, R. F., A. I. Donaldson, and K. A. J. Herniman. 1970. Inhalation, persistence and dispersal of foot-and-mouth disease virus by man. J. Hyg. 68:565–573.

Slade, D. H., ed. 1968. Meteorology and atomic energy. USAEC 56-61, Div. Tech. Inf. no. TID-24190. 445 p.

Solberg, A. N., H. C. Shaffer, and G. A. Kelley. 1956. The collecting of airborne microorganisms. Ohio J. Sci. 56:305–313.

Spendlove, J. C. 1975. Penetration of structures by microbial aerosols. Dev. Ind. Microbiol. 16:427–436.

Sulkin, S. E., and R. M. Pike. 1949. Viral infections contracted in the laboratory. N. Engl. J. Med. 241:205–213.

Sulkin, S. E., and R. M. Pike. 1951a. Survey of laboratory-acquired infections. Am. J. Public Health 41:769–781.

Sulkin, S. E., and R. M. Pike. 1951b. Laboratory-acquired infections. J. Am. Med. Assoc. 147:1740–1745.

Sutnick, A. I., W. T. London, I. Millman, B. J. S. Gerstley, and B. S. Blumberg. 1971. Ergasteric hepatitis: Endemic hepatitis associated with Australia antigen in a research laboratory. Ann. Intern. Med. 75:35–40.

USPHS (U.S. Public Health Serv.). 1966. Clean room and work stations requirements, controlled environment. USPHS Fed. Stand. 209A, 10 August 1966.

Watkins, H. M. S., M. A. Mazzarella, C. E. Meyers, J. P. Hresko, H. Wolochow, M. A. Chatigny, and F. M. Morgan. 1970. Epidemiologic investigations in Polaris submarines. *In* I. H. Silver, ed. Aerobiology: Proceedings of the third international symposium. Academic Press, New York.

Wellock, C. E. 1960. Epidemiology of Q fever in the urban East Bay area. Calif. Health 18:72–76.

Whitfield, W. J., J. C. Mashburn, and W. E. Neitzel. 1962. A new principle for airborne contamination control in clean rooms and work stations. ASTM Spec. Tech. Publ. 342. Am. Soc. Test. Mater., Philadelphia.

Whyte, W., and B. H. Shaw. 1973a. Air flow through doorways. Pages 513–516 *in* J. F. P. Hers and K. C. Winkler, eds. Airborne transmission and airborne infection. Oosthoek Publ. Co., Utrecht, The Netherlands.

Whyte, W., and B. H. Shaw. 1973b. The design and comparative advantages of laminar flow systems. Pages 581–584 *in* J. F. P. Hers and K. C. Winkler, eds. Airborne transmission and airborne infection. Oosthoek Publ. Co., Utrecht, The Netherlands.

Whyte, W., and B. H. Shaw. 1974. The effect of obstructions and thermals in laminar-flow systems. J. Hyg. 72:415–423.

Whyte, W., B. H. Shaw, and M. A. R. Freeman. 1974. An evaluation of a partial-walled laminar-flow operating room. J. Hyg. 73:61–74. SL2 Chapter 4

4. DEPOSITION

M. A. Chatigny, R. L. Dimmick, and J. B. Harrington

PRINCIPLES OF DEPOSITION OF MICROBIOLOGICAL PARTICLES*

Microbiological particles occur over a broad range of sizes extending from viruses having diameters of 0.003–0.05 μm to insects and seeds having dimensions measured in centimeters. This section is limited to a discussion of microbiological particles that can be classed as aerosols; that is, particles that are suspended in air for significant periods. For spherical particles of unit density, the upper size limit for aerosols is about 100 μm diameter. Particles of this size and density settle at a rate of about 20 m/min and therefore do not remain airborne for long.

The size ranges of some common microbiological particles are compared in Figure 4.1, with the ranges used in air pollution and cloud physics. It should be noted that the dividing line between the diameters of aerosols such as cloud and fog droplets and those of precipitation particles such as rain or drizzle lies near 100 μm. Aitken nuclei, having radii of less than 0.1 μm, do not

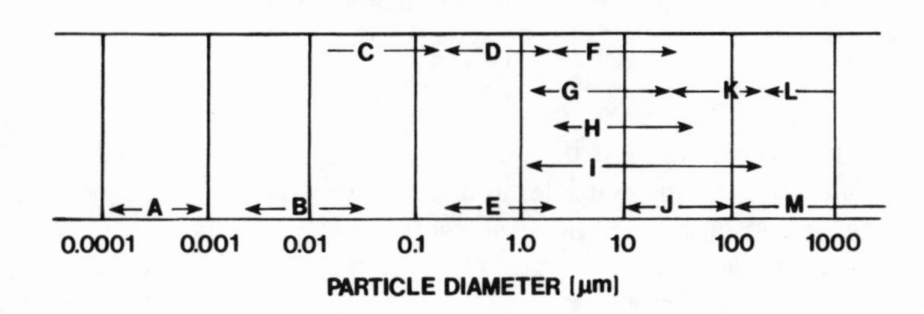

Figure 4.1. Size ranges of some microbiological particles compared with the size classification for atmospheric aerosols: (A) gas molecules, (B) viruses, (C) Aitken nuclei, (D) large nuclei, (E) bacteria, (F) giant nuclei, (G) cloud and fog, (H) milled flour, (I) fungal spores, (J) pollen, (K) drizzle, (L) rain, (M) insects and mites.

---

*J. B. Harrington

play an active role in the precipitation process; large nuclei are chiefly responsible for decreases in visibility; and both large and giant particles act as condensation nuclei.

The deposition of particles occurs either as gravitational settling, molecular diffusion, or impaction onto the earth's surface and vegetation, or, in the precipitation process, through rainout and washout. The aerosol physics required to understand these processes and to compute the aerosol deposition rate is given in the following sections.

SETTLING VELOCITY

Determining the settling velocity of particles is an exceedingly complex problem. The particles often have irregular shapes and difficult-to-measure densities; they may change their densities with changes in atmospheric humidity; they fall in different attitudes. If they are small, they slip between the air molecules and speed up; if they are large, they accelerate the air, create a wake, and slow down; if they fall near a wall, they are slowed; if they fall as a group, they affect the fall speed of others; and they are sensitive to the pressures exerted by thermophoresis, diffusiophoresis, electrostatic forces, and by the smallest air currents. At the Aitken nuclei end of the size range, particles coalesce rapidly to form larger particles. A number of particles gathered by a growing cloud droplet coalesce when the droplet evaporates. Volatile aerobiological substances such as terpenes evaporate. In this section, only the simplest cases are considered.

The drag of air on a moving particle has been found to be proportional to the square of the velocity of the particle, to the air density, and to the cross-sectional area of the particle. Expressed mathematically,

$$D = \frac{1}{2}\rho_a v^2 A C_d \tag{4.1}$$

where ρ_a is the air density, v is the air velocity relative to the particle, A is the cross-sectional area of the particle, and C_d is an experimentally determined proportionality constant called the *drag coefficient*.

For spherical particles

$$A = \frac{\Pi d^2}{4} \quad \text{and} \quad C_d = \frac{24}{\text{Re}} \quad (\text{Re} < 0.05) \tag{4.2}$$

where d is the particle diameter and Re is the Reynolds number. The Reynolds number is a dimensionless number representing the ratio of inertial to viscous forces. In cases where these are the most significant forces, the Reynolds number determines the nature of the flow. The Reynolds number may be written

$$\text{Re} = \frac{vd}{\mu}\rho_a = \frac{vd}{\nu} \tag{4.3}$$

where μ is the dynamic viscosity and ν the kinematic viscosity of the air. Substituting for A, C_d, and Re in Equation (4.1), we obtain the drag force

$$D = 3\Pi\mu v d \tag{4.4}$$

The downward force on the particle is the pull of gravity or Archimedes force. It is proportional to the difference in weights of the particle and an equivalent volume of air and to the gravitational acceleration ($F = ma$),

$$G = \frac{\Pi d^3}{6} g(\rho_p - \rho_a) \simeq \frac{\Pi d^3}{6} g\rho_p \tag{4.5}$$

where g is the acceleration of gravity.

The steady-state gravitational settling speed (Stokes' law) is obtained by equating the gravitational accelerations to the viscous drag and rearranging terms,

$$v = \frac{d^2 g}{18\mu}\rho_p \qquad \text{Stokes' law} \tag{4.6}$$

For particles of unit density, Stokes' law holds within 10% over the particle diameter range 1.6–70 μm.

When the particle is less than 1.6 μm in diameter, it slips between the air molecules. The drag is reduced by a factor called *Cunningham's correction,*

$$D = \frac{3\Pi\mu v d}{1 + \frac{C\ell}{d}} \tag{4.7}$$

where C is 1.72 for liquid drops and smooth spheres and 1.40 for rough spheres, and ℓ is the mean free path of the air molecules (ℓ is 9.42 $\times$ 10^{-6} cm at HTP [25°C] and 7.0 $\times$ 10^{-6} cm at STP [0°C]). In general, Cunningham's correction is applied when the Knudsen number, ℓ/r, is of order unity, and Stokes' law is applied when the Knudsen number is much less than unity. With Cunningham's correction, Stokes' law is written

$$v = \frac{d^2 g\rho_p}{18\mu}\left(1 + \frac{C\ell}{d}\right) \tag{4.8}$$

Particles larger than those in the Stokes range cause acceleration of the air and produce a wake. The resulting increased drag force has been described by Oseen and others by means of the following approximate equation:

$$D = 3\Pi \mu v d \left(1 + \frac{3}{16}\text{Re} - \frac{19}{1280}\text{Re}^2 + \cdot \cdot \cdot\right) \tag{4.9}$$

For liquid drops that change their shape when they fall and for irregular particles, the drag force is written

$$D = 3\Pi \mu v d_s \tag{4.10}$$

where d_s is the equivalent spherical diameter determined experimentally.

An example of the calculation of settling speeds is shown below.

Example: Assuming a pressure of 1000 mbar and a temperature of 20°C, compute the gravitational settling speeds of smooth particles of unit density and diameters of 100, 10, 1, and 0.1 μm using values shown in Table 4.1.

Table 4.1. Air density and viscosity at three temperatures.

Temp (°C)	Dynamic viscosity μ ($g\ cm^{-1}\ sec^{-1}$)	Kinematic viscosity ν (cm^2/sec)	Density ρ_a (g/cm^3 at 1000 mbar)
0	1.71×10^{-4}	0.132	1.275×10^{-3}
20	1.81×10^{-4}	0.150	1.188×10^{-3}
40	1.90×10^{-4}	0.169	1.113×10^{-3}

Answer: As a first approximation, we can estimate the settling speed using Stokes' law.

$$v = \frac{d^2 g \rho_p}{18\ \mu m} = \frac{d^2 \times 981 \times 1}{18 \times 1.8 \times 10^{-4}} \tag{4.11}$$

d (μm)	100	10	1	0.1
v (cm/s)	60	0.6	0.6×10^{-2}	0.6×10^{-4}

The large particle requires Oseen's correction and the two smallest particles require Cunningham's correction.

For the 100-μm particle,

$$\text{Re} = \frac{vd}{\nu} \simeq \frac{60 \times 100 \times 10^{-4}}{0.150} = 4.0 \tag{4.12}$$

$$\frac{C\ell}{d} = \frac{1.72 \times 9 \times 10^{-6}}{100 \times 10^{-4}} = 1.55 \times 10^{-3} \tag{4.13}$$

Oseen's correction = 1 + 0.75 − 0.25 = 1.50

Cunningham's correction = negligible

$$v = \frac{100^2 \times 10^{-8} \times 981 \times 1}{18 \times 1.8 \times 10^{-4}} \times \frac{1}{1.5} = 40 \text{ cm/s} \qquad (4.14)$$

For the 10-μm particle,

$$\text{Re} = 0.004$$

$$\frac{C\ell}{d} = 1.55 \times 10^{-2}$$

Oseen's correction = negligible

Cunningham's correction = 1.55%

$$v = 0.6 \text{ cm/s}$$

For the 1-μm particle,

$$\text{Re} = 4.0 \times 10^{-6}$$

$$\frac{C\ell}{d} = 1.55 \times 10^{-1}$$

Oseen's correction = negligible

Cunningham's correction = 15.5%

$$v = 0.6 \times 10^{-2} \times 1.155 = 0.69 \times 10^{-2} \text{ cm/s} \qquad (4.15)$$

For the 0.1-μm particle,

$$\text{Re} = 4.0 \times 10^{-9}$$

$$\frac{C\ell}{d} = 1.55$$

Oseen's correction = negligible

$$\text{Cunningham's correction} = 155\%$$

$$v = 0.6 \times 10^{-4} \times 2.55 = 1.53 \times 10^{-4} \text{ cm/s} \qquad (4.16)$$

It is interesting to observe the effect of gravitational settling on sampling efficiency when the air to be sampled is drawn through a tube. Suppose air is drawn into a standard tape sampler at the rate of 1 m/s through a tube 1.25 cm in diameter and 7 m long. Before the air reaches the sampler, all of the 100-μm and 10-μm particles, and 5% of the 1-μm particles will have settled out onto the walls of the tube by gravitational settling. As will be seen later, some of the smallest particles will also be lost to the sides of the tube through molecular diffusion.

SIZE DISTRIBUTION

Aerosols that form through the coagulation of numerous small particles have distributions that, over the large and giant nucleus sizes, follow the Junge (1963) power law,

$$f(r) = cr^{-4} \qquad (4.17)$$

where $f(r)$ is the slope of the distribution curve, dN/dr, approximated by the number of particles per unit increment in radius, and c is a constant.

Pollens, fungal spores, and most microbiological particles follow a Gaussian distribution,

$$f(r) = \frac{1}{\sqrt{2\Pi}\sigma_r} \exp - \frac{1}{2}\left(\frac{r - \bar{r}}{\sigma_r}\right)^2 \qquad (4.18)$$

where $\bar{r}$ is the mean radius and σ_r the standard deviation of the distribution. Some spores having highly specialized targets are designed by nature with a uniform particle size distribution and therefore a very small standard deviation. Ragweed pollen is one of these, with a standard deviation of <2 μm. *Alternaria* spores, on the other hand, with a diversity of hosts, are much less uniform in size and therefore have a large standard deviation.

Four definitions that occur frequently in the literature are worth incorporating here, if only to facilitate later reading. These are the log-radius-number distribution,

$$n(r) = rf(r) = r\frac{dN}{dr} = \frac{dN}{d \ln r} \qquad (4.19)$$

the log-radius-surface distribution,

$$S(r) = \frac{dS}{d \ln r} = 4\Pi r^2 \frac{dN}{d \ln r} \tag{4.20}$$

the log-radius-volume distribution,

$$v(r) = \frac{dV}{d \ln r} = \frac{4\Pi r^3}{3} \quad \frac{dN}{d \ln r} \tag{4.21}$$

and the log-radius-mass distribution,

$$m(r) = \rho v(r) \tag{4.22}$$

where ρ is the density of the particle. All these distributioms are derived mainly for purposes of graphical presentation in order to conserve the area as representative of numbers, areas, volumes, or masses, respectively, when they are represented on a linear scale and the radius on a logarithmic scale.

DENSITY

The density of most microbiological particles remains only vaguely known because the particles are often of irregular shape and incorporate unknown quantities of air or water. Ragweed pollen provides a good example. When measured by means of a gas pycnometer (Harrington and Metzger 1963), the density of the dried pollen is found to be 0.83 g/cm^3, whereas the density of the solid material in the pollen is 1.30 g/cm^3. The difference between the two figures is accounted for by the volume of interior air spaces, which occupy about one-third of the volume of the pollen grain. The density of the grain increases as the relative humidity rises above 75% because the intine or protoplasm expands, reducing the size of the air space. Kept at 100% relative humidity (RH) for a long period of time, the protoplasm completely fills the interior space and the pollen density approaches 1.2.

It is interesting to note that ragweed pollen in its sac in the floret is completely saturated and contains no air space. After dehiscence of the pollen sac, water evaporates from the pollen grain extremely rapidly, the rate of evaporation approaching that of a pure liquid drop of the same size. In seconds all the excess water in the pollen grain has evaporated and the grain has become fully buoyant.

When pollen air sacs are external to the spore or when they are connected to the exterior air by large passageways, the effective density cannot be determined using gas pycnometric or weighing techniques. The density of spherical particles can be obtained by measuring the settling speed. For non-spherical particles of diameter $\leqslant 20\ \mu m$, the Millikan oil-drop technique can be used to determine both the effective density and the equivalent spherical radius of the particle. The settling speed of nonspherical particles $> 20\ \mu m$ in diameter must be measured and the equivalent spherical radius must be estimated to determine the effective density.

COAGULATION

Coagulation of particles depends on their collision and adhesion. The former depends upon the particle concentration and mobility. The latter depends upon the nature of the particles and particularly upon surface electrical forces.

The *mobility* of a particle is defined as its terminal velocity per unit of applied force,

$$B = \frac{V}{F} \tag{4.23}$$

Equations (4.24) and (4.25) derived from Equations (4.4) and (4.7) express the mobility of particles in the Stokes range and in the Cunningham correction range, respectively,

$$B = 1/3\pi\mu d \qquad \text{Kn} << 1 \tag{4.24}$$

$$B = \frac{1 + C\ell/d}{3\pi\mu d} \qquad \text{Kn} \simeq 1 \tag{4.25}$$

where the Knudsen number $\text{Kn} = \ell/r$. The mobility B is small for most pollens and spores, moderate for bacteria, and quite large for particles of virus size.

The *molecular diffusivity* of particles is directly proportional to their mobility,

$$D = kTB \tag{4.26}$$

where k is the Boltzmann constant (1.38×10^{-16} erg/°K) and T is the temperature on the Kelvin scale.

The rate of coagulation of particles is directly related to motions on a molecular scale, that is, to the molecular diffusivity. Turbulent diffusion has no appreciable effect on coagulation because turbulent eddies generally exceed 1 cm in size and therefore are much too large to affect particle separations of the order of fractions of micrometers. The rate of coagulation of uniformly sized particles can be expressed as the rate of change of the number of particles per unit time and is given by the equation

$$\frac{dn}{dt} = -KN^2 \tag{4.27}$$

where K is a *coagulation constant* given by the expression

$$K = 4\Pi dD = 4\Pi dkTB \tag{4.28}$$

The number of particles in any size range decreases because of coagulation, but increases because of coagulation of particles in smaller size ranges. The net change is given in a complex equation derived by Smoluchowski (see Junge 1963, p. 129). His results agree well with experimental data.

The net result of coagulation is to rapidly reduce the number of Aitken nuclei. Without constant replenishment they disappear, as they do, for example, over the oceans. Near major sources of Aitken nuclei, such as industrial centers, concentrations as high as 300,000/cm^3 are not uncommon. Over the oceans, far from anthropogenic sources of pollution, the concentration drops to a few hundred per cm^3.

Pollens, fungal spores, bacteria, and most aerobiological particles are too large to be directly affected by coagulation. Many air pollutants affecting biological systems and a few particles of natural origin such as organic hazes and smoke, however, are in the Aitken size range. Viruses, although small, usually exist as an aerosol in a matrix of other material.

During the formation of cloud or fog droplets a number of small particles, propelled by Brownian motion and aided by the flow of water molecules toward a center of condensation (Stephan flow[1]), may diffuse onto the growing drops. Subsequent evaporation of the drops leads to coagulation of all collected particles. Good evidence for the efficiency of this process is demonstrated by the rapid reduction of Aitken's nuclei over calm oceans culminating in a concentration of nuclei closely approximating the average concentration of cloud droplets in cumulus clouds.

IMPACTION

The study of aerosol impaction has important applications in rain scavenging, in dry deposition, and in aerosol sampling. In each case, the dimensionless Stokes number is an important scaling parameter.

$$\mathrm{Stk} = \frac{L}{R} \tag{4.29}$$

where L is the stopping distance of the particle in air and R is the radius of curvature of the air streamlines.

Suppose that a particle is projected with speed V_O into still air. How will its velocity change with time and how far will it travel? Newton's second law states that the force on an object is equal to the product of mass and acceleration. In this case the acceleration, the rate of change of velocity with time, is dV/dT and Newton's law is expressed as

$$F_{drag} = m\frac{dV}{dT} = \frac{-V}{B} \tag{4.30}$$

where B is the mobility discussed previously. Here B is a function of particle size and air viscosity, but is not a function of particle velocity or time. Therefore the equation

$$\frac{dV}{V} = \frac{-dT}{mB} \tag{4.31}$$

can be integrated with respect to time,

$$\int_{t=0}^{t=t} \frac{dV}{V} = -\int_{t=0}^{t=t} \frac{dt}{mB} \tag{4.32}$$

yielding

$$\ln V(t) - \ln V_o = \frac{-t}{mB} \tag{4.33}$$

or

$$V(t) = V_o \exp -\left(\frac{t}{mB}\right) = V_o \exp -\left(\frac{t}{\tau}\right) \tag{4.34}$$

where $\tau = mB$ is called the *inertial period,* that is, the period required for the velocity to be reduced to V_O/e. The distance traveled by the particle can be found by expressing the velocity V by dx/dt and performing a second integration.

$$\int_{t=0}^{t=\infty} dx = \int_{t=0}^{t=\infty} V_o \exp -\left(\frac{t}{mB}\right) dt \tag{4.35}$$

If $x = 0$ at $t = 0$ and $x = L$ at $t = \infty$, then

$$L = -V_o mB \exp -\left(\frac{t}{mB}\right)\Big]_0^\infty = V_o mB = V_o \tau \qquad (4.36)$$

The total distance a particle projected into still air will travel is equal to the distance it would have traveled at its initial velocity in the inertial period τ.

Several aerosol samplers have been designed using impingement on a coated surface as the method of particle capture. One such sampler draws aerosol at high velocity through a tube and impinges it on a flat plate. Whether the particles follow the air trajectory and escape or are carried onto the coated surface and are captured depends upon the size, shape, and density of the particles, the pressure difference across the inlet tube, the dimensions of the nozzle, the separation between the nozzle exit and the plate, and the temperature and pressure of the air. These variables determine the stopping distance of the particle and the radius of curvature of the air streamlines. For situations in which the Stokes number is the same, the sampling efficiency is the same.

A similar dimensionless ratio called the *inertia parameter* K_i has been used to estimate the efficiency of impaction of spherical particles on cylindrical objects.

$$K_i = \frac{2}{9} \cdot \frac{\rho_p r^2 V}{\mu r_c} \qquad (4.37)$$

where r_c is the radius of the cylindrical collector and the other terms are as previously defined. A curve of efficiency versus K_i is determined by solving the equations of motion of a particle moving around a cylinder immersed in a viscous fluid. It is assumed that all particles whose centers pass within one particle radius of the cylinder will be captured. In practice, no adhesive surface is a perfect collector, so only a fraction of the theoretical efficiency is actually attained. For example, numerous tests of the sampling efficiency of dilute rubber-cement-coated flag samplers (Harrington et al. 1959) gave efficiencies of only 72% of the theoretical values. Figure 4.2 may be used to compute the efficiency with which spherical particles are impacted on a cylindrical sampler. A second dimensionless parameter used in the construction of this curve has been omitted because it has a negligible variability under normal atmospheric conditions.

Similar curves for other sampling surfaces have been computed and may be found in the literature. Whereas sampling instruments are designed with specific aerodynamic properties, nature produces a variety of shapes. To find the sampling efficiency of a leaf or stem is not a simple problem. It has been a

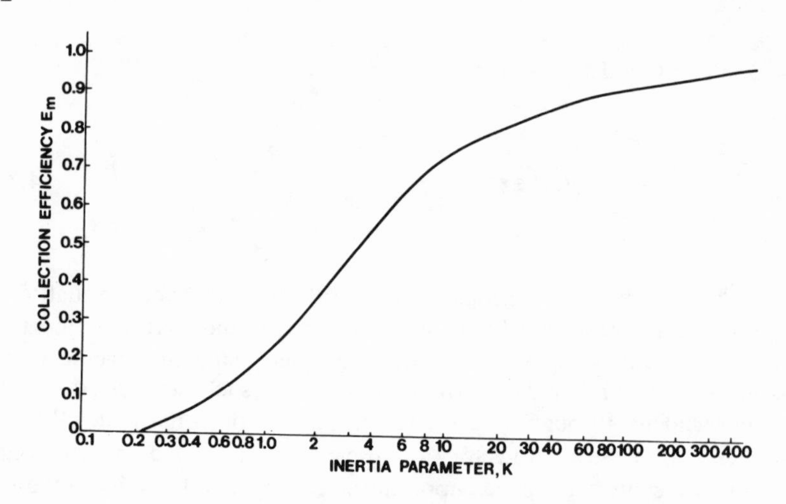

Figure 4.2. Cylinder impingement efficiency (after Brun and Mergler 1953).

common practice to approximate leaves by flat plates and twigs, branches, and trunks by infinite cylinders. Although this practice produces theoretical inaccuracies, they are usually small compared with errors in the estimation of the adhesive efficiency of the surfaces and the wind speed near the object.

DRY DEPOSITION

In the absence of particle sources within a layer of vegetation, the rate of dry deposition is equal to the downward flux of particles into the top of the plant canopy or, in the absence of vegetation, onto the earth or ocean surface. The vertical flux into the vegetative layer (number of particles passing through unit horizontal area in unit time) is composed of two parts: that due to gravitational settling and that due to atmospheric turbulence. They are written:

$$\text{flux} = -v_g \chi \tag{4.38}$$

and

$$\text{flux} = -K_z \frac{\partial \chi}{\partial z} \tag{4.39}$$

respectively, where K_z is the vertical component of the eddy diffusivity, χ is the particle concentration, and v_g is the gravitational settling speed of the particles. Upward flux is taken to be positive. The total flux into a vegetative canopy is equal to the sum of these terms,

$$F(h) = -\left(K_z \frac{\partial \chi}{\partial z} + v_g \chi\right)_h \tag{4.40}$$

where h is the height above ground of the top of the plant layer.

When there is a source of particles below the level h, the flux $F(h)$ could be positive whenever the upward turbulent flux at h exceeds the gravitational flux. This is true during periods of emission of plant pathogens or of pollen, for example.

When there is neither a strong horizontal gradient of particles nor a source of particles below the level h, the concentration profile above the canopy rapidly adjusts to either a constant concentration with height when the particles are deposited primarily by gravitational settling, or a small increase in concentration with height when the particles are deposited primarily by molecular diffusion. In the former instance, if the settling speed of the particle is known, the rate of deposition can be determined from a measurement of particle concentration at a single height. In practice over low vegetation at distances of a few tens of meters or more from active sources, the concentration profile close to the canopy is frequently found to be nearly constant with height (Harrington 1965), and therefore the rate of deposition is easily estimated.

Extremely small particles behave somewhat like gases whose vapor pressure is zero on the sorption surface. They are mixed rapidly by turbulence in the free air but are carried onto the surface by molecular diffusion through the laminar boundary layer. Most of the gradient of concentration occurs through the boundary layer where the movement of particles is slow. It has become customary therefore to compute the rate of deposition as the product of the concentration at some standard distance above the sorbing surface (1 m) and a deposition velocity. This practice makes it unnecessary to measure the weak gradients in the free air or the large gradient through the very thin boundary layer. The inverse of deposition velocity is a resistance. Chamberlain (1975) gives the resistance for perfectly sorbed gases as:

$$r_g = \frac{1.7}{U_*}\left(\frac{U_* Z_o}{\nu}\right)^{0.45}\left(\frac{\nu}{D}\right)^{0.8} \tag{4.41}$$

where the symbols have their usual meanings. It will be noted that the deposition velocity is roughly proportional to $u*^{1/2}$ or the square root of the wind speed, because the friction velocity $U*$ and the wind speed are proportional. An equivalent equation given for extremely small particles ($<0.1\ \mu$m) is,

$$r_p = \beta\left(\frac{d_o}{u}\right)^{1/2}\left(\frac{\nu}{D}\right)^{2/3} \tag{4.42}$$

where the parameter β is a function of leaf type and orientation, d_o is the leaf chord diameter and the wind speed u is measured 1 m above the leaf surface.

The rate of deposition, it should be noted, is much slower for particles than for gases, because the molecular diffusivity is much smaller. Sehmel (1971) has confirmed this conclusion and shows that the rate of deposition for particles of this size is approximately three orders of magnitude smaller than the rate for sorbing gases. For surfaces roughened by plant hairs or similar protuberances, the deposition velocity may be increased by as much as an order of magnitude (Chamberlain 1966). Wesley et al. (1977) find a deposition velocity of approximately 1.0 cm/s for particles having diameters between 0.05 and 0.1 μm.

Particles having diameters greater than 0.1 μm have a decreasing deposition velocity reaching a minimum near 1 μm because of their decreasing mobility in air. As the particle size increases, however, its inertia increases and the particle deposition rate is enhanced by a partial free flight into the laminar boundary layer (Sehmel 1970; Caporaloni et al. 1975). Beyond 1 μm, the deposition velocity increases rapidly (Moller and Schumann 1970) and deposition on horizontal surfaces is determined almost entirely by the gravitational settling speed of the particles. Deposition on irregular surfaces or on vegetation may be greatly enhanced by impaction if the surface is sufficiently moist, hairy (Wedding et al. 1975), or sticky (Belot et al. 1976) to retain the particles.

Within vegetation, if particles are removed from the air by impaction on plant parts, concentration in the plant layer is decreased and the concentration gradient at the top of the layer is not zero. Because K_z is frequently large at h, even a small gradient in concentration can lead to a large increase in the downward flux of particles. A flux of particles into vegetation exceeding the gravitational rate can be shown to create a particle profile in the atmosphere above in which concentration increases with height. This, it should be noted, is not the only cause of a positive concentration gradient in the free air above the canopy; advective effects can lead to the same profile shape.

Within vegetation, the flux of particles is modified by the impaction of particles on vegetation. At any level z within vegetation, the particle conservation equation can be written

$$\frac{\partial}{\partial Z}[K_z(z)\frac{\partial \chi}{\partial \chi} + v_g \chi] = \Upsilon(z)\chi \tag{4.43}$$

where $\Upsilon(z)$ is the particle scavenging rate by vegetation. Equation (4.43) states that the change in the downward flux of particles from one level to the next is exactly equal to the loss of particles to the vegetation in that layer. The scavenging rate Υ depends upon the size, shape, and leaf and stem area densities of the vegetation, upon the wind speed in the vegetation, on various properties including the settling speed of the particle, and upon the efficiency with which the particle is retained by the vegetation once it alights (adhesive efficiency).

The particle scavenging rate by vegetation is composed of two terms, Υ_1, the fractional scavenging rate by inertial impaction, and Υ_2, the fractional scavenging rate by gravitational settling. These may be written as follows:

$$\Upsilon_1(z) = u(z)A_v(z)E_i(z)E_a(z) \tag{4.44}$$

and

$$\Upsilon_2(z) = qA_h(z)E_h(z) \tag{4.45}$$

where A_v and A_h are the vertically and horizontally exposed areas of vegetation per unit volume and E_i, E_a, and E_h are the impingement, adhesive, and horizontal sampling efficiencies, respectively. As noted earlier, E_i can be computed by approximating the shapes of vegetation elements with cylinders, flat plates, and so on, but little is known about the effects of bark, plant hairs, and all of the other irregularities of real vegetation. Similarly, little is known concerning the proper values for E_a and E_h. An estimation of the magnitude of the product E_iE_a can be made by solving Equation (4.43) using reasonable values for each of the variables. When this is done, it is found that observed particle profiles in and above many types of vegetation may be duplicated by the model only if the scavenging rate by inertial impaction is made small (Belot and Gauthier 1975, Harrington 1965, Harrington and Mansell 1973). It is possible that types of vegetation with a high retention capacity exist (Belot and Gauthier 1975, Clough 1975). If so they may prove useful as natural filters for environmental pollutants. The most likely candidate for such a role are the needle-bearing evergreens.

Neither the horizontal wind speed u nor the vertical component of the diffusivity K_z is routinely observed in and above plant layers. They can be computed if the wind speed above the canopy is known, if the drag coefficient as a function of height can be estimated from information on the type and density of vegetation, if the diffusivity above the canopy is known, and if a reasonable model is used for the profile of diffusivity within the canopy.

The first theoretical model for the wind profile under a plant canopy was published by Tan and Ling (1963). Improvements since that time have refined the model but have left the exponential shape of the profile under a plant canopy more or less unchanged. Tan and Ling assumed that the momentum flux, which is constant in the lower few meters of the free atmosphere, decreases downward through the vegetative layer because of friction between the wind and the vegetation. The rate of decrease of momentum flux experienced in traversing a layer beneath the canopy is:

$$\frac{\partial \tau}{\partial z} = C_D A(z) \frac{\rho u^2(z)}{2} \tag{4.46}$$

where τ is the horizontal shearing stress, C_D is the drag coefficient of the plants, and A is the surface area of vegetation per unit volume.

The shearing stress, which transports momentum through the plant layer, may be expressed as

$$\frac{\tau(z)}{\rho} = K_z(z)\frac{\partial u}{\partial z} \tag{4.47}$$

Equations (4.46) and (4.47) can be combined to give a slightly nonlinear expression from which the wind profile can be numerically computed.

$$\frac{\partial}{\partial z}\left[K_z(z)\frac{\partial u}{\partial z}\right] = \frac{C_D A(z)u^2(z)}{2} \tag{4.48}$$

By letting $Z = z/h$ and $U = u(z)/u(h)$, Equation (4.46) can be written in the nondimensional form[2]

$$U'' + f(Z)U' - Sg(Z)U^2 = 0 \tag{4.49}$$

where the functions of height are contained in the dimensionless variables $f(Z)$ and $g(Z)$ and the constants are lumped into S, the so-called shape factor of canopy flow, and where S is a parameter consisting of measured values such as the wind speed and diffusivity at the top of the canopy, the density of vegetation at ground level, and the drag coefficient of the vegetation. Here $f(Z)$ and $g(Z)$ depend upon the distribution of diffusivity and vegetation in the plant layer.

The literature on wind profiles in vegetation is so extensive that only a representative sample can be given here. The pioneering work of Inoue (1963) and Cionoco (1965) in agricultural crops, Cionoco (1972a,b) in both agricultural crops and outdoor artificial canopies, and Plate and Quraishi (1965) in wind tunnel studies of artificial canopies set the stage for a rapid advance during the 1970s. Excellent reviews of the current status of work on wind profiles in vegetation are given by Businger (1975) and Thom (1971). The most thorough research (and collection of the best data) has been conducted in the Thetford Forest, Norfolk, England, and was reported in a series of papers by Oliver and others (Oliver 1971, 1975a,b; Oliver and Mayhead 1974, Smith et al. 1972). Within forests, the complications provided by the presence of plants are still further complicated by problems of scale, of heterogeneity, of the effects of local pressure gradient beneath the canopy, and of the effects of random thermals, which may generate much of the trunk space flow.

Little is known about either $K(Z)$ or $A(Z)$, but both can be approximated on the basis of the best current knowledge (Kinerson and Fritschen 1971, Leonard and Federer 1973). A convenient empirical expression for $A(Z)$ if the actual distribution is unknown is

$$A(Z) = (1 - Z)^{C_1} \exp C_2 Z \tag{4.50}$$

where C_1 and C_2 can be evaluated from the height of the level of maximum vegetation Z_{max} and the ratio of the density at that level to the density at ground level. An example of the distribution of vegetation in a hypothetical deciduous forest is shown in Figure 4.3, where C_1 and C_2 are 1.64 and 8.18, respectively, obtained by letting $A(Z_{max})/A(0)$ equal 50 and Z_{max}/h equal 0.8. A more detailed description of the effect of the distribution of vegetative elements is given by Seginer (1974) and Landsberg and Thom (1971).

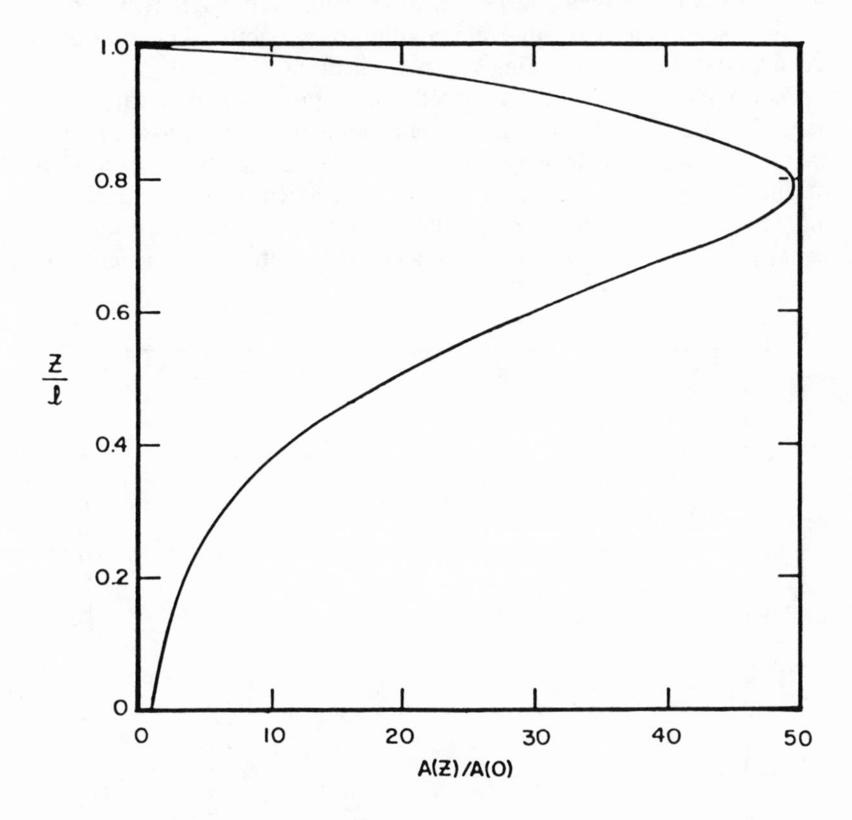

Figure 4.3. Hypothetical distribution of vegetation density in a midlatitude deciduous forest according to the equation: $A(Z)/A(0) = (1 - [Z/l])C_1 \exp (C_2Z/l)$, where $C_1 = 1.64$ and $C_2 = 8.18$.

There has been considerable debate concerning the form of $K_z(Z)$ (see Legg and Monteith 1975), with estimates including an exponential increase with height, a linear increase with height, and a constant value beneath the canopy. It is likely that K_z varies widely depending upon the density and distribution of vegetation. In the absence of definitive information, the best estimate for distribution of the diffusivity beneath the canopy is a constant value.

In terms of the deposition of particles, the shape of K_Z has little significance. The magnitude of K_Z is proportional to the product of the dimension of the eddies and their wind speed. The former can be approximated by the average size of the openings between vegetation elements and the latter by the average standard deviation of the vertical or horizontal wind speeds (Shaw et al. 1974). Having determined values for S, $f(Z)$, and $g(Z)$, Equation (4.49) can be solved for the wind speed profile without difficulty using relaxation methods.

The various terms in Equation (4.43) are now either computed or estimated so that X can be found as a function of height within the plant layer. The rate of deposition is $\Upsilon(z)$, and the total deposition during any time period is the integral of $\Upsilon(z)$ over the period in question. Although $\Upsilon(z)$ is derived under the assumption of steady-state conditions, variability in conditions can be accommodated by recomputing $\Upsilon(z)$ periodically.

An example of computed ragweed pollen profiles within various plant layers is shown in Figure 4.4. These were computed under the assumption that K_Z increases linearly with height in the plant layer. With the exception of curve one, these profiles all display a strong gradient of concentration with height. This implies a rapid flux of pollen into the canopy and a profile above the canopy exhibiting an increase of concentration with height. In my experi-

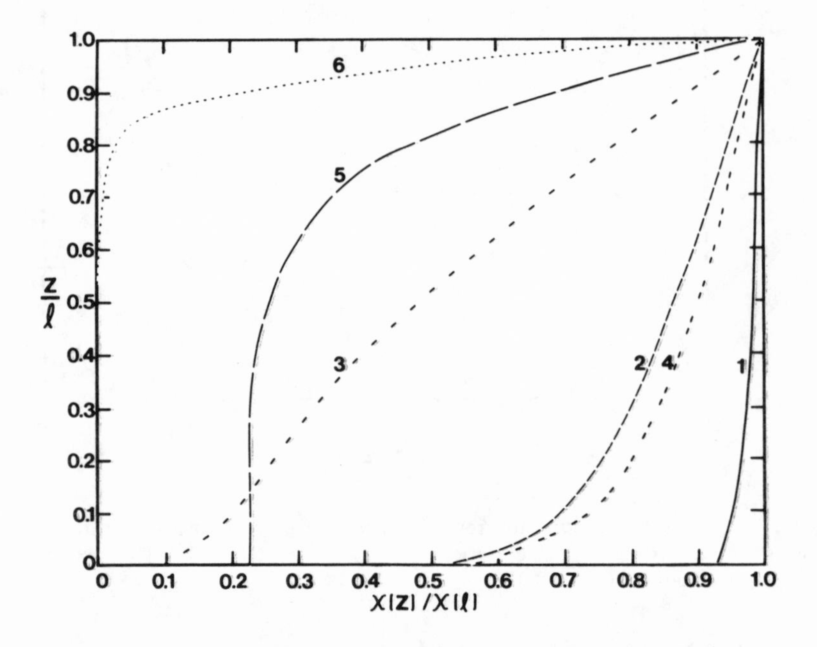

Figure 4.4. Computed ragweed pollen grain profiles within a plant canopy: (1) sparse grass, E_a = 0.1; (2) moderate grass, E_a = 0.1; (3) dense grass, E_a = 0.1; (4) moderate grass, E_a = 0.01; (5) deciduous forest with peak vegetation density of 500 leaves/m^3 of dimension 5 by 10 cm; (6) coniferous forest with peak vegetation density of 100,000 needles/m^3 of dimension 0.02 by 2 cm.

ence, measured profiles in and above plant layers tend to become constant with height as exemplified by measured profiles of naturally occurring pollen and spores over a grassy field (Figure 4.5) or in a moderately dense deciduous forest (Figure 4.6). To reproduce profiles such as these with the model of Equation (4.43) would require extremely low scavenging rates and therefore extremely low collection efficiencies. Definitive studies of the impingement and adhesive efficiencies of various plant parts of various plant species are needed.

At present the best estimate of the dry deposition of most pollens and fungal spores is obtained by assuming that particles deposit at the gravitational rate, that is, at a rate equal to the product of the concentration and settling speed of the particle. Evidence supporting this hypothesis is provided by a 1960 experiment (Harrington 1965) designed to find the relationship between the rate of particle deposition and various meteorological variables.

A sampling array consisting of three semicircular grids located at 6, 12, and 18 m from a central point was laid out on a leveled, fine-sand surface lo-

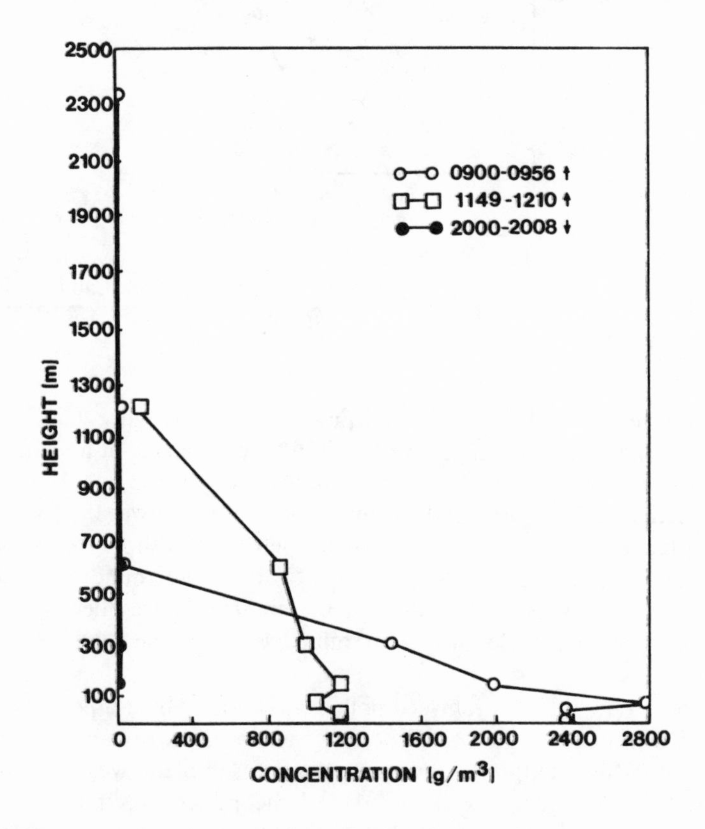

Figure 4.5. Measured ragweed pollen grain concentration profiles over a 4.8-km flight path near Willow Run Airport, 25 August 1962.

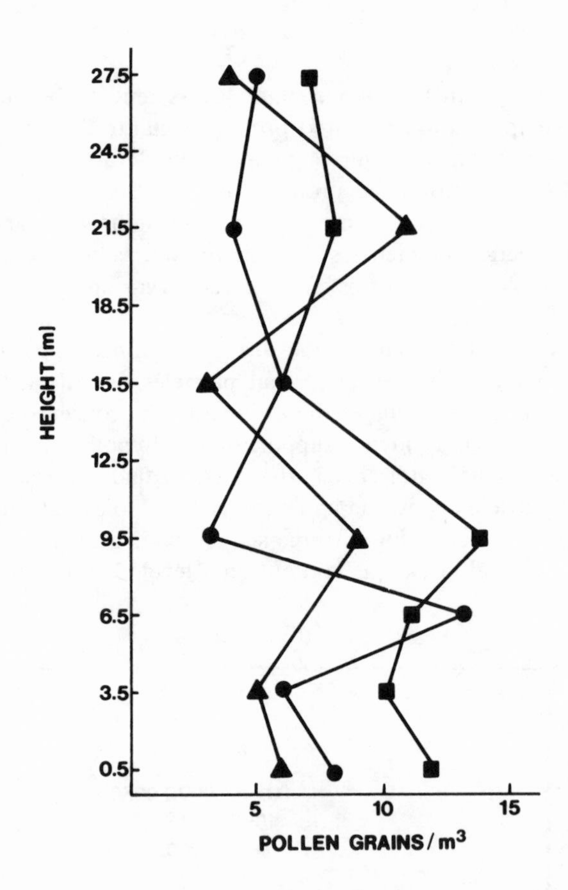

Figure 4.6. Ragweed pollen grain profiles 40 m upwind of a forest and at 40 m and 120 m into the forest: (▲) 40 m upwind, (●) 40 m into forest, (■) 120 m into forest. Tree height 15.5 m.

cated on flat terrain at Willow Run Airport, Michigan (Figure 4.7). An array of 2128 flag samplers (Harrington et al. 1959) was arranged at heights of 0.25 and 0.5 m, and at 0.5-m increments to 3, 4, and 5 m on the 6-, 12-, and 18-m arcs, respectively. The horizontal spacing of the samplers was 0.5 m. Ground-level samplers consisting of 10-cm-diameter petri dishes supported within 26-cm-diameter microfilm cans were sunk level with the surrounding sand along three radii labeled A, B, and C. Both dishes and cans were filled with refined sugar. The sugar was leveled with a straightedge so that the petri dish rim was just visible beneath it.

A weighed quantity of ragweed pollen was emitted over a period of 1 h in single grains from a source located at the center of the array and 0.5 m above the ground. After sampling, the ground-level samplers were capped and sealed. Later the sugar was dissolved and the pollen was filtered out and counted. The flag samples were placed in protective boxes, and later were mounted on microscope slides and counted. As a first estimate it was assumed

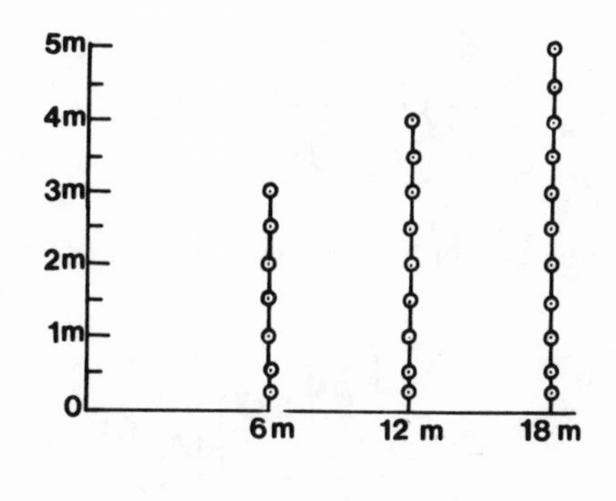

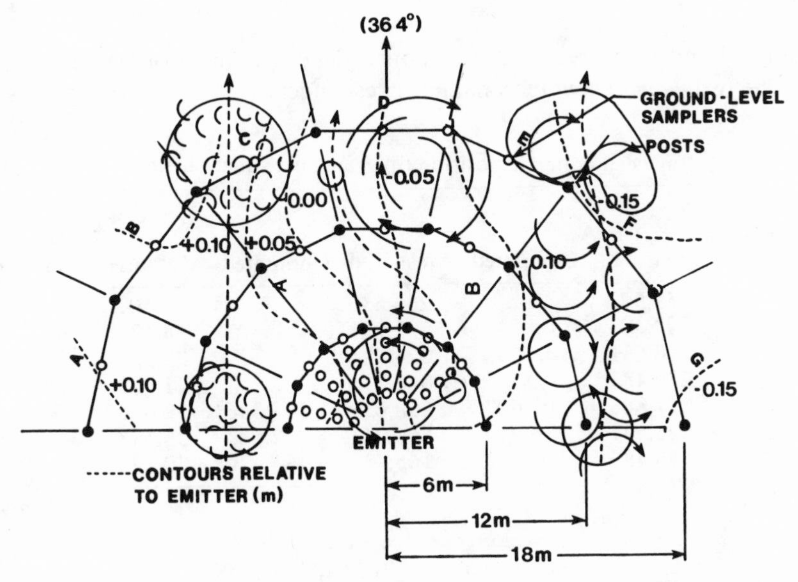

Figure 4.7. Plan view and vertical section through the 1960 out-of-season pollen sampling array. Flag samplers are mounted at 0.5-m intervals on wires located at three radial distances at the heights shown in the vertical section. Land contours relative to the emission point are also shown.

that the rate of deposition was equal to the gravitational fall speed of the particles multiplied by their concentration. The total deposition would then be equal to the product of the deposition rate, the time, and the area of the samplers. Expressed mathematically,

$$\text{dep} = \chi v_g A_s t \tag{4.51}$$

The average concentration at the ground level was assumed to be equal to that at 0.25 m. The concentration is given by the flag count divided by the product of its efficiency, the flag area, the wind speed, and the time.

$$\chi = \frac{N}{E_f A_f u t} \tag{4.52}$$

The deposition is obtained from Equations (4.51) and (4.52) as:

$$\text{dep} = \left(\frac{N}{E_f}\right)\left(\frac{v_g}{U}\right)\left(\frac{A_s}{A_f}\right) \tag{4.53}$$

Computed and measured values of pollen deposition are compared in Table 4.2. The two values show reasonable correspondence.

Table 4.2. Ragweed pollen deposition (in grains per square meter).

Sector Distance (m)	A		B		C	
	Measured	Computed	Measured	Computed	Measured	Computed
2	52,501		264,550		6,601	
3	123,394		503,951		8,173	
4	218,651		1,024,721		29,237	
5	279,483		1,044,213		25,779	
6	217,865	307,692	608,639	1,302,386	32,066	32,206
12	125,437	116,929	566,041	432,583	38,197	3,326
18	58,789	69,199	165,735	185,170	20,120	907

DEPOSITION IN PRECIPITATION

There are two processes at work in rain scavenging of atmospheric aerosols. The first is called *rainout* and occurs when extremely small aerosol particles are carried by molecular diffusion onto cloud droplets or ice crystals. The second process is called *washout* and occurs when falling raindrops or snowflakes collide with and retain large aerosol particles.

The number of particles removed from the air in both cases is proportional to the number of particles present in the appropriate size range. Therefore rain scavenging constitutes an exponential decay process, which can be described by the equations

$$\chi = \chi(0) \exp -\psi t \quad \text{rainout} \tag{4.54}$$

$$\chi = \chi(0) \exp -\Lambda t \quad \text{washout} \tag{4.55}$$

where ψ and Λ are called, respectively, the *rainout* and *washout coefficients*.

The rainout coefficient ψ is determined by three and possibly four different processes. First is the condensation of water vapor on a condensation nucleus, second is the attachment of Aitken particles by Brownian motion, and third is the attachment of Aitken nuclei by Stephan flow. A fourth may be the attachment of Aitken nuclei through their attraction to charged droplets or snowflakes.

The concentration of an aerosol in cloud water k_1 is given by the expression

$$k_1 = \frac{\chi\varepsilon}{\chi_c} \tag{4.56}$$

where χ is the concentration of the aerosol in the air, χ_c is the concentration of condensed water in the cloud, and ε is the fraction of χ which enters the cloud water ($0 \leqslant \varepsilon \leqslant 1$). The quantity ε refers to all four processes. Ignoring possible electrical effects,

$$\varepsilon = \varepsilon_n + \varepsilon_b + \varepsilon_s \tag{4.57}$$

for condensation nuclei, Brownian motion, and Stephan flow, respectively. No measurements of the ε's are available, but estimates of their relative magnitudes can be made. Active condensation nuclei are generally in the size range 0.1–10 μm. The value taken by ε_n depends upon the concentration of condensation nuclei present. In highly polluted air ε_n equals about 0.5, in light pollution 0.8, and in the cleanest air where the number of nuclei is about 200–300/cm^3 ε_n becomes equal to unity. Values of ε_n are much lower if the aerosol is hygrophobic.

Smoluchowsky's theory may be used to estimate the rate of attachment of particles to drops due to Brownian motion. According to this theory the change of concentration of particles with time is

$$\frac{d\chi}{dt} = -4\Pi(\kappa + \kappa_c)(r + r_c)\chi\chi_c \tag{4.58}$$

where the subscript c refers to the cloud droplet, r is the radius, k the diffusion coefficient, and χ the concentration. For most particles

$$\kappa_c << \kappa, \quad r << r_c$$

and therefore

$$\frac{\partial\chi}{\chi} = -4\Pi\kappa r_c\chi_c \; dt \tag{4.59}$$

and the lifetime[3] of the particles before they are captured is

$$\gamma = \frac{1}{\psi} = \frac{1}{4\Pi \kappa r_c \chi_c} \tag{4.60}$$

For $r_c = 10\ \mu m$, $\chi_c = 200\ cm^{-3}$, and k computed from Equation (4.26), we obtain the following lifetimes:

r (μm)	0.01	0.03	0.10
γ (h)	0.64	3.8	38

The average cloud exists for approximately one-half hour. It can be seen from the table above that ε_b is insignificant except for the smallest particles.

The effect of Stephan flow can be shown theoretically and in laboratory tests to be equally small for atmospheric conditions, but significant when large temperature differences between a drop and its environment exist (Pilat and Prem 1976).

Observations of snowflakes under an electron microscope show numerous Aitken particles attached (about $1/\mu m^3$), which aggregate to a mass equal to that of the large condensation nucleus. It is possible that electrical effects are involved. No estimate of the magnitude of the fraction of ε associated with this effect is available.

The *washout* coefficient Λ is determined by the rate at which falling drops capture particles in the air. Small particles evade capture by moving in the airflow around the falling drops. Larger particles have more inertia, move aside less rapidly, and therefore are more easily captured.

The concentration of aerosol in a raindrop at ground level can be computed as follows: the mass of the aerosol collected is given by

$$\frac{4\Pi r^3_d}{3} k_2 \tag{4.61}$$

where r_d is the radius of the falling drop and k_2 is the final concentration of aerosol in the drop. The mass of aerosol collected must be equal to the total mass of aerosol in the volume cut out by the falling drop multiplied by the collection efficiency of the falling drop. The total mass collected in all size categories is

$$\int_{r=0}^{r=\infty} \eta(r)\pi r_d^2 H\bar{\chi}(r)\frac{4}{3}\pi r^3\ dr \tag{4.62}$$

where $\eta(r)$ is the collection efficiency, πr_d^2 is the areal cross section of the falling drop, H is the depth of the aerosol-containing layer, $\bar{\chi}(r)$ is the average concentration of particles in the size range r to $r + dr$ and $4/3\pi r^3$ is the volume of a particle of radius r. Hence

$$k_2 = \frac{\pi H}{r_d} \int_{r=0}^{r=\infty} \eta(r) r^3 \overline{\chi}(r)\ dr \tag{4.63}$$

If $\chi(r)$ is known, then from a graph of $\eta(r)$ versus r similar to Figure 4.2, the value of k_2 may be computed.

For aerosol particles of 10 μm radius and drops ranging from 50 to 2000 μm radius, the efficiency η is practically equal to unity. As the radius of the particle decreases, the efficiency of capture decreases rapidly. It is approximately 0.5 for 5-μm particles, 0.1 for 2-μm particles, and negligible for particles smaller than 1 μm radius.

Washout, therefore, is effective only in removing giant particles. Rainout is effective for condensation nuclei larger than 0.2 μm and for Aitken particles smaller than 0.02 μm radius. There is a gap between 0.02 μm and 0.2 μm in which particles are not easily removed by the precipitation process.

The rapidity of removal of microbiological particles from the air depends on the gravitational settling speed, on the presence or absence of precipitation, possibly on the nature of the vegetation, and also on the degree of turbulence in the lower atmosphere. As an example of the last point, consider a layer of depth H filled with a uniform cloud of spores. If there is no turbulence, all of the spores will settle out in time H/v_g. With uniform mixing, on the other hand, the lifetime of the particles will be H/v_g and some particles will remain airborne for a considerably greater length of time. An example of the lifetime of particles of unit density uniformly mixed in a turbulent layer 5 km in depth is shown in Table 4.3. The lifetime of particles scavenged by precipitation

Table 4.3. Lifetimes of particles of density 1 g/cm³ uniformly mixed to 1.5 km, assuming dry deposition.

Particle radius (μm)	Settling speed (cm/sec)	Lifetime (days)
1	0.013	147
2	0.05	39
4	0.19	10
6	0.44	4.3
8	0.8	2.4
10	1.3	0.6
15	3.0	0.6
20	5.0	0.37
50	33	0.06

may be estimated using a simple model. Let the number of particles removed by a falling drop be equal to the product of a washout scavenging efficiency $\eta(r)$, the average concentration of the aerosol χ, and the volume cut out by the falling drop $\pi r_d^2 H$, where H is the depth of the aerosol-containing layer. Then the change in total number of particles in a column of unit area will be $\pi \bar{\chi} \eta r_d^2 H \cdot N_r$, where N is the number of drops that fall. The change in concentration per unit time will be equal to this value divided by the total volume and by the duration of the precipitation. The latter may be converted into a rainfall rate if the drop size is known. The final expression is

$$\frac{d\bar{\chi}}{dt} = \left(\frac{-\bar{\chi}\eta\pi r_d^2 H}{HA}\right)\left(\frac{RA}{\frac{4}{3}\pi r_d^3}\right) \tag{4.64}$$

$$\frac{d\bar{\chi}}{\bar{\chi}} = -\left(\frac{3}{4}\right)\left(\frac{\eta R}{r_d}\right) dt \tag{4.65}$$

where R is the rainfall rate and A the area. The solution of Equation 4.65 leads to a lifetime equal to $4r_d/3\eta R$. Given a rainfall rate, the lifetime of a particle depends upon its size and the size of the raindrops. An example is shown in Table 4.4. By doubling the rainfall rate the lifetime is halved. It is apparent that, even in light rains, particles of pollen size are rapidly removed from the air. The efficiency of this removal mechanism should be borne in mind by those concerned with long-range transport of pollen or spores (Rodhe and Grandell 1972).

Table 4.4 Lifetimes of particles during a rain consisting of 2-mm drops falling at a rate of 1 mm/h.

Particle radius (μm)	Washout efficiency η	Lifetime (h)
2	0.1	1.3
5	0.5	0.27
10	1.0	0.13

Both dry deposition and rain scavenging are important processes in the removal of aerobiological particles from the atmosphere and their deposition in vegetation. The simple models presented in this chapter are certainly not complete, but are intended to give some indication of the major processes taking place in aerosol deposition. There are many areas where the models are inadequate and where research is needed. Among them, the most needed is the measurement of the collection efficiencies of plant parts for various aerosols. An important subject not treated in this chapter is gaseous deposition. A recent description of gaseous deposition and a slight literature review may be found in articles by Scriven and Fisher (1975a,b).

There are a number of excellent books that cover various aspects of particle deposition including those by Gregory (1973), Gregory and Monteith (1967), deVries and Afgan (1975), Rasool (1973), Junge (1963), and Monteith (1975).

INDOOR DEPOSITION*

Indoor deposition of airborne particulate matter is governed by several factors including number, shape, and size of particles (Lippman and Albert 1969); their density, surface texture, and quality; and the air velocity, humidity, and temperature gradients; in fact, the same factors that govern deposition in other environments (Buchbinder et al. 1945).

The importance of indoor deposition of particles must be judged by whether one is concerned with material effects (Cheng 1973) or with biological effects (see Edmonds, chap. 2, this volume). The average housewife is probably as informed on problems of indoor deposition as are the authors, for the obvious results of deposition of airborne material indoors are dusty horizontal surfaces, cobwebs, and cloudy windows, the result of not only carbon or hydrocarbon deposition (from acts of cooking or smoking) but also from smog and from naturally formed plant and animal material that enters habitats (Winkler 1974, Morrison 1974, Forney and Spielman 1974). These effects are annoying but are not usually considered to be part of the aerobiological domain and are not considered further here except as mechanistic examples of at least one fate of biologically active particles (Leadbetter and Corn 1972). Of course, deposition of this type becomes an important economic factor in certain assembly and food processing plants and clean rooms, but again, the problem is not one of aerobiology. From the biological viewpoint, we are concerned with deposition in or on domestic animals (Rosebury 1962) or man (Fulwiler et al. 1972), or the contamination of food products.

MECHANISMS OF PARTICLE DEPOSITION IN HABITATS

Most of man's habitats, on first approximation, resemble stirred settling chambers. A stirred settling chamber is an enclosed space wherein air is caused to be stirred, or mixed, to the extent that a sample of air taken from any position should contain the same number of suspended, airborne particles (an aerosol) as a sample taken from any other position. Within such enclosures the behavior of the aerosol is so predictable that the expression

$$N_t/N_0 = \exp -kt \tag{4.66}$$

---

*R. L. Dimmick and M. A. Chatigny

(where N_t = number of airborne particles at time *t*, N_0 - number of airborne particles at time 0, *k* = Stokes' velocity/chamber height; see Harrington, chap. 4, this volume) defines the settling loss of particles of a given size on horizontal surfaces. As an example, in a closed room 3 m in height, particles 1 μm in diameter, if created and mixed as a "burst," would disappear from the air at a rate such that 50% would settle in 1150 min, 75% in 2300 min, and so on.

A more rigorous treatment of habitats would include effects of dilution of the aerosol by ventilation and thermal, electrostatic, or turbulent deposition and the increase of the aerosol by activities within the enclosure or the inward leakage of unfiltered, outside air. These factors are more fully discussed by Mason and by Chatigny and Dimmick (chap. 3, this volume). The net result is that mankind is constantly exposed to aerosols of continually varying quantity and quality, and the primary point of deposition is the respiratory tract.

RESPIRATORY DEPOSITION

The factors that influence the deposition of particles in the respiratory tree are those that govern impaction under any situation. The physics of particles are discussed by Harrington (chap. 4, this volume), but briefly, every particle has an inertial property influenced by its size, shape, and density. This property can be defined by a dimensionless number (Ranz and Wong 1952), sometimes called the *stopping distance* (Sinclair 1950) or *inertial parameter* (Harrington, chap. 4, this volume). Since the inertial parameter is related to particle shape and density, the aerodynamic diameter may differ from the diameter measured by microscopy.

When air with contained particles is transmitted through bends in tubing, a given particle in one portion of air (its own airstream) is moved into another stream where it travels for some stopping distance and, if this distance extends as far as the surface of the tubing, the particle then impacts on the wall. If the bend is made to be a jet directed onto a nearby surface, then we have a situation of maximal impaction—a practical right-angle bend. Collectors with this configuration are called *impactors*.

In any given impactor configuration, where air contains particles of varied sizes, we recognize that the larger particles may be removed completely, the smallest ones, with very short stopping distances, may all pass, and the midrange sizes are removed in portions that depend on the positions of the particles in the airstream and their stopping distances. The respiratory tree acts just that way, and the noun *tree* is used to emphasize the similarity to trees that branch again and again and again.

The alveoli, bubblelike structures, are analogous to leaves on a tree, bronchioles are small branches that connect to larger and larger tubular structures (the bronchi), which in turn feed eventually into the trachea (trunk). Inhaled air must pass the nose, the sinus labyrinth, and the glottal opening (which includes the vocal chords) before entering the trachea to be inspired into the multibranched structure of the respiratory tree.

Each of these structures acts as an (albeit imperfect) impactor. The result is that in ordinary breathing, few particles larger than about 5 μm in diameter ever reach the glottal area; those that do seldom penetrate as far as the bronchioles. Particles in the 1- to 3-μm range penetrate and are retained to a maximum extent. Smaller particles are inhaled and then exhaled before they can settle (Stokes velocity), whereas the very small ones (<0.1 μm) behave almost like gas molecules and are retained in large numbers. Particles larger than about 5 μm are collected by the nasal sinuses and throat, and are eventually swallowed. Morrow (1974) has written a particularly well-documented paper describing lung deposition in more detail.

Fate of the Inspired Particle

What happens to an inspired particle depends on a number of factors: the point of deposition, composition and size of the particle (Heyder and Davies 1971), whether the particle is alive or dead, host response (quantitative and qualitative) and physiological state, and synergistic effects of other air contaminants (Morrow 1970). The literature on respiratory retention, clearance, and distribution of inhaled material is voluminous. Though studies have been based on attempts to quantify distribution and retention in the body (Lippmann and Chan 1974), there is no predictive theory that can be generally applied except in a qualitative manner.

A particle that lands in the nasal-pharyngeal area (the glottis, the trachea, the bronchus, in fact, any point above the general bronchiolar branches) adheres to the ubiquitous mucus, which, as a result of ciliary action, is transported into the area of the throat and is swallowed. If the particle lands below this point (particles about 5 μm or less in diameter) then other host mechanisms can serve to isolate or dispose of the particle.

If the particle is not soluble in alveolar fluids, it may be picked up by phagocytes and transported into the bloodstream, where it then behaves as any other insoluble material inoculated intravenously. The particle might cause disruption of the phagocyte and could be transported again, via blood flow, to the lung, or could enter the lymphatic fluid, lodging in the nodes, or, from the blood, could be transported through interstitial walls of the intestine to be excreted in feces. In some cases the particles become immobilized in the lung, causing, for example, such chronic diseases as silicosis (Phibbs et al. 1971), arsenical and other kinds of carcinogenesis (Lee and Fraumeni 1969, Stanton et al. 1969) or pulmonary impairment in miners (Krivanek and Reeves 1972), berylliosis in factory workers (Kanarek et al. 1973), and various other types of allergies or diseases. Radioactive particles are of particular concern (Hanna et al. 1970).

If the particle is soluble, the material simply enters the bloodstream and is diluted throughout the body; subsequently the material is excreted via feces or urine, or it may be metabolized by the liver to other products. If the material is toxic or allergenic (Wittels et al. 1974), it may produce symptoms depending on dosage, excretory rate, or both (Laskin et al. 1970). The mucus provides a solubility barrier that can decrease the immediate concentration effect (for example, with corrosive substances) and limit the rate of absorption.

Typical intramural exposure of the human to both of these types of particles arises from the use of spray cans containing a variety of materials other than the propellant. The authors are not aware of any quantitative studies of the potential chronic toxicity or allergenic reaction from inhalation of these materials, although limited studies of the particle sizes have been reported (Hathaway 1973, Sciarra et al. 1969, Vos and Thomson 1974).

The methods that nature has arranged for expulsion or dilution of inhaled particles are usually effective in the healthy individual (Medici and Buergi 1971, Rassmussen et al. 1968), but persons who are ill often do not have all of the clearance mechanisms available, or the rates of action are impaired, hence these individuals are more susceptible to potential toxic effects (Dadaian et al. 1971). Ciliary action is inhibited to various degrees in the chronic smoker (Boren 1970, Woolf and Suero 1971); the positive-negative air-ion ratio influences ciliary motion; negative ions may stimulate the clearance rate (Krueger and Smith 1959).

Synergistic effects of inhaled gases and particles also influence rates of lung clearance as well as the apparent degree of toxicity (Coffin 1970, Fairchild et al. 1975). Because of the increased surface tension, size, or both of small, moist, particles (Tillery et al. 1973), certain vapors can be more soluble in the particle than predicted by typical equilibrium studies with surface-air interfaces. Whereas inhaling a gas produces a diffuse concentration at any point in the lung (essentially that of the partial pressure of the gas), particles that are inhaled simultaneously with a gas can entrap molecules of the gas that are then concentrated at the point where the particle lands. Further, some gases (e.g., nitrogen dioxide) can irritate the lung mucosa, causing it to be less capable of sustaining all clearance mechanisms (Henry et al. 1973).

Infections

If a particle contains a living microbe(s) in or on it, then it is subjected to essentially the same mechanisms of clearance as above. The exception is whether or not the microbe can propagate in the host, which is simply another way of expressing the quality of infectiousness (Stallybrass 1931). The healthy body contains powerful systems for preventing the growth of microbes; phagocytosis usually destroys the microbe; specific proteins (antibodies) in the blood can lyse or coagulate bacteria, thus preventing their spread; microbial clusters can be walled off, thus preventing access to oxygen or nutrients; interferon can block viral multiplication. Fortunately, only a few microbial species can overcome all these barriers, yet those that do create many problems (Hers et al. 1969), and it is worthwhile to cite some typical examples.

Stress is a factor that can decrease resistance to microbial invasion (Mogabgab 1968). In an epidemiological study of marine recruits (Voors et al. 1968), it was found that having an advanced education, not being promoted at an expected rate, "white ethnic status," or all of these factors caused the recruit to be more susceptible to the common cold. The generally accepted conclusion is that psychological stress (in this case, having to live below one's expected level of achievement or stature) can influence physiological mechanisms that tend to protect the body.

It is generally accepted that the physiological stress per se lowers resistance (Wittels et al. 1974, Vassallo et al. 1970). It is easily demonstrated with laboratory animals; hyperbaric atmospheres and cold stress lowered the 50 percentile lethal respiratory dosage (LD_{50}) of *Klebsiella pneumoniae* in mice from 300 $\pm$ 20 to 200 $\pm$ 20 inhaled cells (Ross and Won 1973). There was no change in the LD_{50} from either one of the stresses alone, so the effect of stress is complex and not fully understood.

The extent and effect of respiratory deposition occurring in man's habitat are influenced by the source of the microbe and the architecture of the habitable structure (Wright et al. 1968, Hammad and Corn 1971). We have already mentioned mines and factories as examples of indoor spaces where problems occur with inorganic aerosols; we now consider some examples of environments where microbes are our reason for concern. In these instances the source is important, and the examples are used to derive some essential principles of air as a vector for infection.

Sources

Hospitals are particularly good examples, for within their walls are contained a greater than normal population of the ill and injured, who serve as reservoirs for many infectious diseases (Fincher 1969, Rosebury 1969). The problems involved with hospitals, in terms of both sepsis and ventilation, are discussed elsewhere (Chatigny and Dimmick, chap. 3, Leedom and Loosli, chap. 6, both this volume), but one classic example of epidemiological investigation is cited here, specifically the work of Riley et al. (1962), who confined guinea pigs in a specially constructed chamber within a duct transporting air from a tuberculosis (TB) ward. Even with appropriate controls and the fact that guinea pigs are especially sensitive to the TB bacillus, a 2-yr exposure of animals was required to obtain undisputed proof that TB was indeed spread via the air. The rate of animal infection was highest when highly active (extensively coughing) patients were in the ward.

Even more elusive to prove is the spread of airborne microbes by industrial operations; two cases are cited. In 1958 a student was studying blood sera of patients with respiratory illness (colds) who were visiting local clinics. He happened to find a sample that contained antibodies for Q fever (Wellock 1960). The records indicated that the donor had not been in areas where the combination of infected host (sheep, goats) and the tick vector was indigenous. On the basis of this finding Wellock sought more samples positive for Q fever and he found them! One donor was an invalid confined for two years in an upstairs apartment. When the dwelling places of positive donors were plotted on a city map of Oakland, California, a large triangle was formed with the apex pointing in the direction of the generally prevailing winds. The neighborhood near the apex contained a tanning plant where sheep hides were scrubbed with a motor-driven brush in combination with steam. Some hides were found to contain the Q fever microbe, and it was also isolated from air within the plant. The finding was essentially accidental. Routine checks on the blood of patients with colds would never have revealed proof of past exposure to Q fever.

Only about 40 cases of Q fever in the whole city of Oakland were found. If only the area of the triangle were included, the probability of infection is approximately 4×10^{-5}—a small number indeed. It is obvious that magnificent efforts would have to be employed to demonstrate the presence of other possible airborne diseases, especially by standard air-sampling methods.

Another incident was an outbreak of brucellosis in a packing plant in Iowa (Hendricks et al. 1962). Although not so dramatic as the previously described case, this outbreak was first noted when several cases of brucellosis were reported within one county. The packing plant appeared to be an obvious source, and indeed the brucellosis microbe (*Brucella melitensis, Br. abortus,* and *Br. suis*) was isolated from animal parts and was found in the air. Additional investigation showed that altogether 128 persons had been infected, most of them workers in the plant.

In 1972 a rabies outbreak that occurred in 1967–1968 in an animal colony in New Mexico was reported (Winkler et al. 1972, 1973). Altogether 64 animals died. This work was one of the most comprehensive epidemiological investigations that has come to our attention. The authors concluded that animals that had previously been experimentally exposed shed the virus, and that the epidemic in the "clean" colony was the result of aerosol transmission of infection although no rabies virus was isolated from the air by mechanical samplers. This tends to illustrate the principle that animals are more sensitive samplers than mechanical samplers; other instances have also shown this to be true (Converse and Reed 1966).

We may conclude from these examples that if operations that produce aerosols are done with infective material, then infective particles will be found in or on people and respiratory outbreaks will occur. If this is so, then we can point to two sources of possible reservoirs of infection that could lead to deposition in the human lung. One is the dental operatory.

An elegant experiment was conducted by Miller et al. (1963). They seeded a patient with a tracer microbe and were able to recover viable bacteria from air in an adjoining waiting room within 3 min after the start of typical cleaning or drilling operations. The obvious conclusion is that if a patient is carrying an infectious microbe in the oral cavity, the persons in the waiting room will be exposed to some probability of infection, and the dentist of course will obtain a moderately high dose. That this has not been named as a focus for epidemics may be the result of the low probability of infection, as noted with Q fever, or it may be that insufficient effort has been made to study the epidemiological picture. There is indeed presumptive evidence that the dental operatory is a public health problem, but dentists do not suffer greater incidence of respiratory disease than other medical doctors. Dental students, however, have nearly twice the number of respiratory illnesses as do medical students in the same institution (Burton and Miller 1963).

Another source for potential deposition of particles in the respiratory tract is the flush toilet. It is known that bursting bubbles and the splash of droplets produce aerosols (Blanchard and Syzdek 1974). Darlow and Bale (1959) studied this problem and found as many as 10^5 airborne microbes per m^3 of air at toilet seat level. They conclude that both the flush toilet and the urinal represent important public health problems. Gerba et al. (1975) recently arrived at the same conclusions.

An example of the importance of "man-made" airborne particles as irritants or allergens of man in an enclosed habitat can be cited—i.e., hypersensitivity lung disease in the turkey-raising industry (Boyer et al. 1974). Of 205 subjects in the turkey industry, including those in processing plants, 142 reported having respiratory symptoms within 1 h after working with the birds. Eighteen percent had positive skin tests to turkey antigens, and 11% had positive sera. The syndrome resembles that frequently observed in pigeon (Riley and Saldana 1973) and parakeet breeders.

Examples of smog-induced or smoke-induced irritation are too numerous and generally well known for an extensive list to be appropriate here, but a final example will be cited. A comparison in 1962 of inhabitants of the Tokyo-Yokahama area by Oshima et al. (1964a,b) with the inhabitants of Niigata showed highly significant differences in four defined types of respiratory symptoms between the two areas. Residents in the Tokyo area were more prone to asthmalike symptoms than were residents of rural Niigata area.

DEPOSITION ON FOODS

Deposition of particles on food products that are to be stored is principally an economic problem, although not serious, in the commercial canning industry. It is an annoyance in home canning when jars leak; though the sterilization process may have been proper, the intake of air while leaky jars cool can cause airborne particles to enter. Improper sterilization rather than leakage is the usual cause of infrequent cases of botulism. In the frozen food industry, isolated, unavoidable, and usually undetected instances of excessive contamination by airborne microbes can cause the product to be susceptible to spoilage after thawing; hence the standard warning not to keep thawed food at refrigeration temperatures for extended periods or to refreeze the product.

CONCLUSIONS

From all the examples cited in this chapter, the following is evident: (1) Indoor air carries many airborne particles, and those that remain airborne for appreciable lengths of time are in the respirable range. (2) There is interaction between indoor and outdoor air. (3) There is a correlation between the number and strength of sources and the average concentration of airborne particles. (4) Assessing the impact of airborne particles is difficult because of variations of the quantitative and qualitative nature of the source, the high variability of individual host response, and the usually low probability of retention per individual. (5) Finally, an ironic corollary to (4) above is that, regardless of the elusive nature of the interaction of air with the ecosystem, for the individual who happens to have inhaled and retained a minimal infective, toxic, or allergenic dosage the probability has become 1.00.

The problems that occur as a result of the interaction of airborne particles and man (including his habitats and food chain) are of the same nature as any other ecological problem, and little effort will be devoted to solutions until and after situations arise that do produce economic effects wherein the price of ignorance will have been greater than the worth of the status quo: The ill-conceived reaction in 1975–1976 to a suspected swine flu epidemic of the 1918 variety is, perhaps, a classic example.

FOOTNOTES

[1]Stephan flow (Facy effect) is the excess flow of molecules toward a center of condensation causing a net pressure on particles toward such centers.

[2]The transformation is carried out as follows:

Setting

$$K_z(z) = K_z(h)K_z(Z)$$

and

$$A(z) = A(0)A(Z)$$

then Equation (4.48) can be written:

$$\frac{\partial^2 u}{\partial z^2} + \frac{\partial \ln K_z(z)}{\partial z}\left(\frac{\partial u}{\partial z}\right) - \frac{C_D A(z) u^2(z)}{2K_z(z)} = 0$$

and transformed into:

$$\frac{\partial^2 U}{\partial Z^2} + \frac{\partial \ln K(Z)}{\partial Z} \cdot \frac{\partial U}{\partial Z} - \frac{A(0)u(h)h^2 C_D}{K_z(h)} \cdot \frac{A(Z)}{K_z(Z)}U^2 = 0$$

or in abbreviated form into:

$$U'' + f(Z)U' - Sg(Z)U^2 = 0$$

[3]Time until the concentration is reduced to $1/e$ of its original value.

LITERATURE CITED

Belot, Y., A. Baille, and J. L. Delmas. 1976. Modèle numerique de dispersion des pollutants atmospherique en presence de couverts végétaux. Application aux couverts forestiers. Atmos. Environ. 10:89–98.

Belot, Y., and D. Gauthier. 1975. Transport of micronic particles from atmosphere to foliar surfaces. Chapter 40 *in* D. A. deVries and N. H. Afgan, eds. Heat and mass transfer in the biosphere. I. Transfer processes in the plant environment. Scripta Publ. Co., Washington, D.C.

Blanchard, D. C., and L. D. Syzdek. 1974. Bubble tube: Apparatus for determining rate of collection of bacteria by an air bubble rising in water. Limnol. Oceanogr. 19:133–138.

Boren, H. G. 1970. Pulmonary cell kinetics after exposure to cigarette smoke. Pages 229–242 *in* M. G. Hanna, Jr., P. Nettesheim, and J. R. Gilbert, eds. Inhalation carcinogenesis. Natl. Cancer Inst./USAEC Div. Tech. Inf., Washington, D.C.

Boyer, R. S., L. E. Klock, C. D. Schmidt, L. Hyland, K. Maxwell, R. M. Gardner, and A. D. Renzetti, Jr. 1974. Hypersensitivity lung disease in the turkey raising industry. Am. Rev. Respir. Dis. 109:630–635.

Brun, R. J., and H. W. Mergler. 1953. Impingement of water droplets on a cylinder in an incompressible flow field and evaluation of rotating multicylinder methods for measurement of droplet-size distribution, volume-median droplet size, and liquid-water content in clouds. NACA Publ. TN 2904. National Advisory Committee for Aeronautics, Washington, D.C.

Buchbinder, L., M. Solowey, and M. Solotorovsky. 1945. Comparative quantitative studies of bacteria in air of enclosed places. (Part I of air pollution survey report.) J. Am. Soc. Heat. Ventil. Eng. 7:389–397.

Burton, W. E., and R. L. Miller. 1963. The role of aerobiology in dentistry. *In* R. L. Dimmick, ed. Proc. First Int. Symp. Aerobiology. Naval Biol. Lab., U.S. Naval Supply Cent., Oakland, Calif.

Businger, J. A. 1975. Aerodynamics of vegetated surfaces. Chapter 10 *in* D. A. deVries and N. H. Afgan, eds. Heat and mass transfer in the biosphere. I. Transfer processes in the plant environment. Scripta Publ. Co., Washington, D.C.

Caporaloni, M., F. Tampieri, F. Trombetti, and O. Vittori. 1975. Transfer of particles in nonisotropic air turbulence. J. Atmos. Sci. 32:565–568.

Chamberlain, A. C. 1966. Transport of *Lycopodium* spores and other small particles to rough surfaces. Proc. R. Soc. London, Ser. A., 296:45–70.

Chamberlain, A. C. 1975. Pollution in plant canopies. Chapter 39 *in* D. A. deVries and N. H. Afgan, eds. Heat and mass transfer in the biosphere. I. Transfer processes in the plant environment. Scripta Publ. Co., Washington, D.C.

Cheng, L. 1973. Formation of airborne-respirable dust at belt conveyor transfer points. J. Am. Ind. Hyg. Assoc. 34:540–546.

Cionco, R. M. 1965. A mathematical model for air flow in a vegetative canopy. J. Appl. Meteorol. 4:717–722.

Cionco, R. M. 1972a. Wind profile index for canopy flow. Boundary-Layer Meteorol. 3:255–263.

Cionco, R. M. 1972b. Intensity of turbulence within canopies with simple and complex roughness elements. Boundary-Layer Meteorol. 2:453–465.

Clough, W. S. 1975. The deposition of particles on moss and grass surfaces. Atmos. Environ. 9:1113–1120.

Coffin, D. L. 1970. Study of the mechanisms of the alteration of susceptibility to infection conferred by oxidant air pollutants. Pages 259–270 *in* M. G. Hanna, Jr., P. Nettesheim, and J. R. Gilbert, eds. Inhalation carcinogenesis. Natl. Cancer Inst./USAEC Div. Tech. Inf., Washington, D.C.

Converse, J. L., and R. E. Reed. 1966. Experimental epidemiology of coccidioidomycosis. Bacteriol. Rev. 30:678–694.

Dadaian, J. H., S. Yin, and G. A. Laurenzi. 1971. Studies of mucus flow in the mammalian respiratory tract. II. The effects of serotonin and related compounds on respiratory tract mucus flow. Am. Rev. Respir. Dis. 103:808–815.

Darlow, H. M., and W. R. Bale. 1959. Infective hazards of water-closets. Lancet 1:1196–1200.

deVries, D. A., and N. H. Afgan. 1975. Heat and mass transfer in the biosphere. I. Transfer processes in the plant environment. Scripta Publ. Co., Washington, D.C.

Fairchild, G. A., S. Stultz, and D. L. Coffin. 1975. Sulfuric acid effect on the deposition of radioactive aerosol in the respiratory tract of guinea pigs. J. Am. Ind. Hyg. Assoc. 38:584–594.

Fincher, E. L. 1969. Aerobiology and hospital sepsis. Chapter 17 *in* R. L. Dimmick and A. B. Akers, eds. An introduction to experimental aerobiology. Wiley-Interscience, New York.

Forney, L. J., and L. A. Spielman. 1974. Deposition of coarse aerosols from turbulent flow. J. Aerosol Sci. 5:257–274.

Fulwiler, R. D., J. C. Abbott, and F. J. Darcy. 1972. The evaluation of detergent enzymes in air. J. Am. Ind. Hyg. Assoc. 33:231–236.

Gerba, C. F., C. Wallis, and J. L. Melnick. 1975. Microbiological hazards of household toilets: Droplet production and the fate of residual organisms. Appl. Microbiol. 30:229–237.

Gregory, P. H. 1973. The microbiology of the atmosphere, 2nd ed. Halstead Press Div., John Wiley, New York. 377 p.

Gregory, P. H., and J. L. Monteith, eds. 1967. Airborne microbes. Cambridge Univ. Press, Cambridge, England. 385 p.

Hammad, Y. Y., and M. Corn. 1971. Hygienic assessment of airborne cotton dust in a textile manufacturing facility. J. Am. Ind. Hyg. Assoc. 32:662–667.

Hanna, M. G., Jr., P. Nettesheim, and J. R. Gilbert, eds. 1970. Inhalation carcinogenesis. Natl. Cancer Inst./USAEC Div. Tech. Inf., Washington, D.C.

Harrington, J. B. 1965. Atmospheric diffusion of ragweed pollen in urban areas. Volume II *in* E. W. Hewson, ed. Atmospheric pollution by aeroallergens: Meteorological phase. Univ. Michigan Press, Ann Arbor.

Harrington, J. B., G. C. Gill, and B. R. Warr. 1959. High-efficiency pollen samplers for use in clinical allergy. J. Allergy 30:357–375.

Harrington, J. B., and J. W. Mansell. 1973. A model of air pollution penetration into a forest canopy. Bull. Am. Meteorol. Soc. 53:1031. (Abstract.)

Harrington, J. B., and K. Metzger. 1963. Ragweed pollen density. Am. J. Bot. 50:532–539.

Hathaway, D. 1973. Particle size measurement of an aerosol deodorant using laser holographic microscopy. Aerosol Age 18:28.

Hendricks, S. L., I. H. Borts, R. H. Heeren, W. J. Hausler, and H. R. Held. 1962. Brucellosis outbreak in an Iowa packing house. Am. J. Public Health 52: 1166–1178.

Henry, M. C., C. Aranyl, and R. Ehrlich. 1973. Scanning electron microscopy observations of the effects of atmospheric pollutants and infectious agents. Pages 216–219 *in* J. F. P. Hers and K. C. Winkler, eds. Airborne transmission and airborne infection. Oosthoek Publ. Co., Utrecht, The Netherlands.

Hers, J. F. P., N. Masurel, and J. C. Gans. 1969. Acute respiratory disease associated with pulmonary involvement in military servicemen in The Netherlands. A serologic and bacteriologic survey, January 1967 to January 1968. Am. Rev. Respir. Dis. 100:499–506.

Heyder, J., and C. N. Davies. 1971. The breathing of half-micron aerosols. II. Dispersion of particles in the respiratory tract. J. Aerosol Sci. 2:437–452.

Inoue, E. 1963. On the turbulent structure of airflow within crop canopies. J. Meteorol. Soc. Japan 41:317–325.

Junge, C. E. 1963. Air chemistry and radioactivity. Academic Press, New York. 382 p.

Kanarek, D. J., R. A. Wainer, R. I. Chamberlin, A. L. Weber, and H. Kazemi. 1973. Respiratory illness in a population exposed to beryllium. Am. Rev. Respir. Dis. 108:1295–1302.

Kinerson, R., Jr., and L. J. Fritschen. 1971. Modeling a coniferous forest canopy. Agric. Meteorol. 8:439–445.

Krivanek, N., and A. L. Reeves. 1972. The effect of chemical forms of beryllium on the production of the immunologic response. J. Am. Ind. Hyg. Assoc. 33:45–52.

Krueger, A. P., and R. Smith. 1959. An enzymatic basis for the acceleration of ciliary activity by negative air ions. Nature (Lond.) 183:1332–1333.

Landsberg, J. J., and A. S. Thom. 1971. Aerodynamic properties of a plant of complex structure. Q. J. R. Meteorol. Soc. 97:565–570.

Laskin, S., M. Kuschner, and R. T. Drew. 1970. Studies in pulmonary carcinogenesis. Pages 321–352 *in* M. G. Hanna, Jr., P. Nettesheim, and J. R. Gilbert, eds. Inhalation carcinogenesis. Natl. Cancer Inst./USAEC Div. Tech. Inf., Washington, D.C.

Leadbetter, M. R., and M. Corn. 1972. Particle size distribution of rat lung residues after exposure to fiberglass dust clouds. J. Am. Ind. Hyg. Assoc. 33:511–522.

Lee, A. M., and J. F. Fraumeni, Jr. 1969. Arsenic and respiratory cancer in man: An occupational study. J. Natl. Cancer Inst. 42:1045–1052.

Legg, B. J., and J. L. Monteith. 1975. Heat and mass transfer within plant canopies. Chapter 11 *in* D. A. deVries and N. H. Afgan, eds. Heat and mass transfer in the biosphere. I. Transfer processes in the plant environment. Scripta Publ. Co., Washington, D.C.

Leonard, R. E., and C. A. Federer. 1973. Estimated and measured roughness parameters for a pine forest. J. Appl. Meteorol. 12:302–307.

Lippmann, M., and R. E. Albert. 1969. The effect of particle size on the regional deposition of inhaled aerosols in the human respiratory tract. J. Am. Ind. Hyg. Assoc. 30:257–275.

Lippmann, M., and T. L. Chan. 1974. Calibration of dual-inlet cyclines for "respirable" mass sampling. J. Am. Ind. Hyg. Assoc. 35:189–200.

Medici, T. C., and H. Buergi. 1971. The role of immunoglobulin A in endogenous bronchial defense mechanisms in chronic bronchitis. Am. Rev. Respir. Dis. 103:784–791.

Miller, R. L., W. E. Burton, and R. W. Spore. 1963. Aerosols produced by dental instrumentation. Pages 97–120 *in* R. L. Dimmick, ed. Proc. First Int. Symp. Aerobiology. Naval Biol. Lab., U.S. Naval Supply Cent., Oakland, Calif.

Mogabgab, W. J. 1968. Acute respiratory illnesses in university (1962–1966), military and industrial (1962–1963) populations. Am. Rev. Respir. Dis. 98:359–379.

Moller, U., and G. Schumann. 1970. Mechanisms of transport from the atmosphere to the earth's surface. J. Geophys. Res. 75:3013–3019.

Monteith, J. L. 1975. Vegetation and the environment. Academic Press, New York.

Morrison, F. A., Jr. 1974. Inertial impaction in stagnation flow. J. Aerosol Sci. 5:241–250.

Morrow, P. E. 1970. Models for the study of particle retention and elimination in the lung. *In* M. G. Hanna, Jr., P. Nettesheim, and J. R. Gilbert, eds. Inhalation carcinogenesis. Natl. Cancer Inst./USAEC Div. Tech. Inf., Washington, D.C.

Morrow, P. E. 1974. Aerosol characterization and deposition. Am. Rev. Respir. Dis. 110:88–99.

Oliver, H. R. 1971. Wind profiles in and above a forest canopy. Q. J. R. Meteorol. Soc. 97:548–553.

Oliver, H. R. 1975a. Wind speeds within the trunk space of a pine forest. Weather 28:345–347.

Oliver, H. R. 1975b. Ventilation in a forest. Q. J. R. Meteorol. Soc. 101:167–172.

Oliver, H. R., and G. J. Mayhead. 1974. Wind measurements in a pine forest during a destructive gale. Agric. Meteorol. 14:347–355.

Oshima, Y., T. Ishizaki, T. Miyamoto, J. Kabe, and S. Makino. 1964a. A study of Tokyo-Yokohama asthma among Japanese. Am. Rev. Respir. Dis. 90:632–634.

Oshima, Y., T. Ishizaki, T. Miyamoto, T. Shimizu, T. Shida, and J. Kabe. 1964b. Air pollution and respiratory diseases in the Tokyo-Yokohama area. Am. Rev. Respir. Dis. 90:572–581.

Phibbs, B. P., R. E. Sundin, and R. S. Mitchell. 1971. Silicosis in Wyoming bentonite workers. Am. Rev. Respir. Dis. 103:1–17.

Pilat, M. J., and A. Prem. 1976. Calculated particle collection efficiencies of single droplets including inertial impaction. Brownian diffusion, diffusiophoresis and thermophoresis. Atmos. Environ. 10:13–20.

Plate, E. J., and A. A. Quraishi. 1965. Modeling of velocity distributions inside and above tall crops. J. Appl. Meteorol. 4:400–408.

Ranz, W. E., and J. B. Wong. 1952. Jet impactors for determining the particle size distribution of aerosols. AMA Arch. Ind. Hyg. Occup. Med. 5:464–477.

Rasool, S. I. 1973. Chemistry of the lower atmosphere. Plenum Press, New York. 335 p.

Rasmussen, D. L., W. A. Laquer, P. Futterman, H. D. Warren, and C. W. Nelson. 1968. Pulmonary impairment in southern West Virginia coal miners. Am. Rev. Respir. Dis. 98:658–676.

Riley, D. J., and M. Saldana. 1973. Pigeon breeder's lung. Subacute course and the importance of indirect exposure. Am. Rev. Respir. Dis. 107:456–460.

Riley, R. L., C. C. Mills, F. O'Grady, L. U. Sultan, F. Wittstadt, and D. N. Shivpuri. 1962. Infectiousness of air from a tuberculosis ward. Am. Rev. Respir. Dis. 85:511–525.

Rodhe, H. J., and H. Grandell. 1972. On the removal of aerosol particles from the atmosphere by precipitation scavenging. Tellus 24:442–454.

Rosebury, T. 1962. Microorganisms indigenous to man. McGraw-Hill Publ., New York.

Rosebury, T. 1969. Life on man. Viking Press, New York. 239 p.

Ross, H. C., and W. D. Won. 1973. Susceptibility of mice to airborne *Klebsiella pneumoniae* infection in helium-oxygen atmospheres. Aerosp. Med. 44:1009–1012.

Sciarra, J. J., P. McGinley, and L. Izzo. 1969. Determination of particle size distribution of selected aerosol cosmetics. I. Hair sprays. J. Soc. Cosmet. Chem. 20: 385.

Scriven, R. A., and B. E. Fisher. 1975a,b. Long range transport of airborne material and its removal by deposition and washout. 1. General considerations. 2. Effect of turbulent diffusion. Atmos. Environ. 9:49–68.

Seginer, T. 1974. Aerodynamic roughness of vegetated surfaces. Boundary-Layer Meteorol. 5:383–393.

Sehmel, G. A. 1970. Particle deposition for turbulent air flow. J. Geophys. Res. 75:1766–1781.

Sehmel, G. A. 1971. Particle diffusivities and deposition velocities over a horizontal smooth surface. J. Colloid Interface Sci. 37:891–906.

Shaw, R. H., R. H. Silversides, and G. W. Thurtell. 1974. Some observations of turbulence and turbulent transport within and above plant canopies. Boundary-Layer Meteorol. 5:429–449.

Sinclair, D. 1950. Stability of aerosols and behavior of aerosol particles. Chapter 5 *in* Handbook of aerosols. U.S. Gov. Print. Off., Washington, D.C.

Smith, F. B., D. J. Carson, and H. R. Oliver. 1972. Mean wind direction shear through a forest canopy. Boundary-Layer Meteorol. 3:178–180.

Stallybrass, C. O. 1931. The principles of epidemiology and the process of infection. Macmillan Co., New York.

Stanton, M. F., R. Blackwell, and E. Miller. 1969. Experimental pulmonary carcinogenesis with asbestos. J. Am. Ind. Hyg. Assoc. 30:236–244.

Tan, H. S., and S. C. Ling. 1963. A study of atmospheric turbulence and canopy flow. Part II *in* E. R. Lemon, ed. The energy budget at the earth's surface. Prod. Res. Rep. 72, USDA Agric. Res. Serv.

Thom, A. S. 1971. Momentum absorption by vegetation. Q. J. R. Meteorol. Soc. 97:414–428.

Tillery, M. I., O. R. Moss, H. J. Ettinger, and G. W. Royer. 1973. Effect of humidity on the aerodynamic size characteristics of nonhygroscopic aerosols. J. Am. Ind. Hyg. Assoc. 34:440–449.

Vassallo, C. L., Z. A. Zawadzki, and J. R. Simons. 1970. Recurrent respiratory infections in a family with immunoglobulin A deficiency. Am. Rev. Respir. Dis. 101:245–251.

Voors, A. W., G. T. Stewart, R. R. Gutekunst, C. F. Moldow, and C. D. Jenkins. 1968. Respiratory infection in marine recruits. Am. Rev. Respir. Dis. 98:801–809.

Vos, K., and D. B. Thomson. 1974. Particle size measurement of eight commercial pressurized products. Powder Technol. 10:103.

Wedding, J. B., R. W. Carlson, J. J. Stukel, and F. A. Bazzaz. 1975. Aerosol deposition on plant leaves. Environ. Sci. Technol. 9:151–153.

Wellock, C. E. 1960. Epidemiology of Q-fever in the urban East Bay area. Calif. Health 18:72–76.

Wesley, M. L., B. B. Hicks, W. P. Dannevik, S. Frisella, and R. B. Husar. 1977. Measurement of particle deposition from the atmosphere. Atmos. Environ. 11:561–563.

Winkler, P. 1974. Relative humidity and the adhesion of atmospheric particles to the plates of impactors. J. Aerosol Sci. 5:235–240.

Winkler, W. G., E. F. Baker, Jr., and C. C. Hopkins. 1972. An outbreak of non-bite transmitted rabies in a laboratory animal colony. Am. J. Epidemiol. 95: 267–277.

Winkler, W. G., T. R. Fashinell, L. Leffingwell, P. Howard, and J. P. Conomy. 1973. Airborne rabies transmission in a laboratory worker. J. Am. Med. Assoc. 226:1219–1221.

Wittels, E. H., J. J. Coalson, M. H. Welch, and C. A. Guenter. 1974. Pulmonary intravascular leukocyte sequestration: A potential mechanism of lung injury. Am. Rev. Respir. Dis. 109:502–509.

Woolf, C. R., and J. T. Suero. 1971. The respiratory effects of regular cigarette smoking in women. Am. Rev. Respir. Dis. 103:26–37.

Wright, D. N., E. M. K. Vaichulis, and M. A. Chatigny. 1968. Biohazard determination of crowded living-working spaces: Airborne bacteria aboard two naval vessels. J. Am. Ind. Hyg. Assoc. 29:574–581.

5. SAMPLING TECHNIQUES IN AEROBIOLOGY

G. S. Raynor

Sampling methods in aerobiology are nearly as diverse as the disciplines included in the field. This results from the great variety of airborne biological particles and the various reasons for sampling them. Although some particles are of purely scientific interest so far, the economic or medical importance of others has been known for many years. Thus sampling techniques for certain purposes were first devised over a century ago while new and improved instruments have appeared periodically to the present time. Gregory (1973) gives an excellent historical account of early work in the field.

Air sampling is conducted for many purposes and may be qualitative or quantitative. In the former, the investigator may wish to determine which species of microparticles within some selected group are present and how their occurrence changes with time or other variables. Such studies include surveys over space, time, or both. In quantitative sampling, an attempt is made to measure actual concentrations in the sampled air. These may then be related to other variables such as outbreaks of plant disease, occurrence and severity of human allergy, or other known or presumed effects of the particles sampled. Some samplers give quantitative measurements, at least for a restricted size range of particles, while others are useful only for qualitative studies.

Techniques of air sampling must satisfy the purpose of the sampling program, be reasonably efficient at catching the particles of interest, and be compatible with required counting or analytical methods. There is no universal sampler and each discipline has developed its own sampling methods. Therefore a sampling device or method should be selected only after the purpose of sampling has been established, the characteristics of the particles to be sampled are known, and methods for handling the samples have been chosen. No attempt has been made to cover air sampling methods for gaseous air pollutants. tions.

CHARACTERISTICS OF AIRBORNE BIOLOGICAL PARTICLES

Since choice of preferred sampling methods is dependent on characteristics of airborne particles, these characteristics are briefly summarized for several important groups. Size, shape, and surface structure are well known for many of the more common particles but few measurements of density have been made and some reported in the literature may be inaccurate. Also, little information exists on the viability in air or on a sampling surface of most small living organisms.

Characteristics of airborne particles in general were summarized briefly by Edmonds (1972). Anemophilous pollens were described by several authors including Erdtman (1969), Faegri and Iverson (1964), Wodehouse (1935, 1942), and Ogden et al. (1974). Sizes of airborne pollen grains range from about 10 μm in diameter to somewhat over 100 μm in greatest dimension, but most are from 20 to 40 μm. Little range in size occurs within grains from a single species. Most are more or less spherical, often with roughened or sculptured surfaces, but a variety of other shapes occur. The pollens of many conifers, for instance, have large bladderlike structures on either end. Certain thin-walled pollens such as in some grasses are roughly spherical but may partially collapse, resulting in varied shapes.

Many more species of fungus spores than of pollens are found in the atmosphere. They are more variable in both size and shape and usually are more difficult to identify. Many diverse families produce spores that are outwardly similar and not visually separable. On the other hand, an appreciable range of sizes is found among spores of a single species. Many fungus spores are in the 4- to 16-μm range, but some are only 1–2 μm in diameter while others are well over 100 μm. Thus not all can be efficiently sampled by a single device even for visual identification, while many must be cultured to determine their identity. No single source exists that describes most fungus spores, but a good introduction with many references is given in Ogden et al. (1974). Many of the common airborne fungus spores are also described by Southworth (1974).

Spores of mosses and ferns are seldom predominant components of the air spora but may be locally common at times. Moss spores are in the same size range as fungus spores, mostly from 2 to 40 μm, but fern spores average larger, in the pollen size range.

Airborne algae and Protozoa are less well known than pollens and spores. Concentrations average less except in favored locations, but some may be important allergens. Algae may occur as single cells no greater than a few micrometers in size or as larger vegetative fragments of varying size and shape. Protozoa have a wide range in size but those airborne are mostly under 100 μm. The same applies to lichen spores and fragments, which are sometimes taken in air samples.

A number of larger particles must also be considered. Adult insects of small, weak, flying species are often transported more or less passively by the wind as are certain larvae and spiders that spin silken threads to aid in aerial dispersal. Most are in the millimeter size range and require sampling methods quite different from smaller particles. Insect hairs and other fragments as well as various small vegetative debris (Benninghoff 1971) occur in various sizes and are frequently caught while sampling for other particles. Certain seeds are often transported appreciable distances through the air, including those such as the dandelions and milkweeds with attached filaments and those with wing-like appendages. Deposition of seeds is usually of more interest than their concentration in air.

At the small end of the size spectrum are bacteria and viruses, but they are not considered extensively here since sampling for them is specialized and practiced most often in enclosed spaces.

SAMPLING PRINCIPLES

A great number of devices are in use for sampling airborne particles. Each is best suited for a limited particle size range and all operate on only a few basic principles, which are described and illustrated below.

GRAVITATIONAL SETTLING

Exposure of a horizontal surface on which particles can settle by gravity is the simplest method of collecting airborne particles and is frequently used. In theory, particles simply settle at their terminal velocity and are retained by an adhesive on the sampling surface. Terminal velocity of a small, smooth, spherical particle can be computed from the Stokes equation (Harrington, chap. 4, this volume), which is not exact for rough or nonspherical particles but should not be greatly in error for many airborne particles.

In practice, conditions are more complicated as shown in Figure 5.1. Collection efficiency is a complex function of particle size, wind speed, wind direction, and turbulence, as well as particle concentration. It is therefore impossible to define the volume of air sampled or to compute the concentration. Counts are not comparable with time or place unless meteorological conditions are identical; however, such samples give an indication of particles present and a very rough idea of their abundance.

When a horizontal sampling surface is exposed on the ground, it does give a measure of deposition per unit area on that particular surface. Deposition to the surrounding natural surface, however, may be much different.

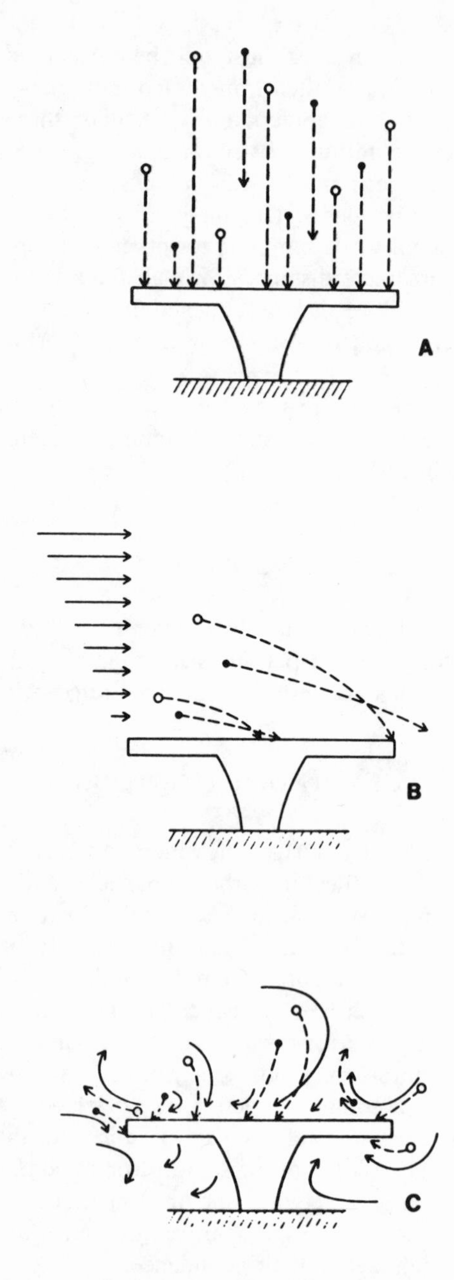

Figure 5.1. Mechanisms of particle collection by "gravity slide" sampler; solid arrows = air trajectories, dashed arrows = particle trajectories. (A) Gravitational settling or sedimentation in calm air—large particles fall more rapidly than small particles, which may not descend to sampling surface during a finite sampling period. (B) Settling in a wind speed increasing with height—trajectories become less horizontal and more vertical as particle descends to layers of decreasing wind speed. Large particles still settle more rapidly than small particles. (C) Settling in turbulent air—particles are collected by turbulent impingement. Particles generally follow eddy motions, but their paths are modified slightly by gravity.

Since wind speeds are generally much greater than gravitational settling rates, most small airborne particles travel a nearly horizontal course. Their mass and velocity give them an inertial force that resists changes in speed and direction. When a particle approaches a physical obstacle, the air molecules surrounding the particle divert and flow around the obstacle. If the particle has sufficient inertia, it will continue on its original course or on a path somewhere between that and the path of the air molecules, and may strike the obstacle. In the atmosphere the efficiency of impaction (the percentage of particles approaching an obstacle that actually strike it) is a direct funcion of the size, mass, and velocity of the particle and an inverse function of the size of the obstacle. Equations describing the impaction process are mathematically difficult and have been solved only for smooth, spherical particles and for simple obstacle shapes such as spheres and cylinders. The mathematics are given by Brun and Mergler (1953) and by Green and Lane (1964).

Although theory deals only with efficiency of impaction, the efficiency of retention is also important. A particle, upon impact, may either stick to the obstacle or rebound from it and reenter the airstream. A sampling surface must be coated with a good adhesive to ensure adequate retention. Sampling efficiency is a product of impaction efficiency and retention efficiency and can be determined experimentally in a wind tunnel.

Since impaction efficiency is a function of wind speed, particle characteristics, and collector size, acceptable efficiency can be obtained only for certain combinations of these variables. Particles may impact on obstacles of any shape, but vertical cylinders are most commonly used as impaction samplers since they are horizontally symmetrical and their impaction efficiency can be calculated.

The simplest form of wind impaction sampler is a small cylinder coated with adhesive and mounted in a fixed position (Figure 5.2). Such samplers are seldom used in the atmosphere since the entire circumference must be examined for collected particles if exposed to variable or shifting winds. The cylinder is usually mounted on the front of a wind vane so that collection occurs on only one side. Choice of cylinder diameter is governed by the wind speeds expected and the size of the particles to be sampled, and usually must be a compromise between high efficiency and length of the sampling period. Small cylinders are more efficient (Figure 5.3) but are more subject to overloading, which causes a progressive decline in efficiency. Cylinders from 1 to 6 mm in diameter have been used for collecting pollens.

Wind impaction samplers are better suited for use in controlled experiments than for routine daily sampling or for the collection of a variety of particulates. Conversion of counts to concentrations requires an efficiency curve for each type of particle sampled and an accurate measure of wind speed.

The basic disadvantage of wind impaction samplers (their change in efficiency with wind speed) was largely overcome, but their advantages were re-

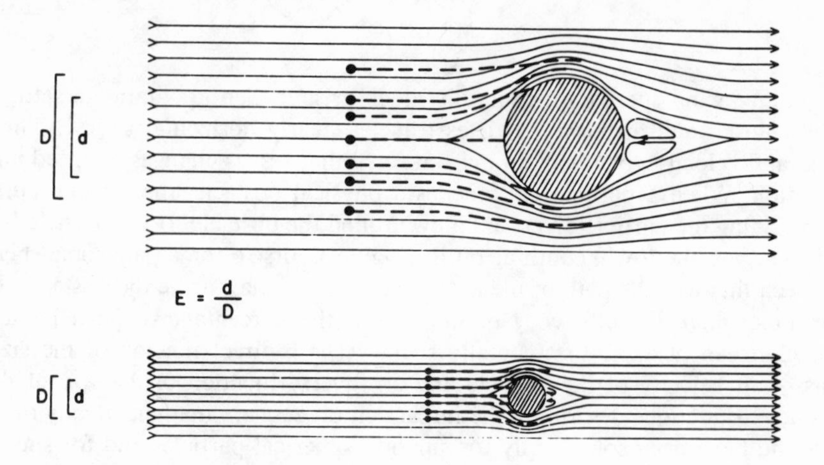

Figure 5.2. Air (▬▬) and particle (▬ ▬) trajectories around large and small cylinders illustrating the greater impaction efficiency (E) of the small cylinder. Here $E = d/D$, where d = partial diameter of the cylinder from which particles impact and D = cylinder diameter.

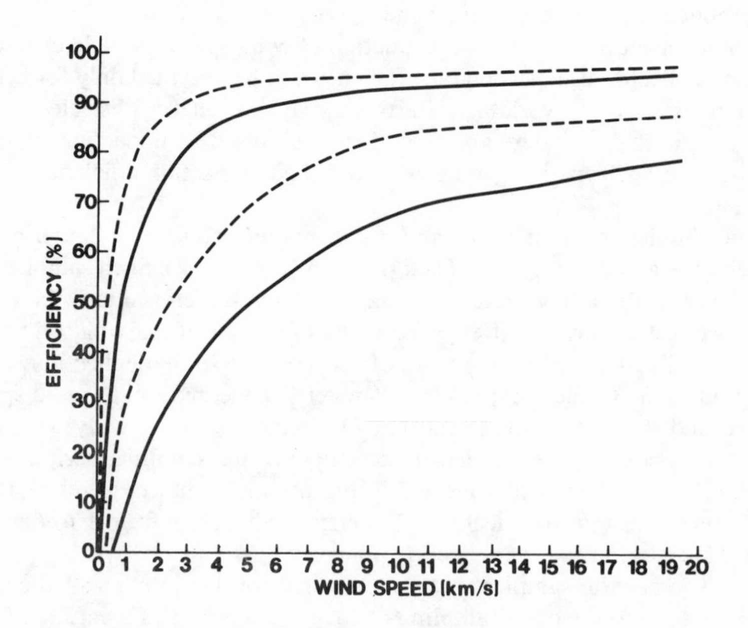

Figure 5.3. Computed curves relating efficiency of impaction of pollen particles of two diameters, ragweed (20 μm, ▬▬) and timothy (33 μm, ▬ ▬), on 1-mm-(upper curves) and 6.3-mm-diameter (lower curves) cylinders as a function of wind speed.

tained by the development of motor-driven, rotating, impaction samplers. In these devices two vertical collector arms are mounted at the ends of a cross-arm centered on a vertical motor shaft. The arms are rotated at speeds from 1500 to 3600 rpm with the radius of rotation usually chosen to give a linear speed of about 10 m/s. As the arms rotate, the leading surfaces impact particles in their path with a high efficiency and, being coated with a suitable adhesive, capture the impacting particles. The first such device was the Rotorod sampler (Perkins 1957), of which several versions have been developed. A number of other rotating impaction samplers were subsequently developed, each with certain advantages, but all operate on the same principle. Since the linear speed of the collecting arms is constant and usually much greater than the ambient wind speed, the latter is normally neglected in calculating the efficiency and the volume sampled. A measure of wind speed is usually not required, although it has been shown that efficiency decreases somewhat as wind speed increases (Ogden and Raynor 1967). The collection efficiency also varies with collector and particle size as in wind impaction samplers, but is normally much higher and can be calculated or measured experimentally in a wind tunnel. Edmonds (1972) calculated the efficiency of collection of Rotorod samplers for various sizes of fungal spores.

If these devices are operated for prolonged periods, their high collection efficiency leads to overloading, and versions have been designed that operate sequentially or intermittently. Some of these instruments actually rotate for only a small percentage of the total exposure time, and it is necessary to shield the collecting surfaces from wind impaction during their idle periods. Rotating impaction samplers are well suited for sampling airborne pollens and particles of similar size, and their use is increasing.

SUCTION

Samplers in which air containing material to be sampled is drawn into an entrance by suction from a vacuum pump or other air-moving device may be classified as suction samplers and are used for many air sampling purposes. Many methods are used within such samplers for collecting the material of interest from the airstream. These methods include filtration, impaction, electrostatic and thermal precipitation, and liquid impingement. Some suction samplers do not collect the material but measure its concentration, usually by optical methods, as it passes through a special viewing or measurement section.

Techniques and devices for measuring and removing materials drawn into suction samplers are generally highly efficient for the specific type or size of particle for which the sampler is designed, but little attention has been given to the problem of getting a representative sample into the entrance. This is partially because suction samplers are most commonly used to sample gases and submicron-sized particles whose entrance efficiency is normally high.

Such samplers are often used to sample larger particles, however, which tend to deviate from the airstream entering the sampler if the air is forced to change direction or speed. In such cases the number collected may be much different (smaller or larger) from the number originally in the air sampled.

Isokinetic sampling (Watson 1954) is the ideal method of taking an accurate sample of large particulates, such as pollen, from the atmosphere. In this method, illustrated by the central diagram of Figure 5.4, air is drawn into a sharp-edged orifice aligned with the airstream. Air within the sampler is drawn away from the entrance at the same velocity (V_S) as ambient air approaches (V_A). The approaching air does not have to change direction or speed, and enters the sampler entrance smoothly carrying all entrained particulates with it. If V_A is greater than V_S as in the upper diagram, not all the approaching air can enter and some must divert around the entrance. Particles in the diverted air may have enough inertia to be carried into the entrance and cause oversampling. If V_A is less than V_S, as in the lower diagram, air is drawn in from outside the stream approaching the entrance, but particles in

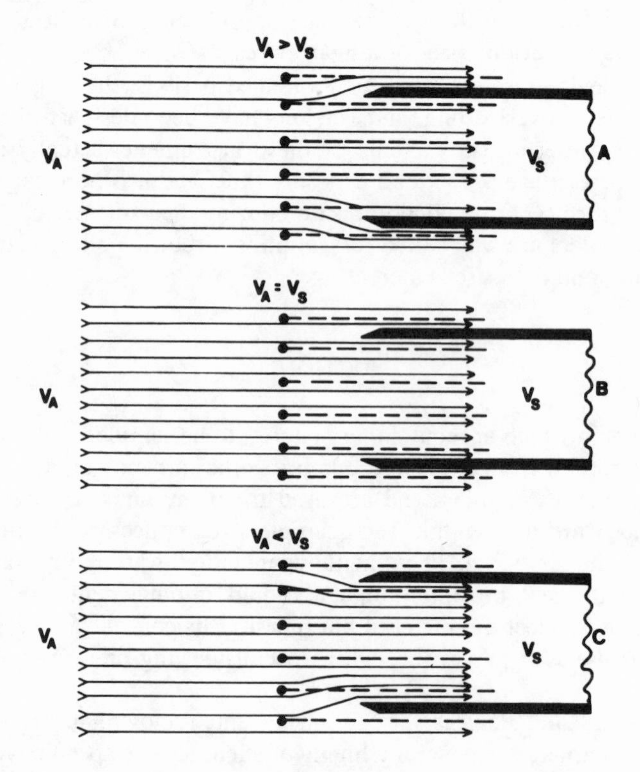

Figure 5.4. Air (▬▬) and particle (▬ ▬) trajectories at the entrance of a suction-type sampler facing the wind: (A) Air speed (V_A) exceeds sampling speed (V_S); particles in air that diverts around entrance enter, resulting in oversampling. (B) $V_A = V_S$; all particles in air sampled enter sampler (isokinetic sampling). (C) V_S; particles in air that enter sampler escape, resulting in undersampling.

this region may have enough inertia to be carried past the entrance and undersampling results. Isokinetic sampling is a standard method of obtaining an accurate sample in wind tunnels, tubes, ducts, and chimneys where the flow is constant, but no samplers have been perfected that can adjust to the rapid fluctuations in speed and direction present in the free atmosphere. If samplers are vane mounted and have an adjustable flow rate, they can be used to approximate isokinetic sampling by matching the flow rate to the average wind speed; but most samplers are designed for a fixed flow rate only. Equations for estimation of the errors involved in nonisokinetic sampling are given by Badzioch (1960).

Most suction samplers are so designed that even an approximation to isokinetic sampling is impossible. In samplers where the entrance does not face the airstream, flow patterns near the entrance are complex and entrance efficiency for large particles is generally low. Air and particle trajectories around a typical filter sampler with the entrance at a right angle to the airstream are illustrated in Figure 5.5. Many of the particles originally in the air that enters the sampler pass by or strike the filter holder and are not captured. The efficiency (E) is given by h/H, where h = extent of the vertical air column from which particles enter the sampler and H = extent of the vertical air column that enters the sampler. Wind tunnel experiments and field experience with filter samplers, oriented as in Figure 5.5, show that entrance efficiency increases with increased flow rate into the entrance and decreases with increasing wind speed and with increasing particle size. It may drop to near

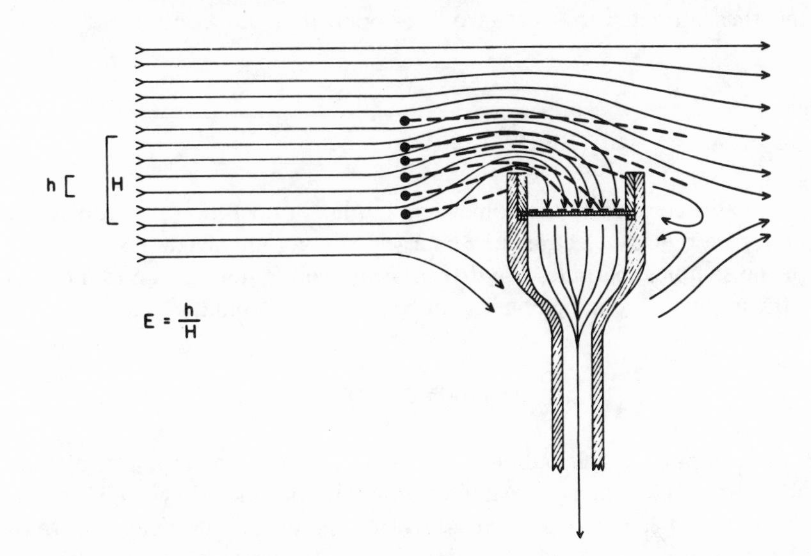

Figure 5.5. Air (——) and particle (— —) trajectories around a filter sampler at right angles to the airflow. Entrance efficiency *(E)* is given by h/H, where h is the height of the air column from which particles enter the sampler and H is the height of the air column entering the sampler.

zero for pollen-sized particles at wind speeds often found in the atmosphere and with flow rates commonly used. These results suggest that suction samplers are not normally suitable for sampling the large airborne particles. Although they may be 100% efficient in removing and retaining particles that enter them, their low and variable entrance efficiency for large particles is often not realized by the user.

FILTRATION

Filtration is the commonest method for removing particles from air drawn into an entrance by suction. The air passes through a fibrous or porous medium that impacts or sieves the particles. Only filters with a smooth surface such as molecular membranes are suitable for direct microscopic examination of captured particles. Particles to be cultured must be removed to a suitable medium. By suitable choice of filter media and flow rates, filtration can be highly effective in capturing particles that enter the sampler, although loss can occur if a lengthy entrance section precedes the filter.

ELECTROSTATIC PRECIPITATION

Electrostatic precipitation is best suited for small particles. As air is drawn through the sampling unit, particles are charged near the entrance region and then attracted to an electrode of opposite charge within the instrument.

THERMAL PRECIPITATION

Thermal precipitators are somewhat similar in operation to electrostatic precipitators although electrostatic charges are not involved. As air flows through the sampler, particles are driven away from a hot surface to a colder one by the more energetic bombardment by molecules from the hotter side.

LIQUID IMPINGMENT

Liquid impingers and bubblers operate by drawing a stream of air into the bottom of a container of water or other liquid and allowing it to rise through the liquid as buoyant bubbles. During the process particles are transferred to the liquid and retained. Liquid impingers are most commonly used for sampling certain gases but have acceptable efficiency for some airborne particles. They are particularly advantageous where dehydration of a viable particle, as on a filter surface, must be avoided. Bubblers have been recommended for sampling delicate organisms such as algae (Schlichting 1971).

SAMPLING INSTRUMENTS

Many sampling instruments operating on the principles described above have been used for collecting biological particles. Most are useful only over a limited range of particle sizes but some are being used for particles that they cannot sample efficiently or representatively. Only a selection from those most frequently used or commericially available is included here. The information and descriptions herein should guide to an appropriate choice of instrument for each sampling need. A previous review of pollen and spore samplers was given by Henderson and Stalker (1966).

DEPOSITION SAMPLERS

The familiar Durham or gravity slide sampler (Durham 1946) is usually classed as a sedimentation sampler, but measures neither actual deposition nor air concentrations. Although used almost exclusively for pollen sampling for many years, and still used extensively, it can no longer be recommended now that more quantitative devices are available. The volume of air sampled is unknown as is its efficiency for any particle type. Its catch is a function of wind speed, turbulence, and orientation of the slide with respect to wind direction. These effects were shown experimentally by Ogden and Raynor (1960).

Deposition to the ground over short or medium sampling periods has been measured by microscope slides coated with an adhesive and by sheets of sticky paper in a flat aluminum frame (Lewis and Ogden 1965). Neither is necessarily representative of surrounding natural surfaces but better methods have not been developed.

The Tauber trap (Tauber 1974) has been used for collecting particles over long periods, usually a pollen season. The trap consists of a cylindrical container covered with an aerodynamically shaped collar with a centered circular orifice through which particles enter. Particles are trapped by a layer of glycerol in the bottom of the container. The trap is operated either unprotected or with a roof to prevent collection of precipitation, and has been used underwater as well as on the ground. Its efficiency is low and varies with particle size and wind speed, but it can collect large quantities without loss and can be left unattended for long periods. Petri dishes containing a nutrient medium are frequently used as deposition collectors for organisms to be cultured.

IMPACTION SAMPLERS

One of the earliest users of wind impaction samplers was Gregory (1951) but his work was carried out in a wind tunnel. More recently, vane-mounted models were used in field research programs at the University of Michigan (the flag sampler; Harrington et al. 1959) and Brookhaven National Laboratory (the slide-edge-cylinder sampler; Raynor et al. 1970). Both were evaluated by Ogden et al. (1974). They are useful for sampling large particles under controlled or carefully monitored conditions but are not recommended for field use under light or variable wind speeds.

Several rotating impaction samplers are in use and have proved the most useful and reliable devices for sampling particles in the pollen size range. All have been recently described and evaluated in detail by Raynor (1972b) and Ogden et al. (1974), so only brief mention is made here.

The first such sampler was the Rotorod (Perkins 1957) developed at Stanford Research Institute. Particles are collected on the surfaces of two upright metal arms mounted on a small motor that rotates them at about 2500 rpm. Several later versions have been produced. The Rotorod fluorescent particle (FP) sampler (Webster 1963) is designed for sampling small particles. A version with plastic inserts on the upright arms is useful for particles that must be examined under a microscope (Webster 1968). A recent version designed for intermittent use has retracting arms that are shielded from wind impaction when not rotating (Metronics Associates 1967).

The Rotobar sampler was developed at the University of Michigan (Harrington et al. 1959). Samples are taken on metal bands held between two crossarms. It was later modified by Solomon et al. (1968), who added hinged shields that protect against wind impaction during idle periods. Magill et al. (1968) described a modified Rotorod sampler similar to the Rotobar.

Several versions of the Rotoslide sampler were developed at Brookhaven National Laboratory and use the edges of two standard microscope slides in upright holders as sampling surfaces. The simplest model consists only of a motor and the slide holder. The first intermittent model (Ogden and Raynor 1967) used a hood to shield the slides from wind impaction when not operating. A timer operated the sampler for 1 min of each 12. Later the swing shield (Raynor and Ogden 1970) was developed to give protection more simply. These samplers have been adopted by numerous allergists, hospitals, and public health departments for sampling pollen and other particles. Several sequential Rotoslide models were developed for use in research studies (Ogden et al. 1974) but have not been generally used.

A rotating sampler of different design, the sequential Rotodisk pollen sampler (Cole and Stohrer 1966) takes sequential samples on discrete segments of the edge of a rotating disk by use of timing and stepping devices.

SUCTION SAMPLERS

The Hirst spore trap (Hirst 1952, 1972) was specifically designed for sampling fungus spores as a function of time. It contains an intake orifice on an upright cylindrical housing kept facing the wind by a vane tail. Flow rate is fixed and provided by an external vacuum pump. Inside the orifice a greased microscope slide is drawn upward at 2 mm/h. Particles drawn into the orifice are deposited by impaction on the slide. Efficiency is reasonably high but varies with wind speed and particle size.

A later model, the Burkard 7-day recording volumetric spore trap, is similar in principle but has a built-in motor and collects for a week on an adhesive-coated transparent tape drawn past the entrance orifice. Samplers of somewhat similar design have been described by Panzer et al. (1957), Voisey and Bassett (1961), and Schenck (1964), but apparently none are commercially available.

A sampler commonly used for sampling plant pathogens was first described by Pady (1959). A newer version was described by Kramer and Pady (1966) and is commercially available as the Kramer-Collins spore sampler. The sampler gives discrete hourly samples over a 24-h period on a single slide. The entrance orifice is in the top so that entrance efficiency is poor for large particles. Another device with 24-h time discrimination and a top entrance is the Marx impinger (Marx et al. 1959).

CASCADE IMPACTOR

Although cascade impactors are also suction samplers, they are discussed separately since their construction, operation, and use differ appreciably from other samplers. The first cascade impactor was described by May (1945), but many models are now available from various sources and numerous papers on their design and calibration have appeared in the literature. The instrument contains a number of orifices in series, each smaller than the preceding. Behind each orifice is a slide or other deposition surface. Air is drawn through the sampler at a constant volumetric rate giving progressively higher speeds through each orifice. The large particles impact on the first stage and progressively smaller ones on succeeding stages, although the size separation has considerable overlap. Some models have moving slides for time discrimination and some have a filter as the first stage. These samplers are subject to the same anisokinetic entrance problems for large particles as other suction samplers but serve a useful purpose where size separation is desired.

A commonly used cascade impactor is the Andersen sampler (Andersen 1958), which is available in several models. In one, particles are deposited on petri dishes of nutrient media for culture of collected particles. In other models, particles are impacted on glass or stainless steel collection plates. A recent model is designed to replace the filter assembly on the entrance to a high-volume sampler and uses a filter as the final stage.

FILTER SAMPLERS

Filter samplers constitute another distinctive group of suction samplers. Most widely used, particularly for sampling nonbiological air pollutants, are the high-volume samplers that are available in several models. Each consists of a vacuum-cleaner-type motor that pulls air through any of several types of large filters. Flow rates are high but visual inspection of the catch is impossible on most of the filters used except the glass fiber filter, through which flow rate is least. Since the high flow rate is attained by use of a large filter area, face velocity through the filter is no higher than through many smaller filter samplers and entrance efficiency for large particles is questionable.

For sampling at low flow rates, filter holders in various sizes can be obtained from several sources and may be used with any of the large variety of filter types available. They differ in physical and chemical composition, in pore size, resistance to flow, and other characteristics. Those with smooth surfaces are preferred for microscopic examination of collected particles, and include the various molecular membrane types and some glass fiber filters.

LIQUID IMPINGERS

Several types of liquid impingers or bubblers are available commercially and similar devices can be fabricated from laboratory glassware. The two most frequently used are the Greenberg-Smith and the midget impingers, both available from various commercial suppliers. Both operate on the same principle as described above but the former is larger and samples at a flow rate of 0.03 m^3/min, while the midget impinger samples at one-tenth of that rate.

AIRBORNE SAMPLERS

Pollen and spore samples have been taken from aircraft for many years. Early investigators used an adhesive-coated slide or other impaction surface held out a window or exposed to a stream of air taken into the aircraft. More recently a number of samplers, mostly with nonisokinetic entrances, have been used (e.g., Kelly et al. 1951, Holzapfel and Gressitt 1964, Harrington 1965, Timmons et al. 1966, Burleigh et al. 1967). Raynor (1972a) developed an isokinetic sampler for use on light aircraft that was used successfully for sampling pollen aloft (Raynor et al. 1974).

OTHER SAMPLERS

Other types of samplers such as electrostatic and thermal precipitators or devices that count and size particles optically as they pass through a viewing section are seldom used for sampling biological particles. Information covering them may be obtained from the literature or from the manufacturers.

BACTERIA AND VIRUSES

Methods for sampling microbiological aerosols were reviewed by Buchanan and Dahlgren (1959).

INSECT SAMPLERS

A recent review of insect sampling methods with many literature references was given by Gara (1972). Devices that may be classed as air samplers include moving or rotating nets, townets, sticky barriers, and suction traps.

USE OF SAMPLERS

CHOICE OF SAMPLER

Although cost, availability, or disciplinary practices often govern the choice of a sampler, better results may be expected if the sampler best adapted for a particular program is chosen. The user must consider the size and other characteristics of the particles to be sampled, methods of handling and analyzing the sample, length of the sampling period, and other pertinent factors. Each sampling program has somewhat different needs but several general guidelines can be given.

Particles less than about 5 μm in diameter and not requiring culture are best sampled by suction samplers, since most impactors are inefficient for particles this small. Small particles follow air motions quite well so that anisokinetic sampling does not result in serious errors. A simple filter sampler is adequate for short-period samples where time discrimination is not required. Liquid impingers may also be used. If changes with time are of interest, the Hirst or Burkard spore traps are recommended. If large volumes of air must be sampled, a high-volume sampler with a glass fiber filter may be suitable. If it is wished to sort collected particles into size classes, one of the cascade impactors is needed. For culture of collected particles, the Anderson sampler is usually the instrument of choice.

Particles between 5 and 10–15 μm may not be sampled very efficiently by either suction or impaction-type samplers. If the efficiency can be determined, however, either type may be adequate.

Particles above 15 μm are best sampled by rotating impactor samplers whose efficiency is good for particles of these sizes. A choice may be made between the several available types of rotating impactors and between continuous, intermittent, and sequential operation.

If deposition to the surface is to be measured rather than air concentration, some deposition sampler must be used. For long-period samples, the Tauber trap has proved useful. For shorter periods, microscope slides or sticky paper sheets are often used. Petri dishes with a suitable culture medium may be used for particles that must be cultured; however it should be realized that no deposition collector necessarily collects at the same rate as an adjoining natural surface and that particles of different types may be collected with different efficiency on the same surface.

Specialists in various fields have recommended specific samplers. Rotating impactor samplers were suggested for pollens by Ogden et al. (1974). Plant pathologists prefer the Kramer-Collins spore trap and the Rotorod sampler (Wallin and Loonan 1974). Fungus spores are frequently sampled by the Hirst and Burkard spore traps, while students of airborne algae use liquid impingers, membrane filters, exposed culture media (Schlichting 1969), and rotating impactors. Filters or impactor surfaces may be placed on nutrient media for culture of collected algae. Palynologists studying modern pollen rain favor the Tauber trap.

CHOICE OF SAMPLING LOCATION

Selection of sampling location is often governed by expediency but attempts should be made to select a site representative of the area to which the data are expected to apply. A rooftop site, for instance, may not be representative of locations in the streets below, while a sampling station near a field of ragweed may give much higher pollen concentrations than occur in the region as a whole.

Other considerations may be equally important. Major contamination from unwanted particles cannot be tolerated. Dust from a road or smoke from a nearby chimney, for instance, may obscure particles of interest and render samples useless. A site must also be easily accessible for changing samples, have electric power if needed, and be secure from theft and vandalism. Ogden et al. (1974) discussed location of pollen samplers in more detail.

SELECTION OF SAMPLING SEASON

Since many airborne particles are seasonal in occurrence, sampling for them at the wrong time of year is a waste of time and effort. Some information on season of occurrence should be incorporated into the presampling planning process. Most pollens are highly seasonal and their periods of occurrence are generally known. Some fungus spores occur over longer portions of the year while others appear in response to specific meteorological conditions and may be largely unpredictable much in advance. Thus sampling over extended periods may be necessary to document their appearance or relate their occurrence to other factors.

SELECTION OF SAMPLING PERIOD

Choice of the length of each basic sampling period will depend on the requirements of the sampling program, the characteristics of the sampler, and the capability for processing samples. Conflicting considerations must often be compromised and no general guidelines can be given. Daily averages may be adequate for studies of seasonal change but tell nothing about peak concentrations during the day, which may be more important, in allergy for instance. Study of diurnal variability requires samples no more than an hour or two in length, but if concentrations are low some counts may be so small that their reliability is questionable. Also, frequent attention or more complex sequential samplers are required and more samples may be obtained than can be processed. Thus sampling periods must be carefully chosen to meet the requirements of the program within the limitations of the available resources. Some considerations in sampling design were discussed by Reagan (1971).

ANALYTICAL INSTRUMENTS AND METHODS

Since most airborne particles are too small to be seen with the unaided eye, microscopes are commonly used for identification and counting. Identification can often be facilitated by selective staining of the particles of interest. Species that cannot be identified visually may be deposited on suitable culture media and the resulting colonies may be identified and counted.

LIGHT MICROSCOPE

The light microscope is the most frequently used tool for identification and counting of particles above about 0.8 μm. For small-particle work a compound, binocular microscope with a condenser, a calibrated mechanical stage, and a selection of power is essential. A Whipple micrometer disk or other scale in one of the oculars is often helpful. For most routine observations, magnifications of 100 × to 400 × are used as appropriate. For critical identification or counting of very small particles, an oil immersion lens is used.

ELECTRON MICROSCOPE

Electron microscopes are seldom needed for routine counting or identification of airborne biological particles but are often useful for examining surface details or for checking for the presence of very small particles below the resolution of the light microscope. A scanning electron microscope is preferred since its image is three-dimensional and the higher magnification of the transmission scope is seldom needed. The cost of acquiring and operating such instruments, however, the need for a trained operator, the small field of view, and the large number of fields necessary to photograph an adequate sample normally restrict their use to large institutions or well-funded research programs.

STAINS

Differential staining of particles to be examined greatly aids identification and counting, especially of particles that would otherwise appear colorless or transparent. Pollens may be readily stained with a variety of dyes but the most commonly used is basic fuchsin. Directions for use were given by Ogden et al. (1974). Fungus spores and inorganic materials are not stained, thus giving good contrast. Wodehouse (1935), Venning (1954), and Brown (1960) described other dyes for staining pollens. The darker fungus spores are not easily stained but can usually be examined adequately as collected. Identification of other small particles may be aided by suitable stains and information on techniques can be obtained in the literature covering appropriate disciplines.

CULTURE TECHNIQUES

Culture techniques are often needed for identification of fungus spores, algae, bacteria, and other biological particles, but media and methods are specialized for specific organisms and details should be sought in the appropriate literature. Rogerson (1958) described culture media for airborne fungi. Schlichting (1971) mentioned a medium for green algae and gave references to other media.

IDENTIFICATION OF PARTICLES

Except for a few common and well-known pollens and spores, identification of airborne particles requires considerable training and experience, or extensive consultation of the literature and reference collections, and the assistance of specialists. Pollen identification is well covered in books by Wodehouse (1935, 1942, 1971), Erdtman (1943, 1952, 1957, 1965, 1969), Hyde and Adams (1958), Kapp (1969), Solomon et al. (1967), and Ogden et al. (1974). A good introduction to fungus spore identification with many references was given by Ogden et al. (1974).

Pollen and spore identification is often aided by a collection of reference slides containing particles of known identity usually taken directly from the source. Many species may be purchased from firms supplying allergens to physicians. Reference slides are on file at several research institutions.

PRESENTATION OF DATA

Aerobiological sampling is carried out for many purposes so data recording and presentation vary with the purpose of the program. If the data are to be made available to other workers, however, certain minimum information should accompany the counts or concentration measurements. These include date, time, duration of sampling, type of sampler, location and height of measurement, and at least a minimum description of weather conditions. Data may be presented in tabular form with appropriate summaries or indexes of variability, but are often better understood if shown graphically.

INTERPRETATION OF DATA

Although accurate measurements taken with a well-exposed sampler may be obtained for one or more airborne particles of interest, the data must be evaluated with an understanding of the many variables that may have influenced the measurement. Concentrations of airborne particles depend on such factors as distance and direction from sources, height above ground, season, time of day, weather conditions, terrain, and nearby obstacles to free airflow. Since concentrations may change quickly with both time and space, each measurement should be regarded as representative only of the location and time period sampled. Extrapolation to either longer or shorter time periods may not be valid nor may assumptions that the measurement is representative of a wider area. Thus extension of the data should be made only if other evidence such as experience or previous results suggest its validity.

Data should be interpreted with respect to other conditions observed or recorded during or prior to the sampling period. Low counts may be correlated with precipitation, for instance, or high counts with a wind direction from a known source area. Collection of data without relation to causative variables has little value.

Data may also be related to the responses of receptors. High pollen concentrations, for instance, may be related to the occurrence or severity of allergic symptoms in hay-fever patients. High concentrations of plant disease spores should serve as a warning to watch for disease outbreaks in susceptible crops. Thus, for maximum usefulness, sampling data should be related to source, transport, and receptor parameters.

LITERATURE CITED

Andersen, A. A. 1958. New sampler for the collection, sizing and enumeration of viable airborne particles. J. Bacteriol. 76:471–484.

Badzioch, S. 1960. Correction for anisokinetic sampling of gasborne dust particles. J. Inst. Fuel 33:106–110.

Benninghoff, W. S. 1971. Dust of biological origin and significance. Pages 14–15 *in* W. S. Benninghoff and R. L. Edmonds, eds. Aerobiology objectives in atmospheric monitoring. US/IBP Aerobiol. Program Handb. 1. Univ. Michigan, Ann Arbor.

Brown, C. A. 1960. Palynological techniques. Published by author, Baton Rouge, La. 188 p.

Brun, R. J., and H. W. Mergler. 1953. Impingement of water droplets on a cylinder in an incompressible flow field and evaluation of rotating multicylinder methods for measurement of droplet-size distribution, volume-median droplet size, and liquid-water content in clouds. NACA Publ. TN 2904. National Advisory Committee for Aeronautics, Washington, D.C.

Buchanan, L. M., and C. M. Dahlgren. 1959. Sampling microbiological aerosols. Public Health Monogr. 6. USDHEW, Washington, D.C. 53 p.

Burleigh, J. R., C. L. Kramer, and T. J. Collins. 1967. A spore sampler for use in aircraft. Phytopathology 57:434–436.

Cole, A. L., and A. W. Stohrer. 1966. Sequential roto-disk pollen sampler. U.S. Public Health Serv. Rep. 81:577–578.

Durham, O. C. 1946. The volumetric incidence of airborne allergens. IV. A proposed standard method of gravity sampling, counting and volumetric interpolation of results. J. Allergy 17:79–86.

Edmonds, R. L. 1972. Collection efficiency of rotorod samplers for sampling fungus spores in the atmosphere. Plant Disease Reporter 56:704–708.

Erdtman, G. 1943. An introduction to pollen analysis. Chronica Bontanica, Waltham, Mass. (Reprinted 1954 by Ronald Press, New York.) 239 p.

Erdtman, G. 1952. An introduction to palynology. I. Pollen morphology and plant taxonomy: Angiosperms. Almqvist & Wiksell, Stockholm, and Chronica Botanica, Waltham, Mass. 539 p.

Erdtman, G. 1957. An introduction to palynology. II. Pollen and spore morphology. Plant taxonomy: Gymnospermae, Pteridophyta, Bryophyta (illustrations). Almqvist & Wiksell, Stockholm, and Ronald Press, New York. 151 p.

Erdtman, G. 1965. An introduction to palynology. III. Pollen and spore morphology. Plant taxonomy: Gymnospermae, Bryophyta (text). Almqvist & Wiksell, Stockholm. 191 p.

Erdtman, G. 1969. Handbook of palynology. Hafner Publ. Co., New York. 486 p.

Faegri, K., and J. Iverson. 1964. Textbook of pollen analysis, 2nd rev. ed. Hafner Publ. Co., New York. 237 p.

Gara, R. I. 1972. Sampling the atmosphere for airborne insects. Pages 109–118 *in* W. S. Benninghoff and R. L. Edmonds, eds. Ecological systems approaches to aerobiology. I. Identification of component elements and their functional relationships. US/IBP Aerobiol. Program Handb. 2. Univ. Michigan, Ann Arbor.

Green, H. L., and W. R. Lane. 1964. Particulate clouds: Dusts, smokes, and mists: Their physics and physical chemistry, and industrial and environmental aspects, 2nd ed. Van Nostrand Co., Princeton, N.J. 471 p.

Gregory, P. H. 1951. Deposition of air-borne *Lycopodium* spores on cylinders. Ann. Appl. Biol. 38:357–376.

Gregory, P. H. 1973. The microbiology of the atmosphere, 2nd rev. ed. John Wiley & Sons, New York. 377 p.

Harrington, J. B. 1965. Atmospheric diffusion of ragweed pollen in urban areas. Volume II *in* E. W. Hewson, ed. Atmospheric pollution by aeroallergens: Meteorological phase. Univ. Michigan Press, Ann Arbor.

Harrington, J. B., G. C. Gill, and B. R. Warr. 1959. High-efficiency pollen samplers for use in clinical allergy. J. Allergy 30:357–375.

Henderson, J. J., and W. W. Stalker. 1966. Pollen and spores: Comparison of sampling and counting methods. USDHEW, Div. Air Pollut., Cincinnati, Ohio.

Hirst, J. M. 1952. An automatic volumetric spore trap. Ann. Appl. Biol. 39: 257–265.

Hirst, J. M. 1972. Sampling atmospheric particulates with suction devices: The Hirst trap. Pages 80–81 *in* W. S. Benninghoff and R. L. Edmonds, eds. Ecological systems approaches to aerobiology. I. Identification of component elements and their functional relationships. US/IBP Aerobiol. Program Handb. 2. Univ. Michigan, Ann Arbor.

Holzapfel, E. P., and J. L. Gressitt. 1964. Airplane trapping of organisms and particles. Pages 151–162 *in* Proc. Atmos. Biol. Conf., Univ. Minnesota, 13–15 April 1964. NASA, Washington, D.C.

Hyde, H. A., and K. F. Adams. 1958. An atlas of airborne pollen grains. Macmillan Co., London, and St. Martin's Press, New York. 112 p.

Kapp, R. O. 1969. How to know pollen and spores. W. C. Brown Co., Dubuque, Iowa. 249 p.

Kelly, C. D., S. M. Pady, and N. Polunin. 1951. Aerobiological sampling methods from aircraft. Can. J. Bot. 29:206–214.

Kramer, C. L., and S. M. Pady. 1966. A new 24-hour spore sampler. Phytopathology 56:517–520.

Lewis, D. M., and E. C. Ogden. 1965. Trapping methods for modern pollen rain studies. Pages 613–626 *in* B. Kummel and D. Raup, eds. Handbook of paleontological techniques. W. H. Freeman, San Francisco.

Magill, P. L., E. D. Lumpkins, and J. S. Arveson. 1968. A system for appraising airborne populations of pollens and spores. J. Am. Ind. Hyg. Assoc. 29:293–298.

Marx, H. P., J. Spiegelman, and G. I. Blumstein. 1959. An improved volumetric impinger for pollen counting. J. Allergy 30:83–89.

May, K. R. 1945. The cascade impactor: An instrument for sampling coarse aerosols. J. Sci. Instrum. 22:187–195.

Metronics Associates. 1967. Prod. Bull. 17-67. Metronics Assoc., Palo Alto, Calif.

Ogden, E. C., and G. S. Raynor. 1960. Field evaluation of ragweed pollen samplers. J. Allergy 31:307–316.

Ogden, E. C., and G. S. Raynor. 1967. A new sampler for airborne pollen: The Rotoslide. J. Allergy 40:1–11.

Ogden, E. C., G. S. Raynor, J. V. Hayes, D. M. Lewis, and J. H. Haines. 1974. Manual for sampling airborne pollen. Hafner Press, New York. 182 p.

Pady, S. M. 1959. A continuous spore sampler. Phytopathology 49:757–760.

Panzer, J. D., E. C. Tullis, and E. P. VanArsdel. 1957. A simple 24-hour slide spore collector. Phytopathology 47:512–514.

Perkins, W. A. 1957. The Rotorod sampler. 2nd Semiannu. Rep. CML 186. Aerosol Lab., Stanford Univ., Stanford, Calif.

Raynor, G. S. 1972a. An isokinetic sampler for use on light aircraft. Atmos. Environ. 6:191–196.

Raynor, G. S. 1972b. Sampling atmospheric particulates with rotating arm impaction samplers. Pages 82–105 *in* W. S. Benninghoff and R. L. Edmonds, eds. Ecological systems approaches to aerobiology. I. Identification of component elements and their functional relationships. US/IBP Aerobiol. Program Handb. 2. Univ. Michigan, Ann Arbor.

Raynor, G. S., J. V. Hayes, and E. C. Ogden. 1970. Experimental data on ragweed pollen dispersion and deposition from point and area sources. Rep. BNL 50224 (T-564). Brookhaven Natl. Lab., Upton, N.Y.

Raynor, G. S., J. V. Hayes, and E. C. Ogden. 1974. Mesoscale transport and dispersion of airborne pollens. J. Appl. Meteorol. 13:87–95.

Raynor, G. S., and E. C. Ogden. 1970. The swingshield: An improved shielding device for the intermittent Rotoslide sampler. J. Allergy 45:329–332.

Reagan, J. A., III. 1971. Sampling design. Pages 44–46 *in* W. S. Benninghoff and R. L. Edmonds, eds. Aerobiology objectives in atmospheric monitoring. US/IBP Aerobiol. Program Handb. 1. Univ. Michigan, Ann Arbor.

Rogerson, C. T. 1958. Kansas aeromycology. I. Comparison of media. Trans. Kans. Acad. Sci. 61:155–162.

Schenck, N. C. 1964. A portable, inexpensive, and continuously sampling spore trap. Phytopathology 54:613–614.

Schlichting, H. E., Jr. 1969. The importance of airborne algae and protozoa. J. Air Pollut. Control Assoc. 19:946–951.

Schlichting, H. E., Jr. 1971. Algae and protozoa in the atmosphere. Pages 16–18 *in* W. S. Benninghoff and R. L. Edmonds, eds. Aerobiology objectives in atmospheric monitoring. US/IBP Aerobiol. Program Handb. 1. Univ. Michigan, Ann Arbor.

Solomon, W. R., O. C. Durham, and F. L. McKay. 1967. Aeroallergens. II. Pollens and the plants that produce them. Pages 340–397 *in* J. M. Sheldon, R. G. Lovell, and K. P. Mathews, eds. A manual of clinical allergy, 2nd ed. Saunders Co., Philadelphia.

Solomon, W. R., A. W. Stohrer, and J. A. Gilliam. 1968. The "fly-shield" Rotobar: A simplified impaction sampler with motion-regulated shielding. J. Allergy 41:290–296.

Southworth, D. 1974. Introduction to the biology of airborne fungal spores. Ann. Allergy 32:1–22.

Tauber, H. 1974. A static non-overload pollen collector. New Phytol. 73:359–369.

Timmons, D. E., J. D. Fulton, and R. B. Mitchell. 1966. Microorganisms of the upper atmosphere. I. Instrumentation for isokinetic air sampling at altitude. Appl. Microbiol. 14:229–231.

Venning, F. D. 1954. Manual of advanced plant microtechnique. W. C. Brown, Dubuque, Iowa. 96 p.

Voisey, P. W., and I. J. Bassett. 1961. A new continuous pollen sampler. Can. J. Plant Sci. 41:849–853.

Wallin, J. R., and D. V. Loonan. 1974. Air sampling to detect spores of *Helminthosporium maydis* race T. Phytopathology 64:41–44.

Watson, H. H. 1954. Errors due to anisokinetic sampling of aerosols. Am. Ind. Hyg. Assoc. Q. 15:21–25.

Webster, F. X. 1963. Collection efficiency of the Rotorod FP sampler. Tech. Rep. 98. Metronics Assoc., Inc., Palo Alto, Calif. 50 p.

Webster, F. X. 1968. The fluorescent particle atmospheric tracer technique. Pages 48–80 *in* J. Y. Wang, ed. Proc. Conf. Air Pollut. Calif. San Jose State College, San Jose, Calif.

Wodehouse, R. P. 1935. Pollen grains. McGraw-Hill, New York. 574 p. (Reprinted 1959 by Hafner Publ. Co., New York.)

Wodehouse, R. P. 1942. Atmospheric pollen. Pages 8–31 *in* S. Moulton, ed. Aerobiology. AAAS Publ. 17. Am. Assoc. Advan. Sci., Washington, D.C.

Wodehouse, R. P. 1971. Hayfever plants, 2nd rev. ed. Hafner Publ. Co., New York. 280 p.

6. IMPACT OF AIRBORNE MATERIALS ON LIVING SYSTEMS

I. L. Bernstein, L. Calpouzos, R. L. Edmonds, H. F. Hasenclever, J. M. Leedom, C. G. Loosli, M. L. McManus, R. S. Safferman, A. M. Solomon, W. R. Solomon, and G. Vali

IMPACT ON PLANT SYSTEMS

IMPACT OF AIRBORNE PLANT PATHOGENS AND AIR POLLUTANTS*

Propagules capable of infecting plants are dispersed in the atmosphere along with plant damaging air pollutants. Finally, they are deposited on susceptible plants. What is the ultimate impact of this process? *Impact* suggests a single swift blow; however, the effect is more often a growing destructive pressure upon the host plant population provided there is a convergence of favorable circumstances for the pathogen and the air pollutant. The major organisms and air pollutants causing impact to plants are shown in Table 6.1.

Le Clerg (1964) estimated that the total average annual loss to all crop groups from plant pathogens from 1951 to 1960 was about $3.25 billion. Actual losses from plant diseases are probably greater than 30 years ago as farm output has increased, but the percentage of losses may be less (Barnes 1964). Total dollar losses to forest crops as a result of disease have not been calculated, but rather estimated as losses of cubic feet of timber. Agricultural losses to air pollutants have been increasing and in 1968 annual losses in the United States were estimated to be $500 million (Heggestad 1968). This chapter will dwell primarily on examples of crop and forest tree losses due to disease and air pollution, with some emphasis on socioeconomic effects.

Impact of Airborne Plant Pathogens

The impact of airborne plant pathogens can be thought of as having three sequential stages: (1) epidemic, (2) biological loss, and (3) socioeconomic loss. Biological propagules settling out of the atmosphere may cause a few infections, but to have an appreciable impact they must multiply their effect on a

*L. Calpouzos and R. L. Edmonds

Table 6.1 Major organisms and air pollutants causing economic impacts to plants.

Causal organism	Disease
Crops	
Puccinia graminis tritici	Wheat rust
Puccinia striiformis	Wheat stripe rust
Mycosphaerella musicola	Sigatoka disease of bananas
Hemileia vastatrix	Coffee rust
Pyricularea oryzae	Rice rust
Trees	
Cronartium ribicola	White pine blister rust
Cronartium fusiforme	Southern fusiform rust
Fomes annosus	Annosus root and butt rot
Ceratocystis ulmi	Dutch elm disease
Air Pollutants	
Sulfur dioxide	
Fluorides	
Ozone	
Peroxyacetyl nitrate	
Oxides of nitrogen	

large scale—hence the epidemic, which leads to a loss of biological material, which in turn may lead to important changes in ecosystems often with socio-economic effects on humans. The present state of the science has been reviewed comprehensively by Kranz (1974) and Van der Plank (1963, 1968, 1975) with the conclusion that there are two major classes of airborne plant diseases: (1) those involving long-range dispersal and (2) those involving short-range dispersal. Specific examples of the impact of these types of diseases are discussed below.

Long-range dispersal

Most airborne plant pathogens are fungi, some of which are assumed to travel from a few kilometers to several hundred kilometers. Ascospores of *Mycosphaerella musicola* Leach were suggested to have been carried by the tropical trade winds from western Africa to the Caribbean area where this fungus was first found in the Western Hemisphere in 1933 (Wardlaw 1934, Stover 1962, 1972). This pathogen causes lesions on leaves of banana plants —the dreaded sigatoka disease. By 1937 a major epidemic was under way and sigatoka had appeared in most of the banana-producing areas of Latin America. It was becoming impossible to produce bananas for shipment to distant markets in the United States or Europe since the fruits from diseased plants showed slow or arrested development and were spoiled by premature ripening. The situation was saved by the introduction of fungicidal copper sprays that controlled the disease reasonably well. The socioeconomic impact was appreciable, and the labor force in banana plantations was increased in order to apply the 15–20 sprays each year on thousands of acres. To control sigatoka, the banana industry became one of the world's largest users of copper. The expense of chemical control by copper sprays was so considerable that only the large companies could afford it. For example, in Jamaica about 25% of the value of the fruit to the farmer was going to controlling this disease (Martyn and McIlwaine 1951 as reported in Wardlaw 1972). Since around 1960, mineral oil low-volume sprays applied by aircraft have replaced the more laborious and more expensive copper sprays. Downward shifts in the labor force occurred; on the other hand, smaller producers could now afford to enter the banana market. The sigatoka disease must be controlled by chemical means; no resistant commercial varieties are available. Without chemical control methods, sigatoka would drastically reduce or eliminate the banana from the human diet in the temperate regions of the world.

Coffee is native to the region encompassing Ethiopia, where in the wild state it has presumably lived in a precarious balance with the wind-disseminated rust fungus *Hemileia vastatrix* Berk. et Br. (Purseglove 1968). Since the Middle Ages the coffee plant has been cultivated and has spread around the world within the tropical belt. The rust, meanwhile, was left behind temporarily and has caught up with the coffee tree in its new surroundings only years later. Of the many episodes, two are worthy of note. Sri Lanka (Ceylon) before 1869 was a major producer of coffee, most of which went to the British. In that fateful year coffee leaf rust was found in Sri Lanka (Large 1940). Within five years all of the coffee plantations were being attacked. By 1878 the monetary losses were estimated at over two million pounds sterling. By then the disease had spread over southern India and Malaya. Many planters were ruined; the Oriental Bank went bankrupt (Large 1940). Sri Lanka was replanted to tea, which is resistant to this fungus, a situation which has kept Britons primarily as tea drinkers. The rust fungus attacks the coffee leaves. Bright orange circular lesions form on the lower surface of each infected leaf. In heavy infestations the rusted leaves drop, the tree produces new flushes of leaves, which again drop off because of further infections, and the process is repeated until the tree is finally exhausted, the harvest is greatly reduced, and often the tree dies.

South America in this century has become the world's major supplier of coffee. Leaf rust was unknown in the Western Hemisphere until January 1970, when a small outbreak was discovered in northeastern Brazil in Bahia. Although no one is certain, it is assumed that spores of the fungus were windblown across the South Atlantic from western Africa to Brazil, a distance greater than 2000 km (Wellman 1972). The fungus spread quickly; by 1972 it was in Parana some 1800 km southwest of the area where it was first discovered. Will the Sri Lanka experience be repeated in Brazil? So far the coffee plantations seem to have survived the pathogen—there are losses, the full extent of which are not yet known. Chemical control methods are being introduced. The battle is in progress; the outcome is uncertain.

Wheat is one of man's basic foods upon which several civilizations have depended for their existence. The central plains of North America are one of several major wheat-producing areas of the world. From Mexico and Texas toward the Canadian prairie provinces each spring there is a wave of wheat development as the cold weather recedes northward before the advancing warm winds. These winds carry spores of the rust fungus *Puccinia graminis tritici,* which can infect susceptible wheat varieties, mainly their stems and to a lesser extent their leaves. Five to ten days later a blister forms at the surface of the infection area; it bursts open and appears brick-red because of the color of thousands of rust spores being produced. This is the infamous stem rust or black rust disease of wheat. When many blisters or pustules appear on the plant early, the plant can die before any grain is formed, and the loss is total. When the pustules appear later in the season, the grain that forms is smaller and often shriveled; loss is partial but still significant.

Data suggest that viable spores of *P. graminis tritici* can travel at least 800 km through the atmosphere (Craigie 1945 as reported in Ingold 1971). Epidemics do not occur each year primarily because research programs in many nations breed new rust-resistant varieties, an effort that costs several millions of dollars annually. It is clear that this effort pays since, when epidemics occur, losses occasionally can be devastating. For example, on the North American continent during the past 50 yr widespread wheat rust epidemic occurred in 10 of those years, the most recent being 1965. The conservative estimate of loss in millions of bushels ranged from 24 (in 1938) to 126 (in 1935; unpublished data from USDA Cereal Rust Laboratory, University of Minnesota, St. Paul). The average loss of those 10 epidemics was 59 million bushels, which at current prices of wheat represents an economic loss of about $220 million per epidemic. In the present era of rapidly increasing population and changing weather patterns, losses from stem rust from one major epidemic could mean widespread starvation somewhere in the world.

Short-range dispersal

The role of transoceanic and transcontinental transport has been illustrated by the three diseases just discussed. The impact of short-range dispersal is the more common situation (it occurs also in the three examples given above), and will be shown for two other disease situations.

Rice, like wheat, is another of man's basic foods. The most economically important disease of rice is probably rice blast, incited by the fungus *Piricularia oryzae* (Ou 1972). This disease is spread primarily by airborne spores, most of which are trapped near the ground when there is no wind; however, they have been collected as high as 24 m above the ground and there is a report of capturing spores from an airplane at about 2150 m (Ou 1972). The disease appears as lesions on the panicles and on the leaves of the rice plant. Seedlings or tillering plants are often killed. On older plants, numerous infections on the panicles reduce yield. Those panicles with one-third or more of the spikelets affected are considered as blasted panicles in the formula $Y = 0.69X + 2.8$, where Y is the percent of loss and X is the percentage of blasted panicles (Kuribayashi and Ichikawa 1952 as reported in Ou 1972). Accurate loss estimates from leaf blast are not available. Losses due to panicle blast were estimated in Japan during 1953–1960 to range from 1.4% to 7.3% of the total yield even though chemical control methods were being extensively used. In epidemic areas of the Philippines, thousands of acres suffered more than 50% yield loss. This disease is widespread in major rice-growing areas; the impact is principally by short-range aerially disseminated spores.

Yield losses are important not only to food crops but to other plant species as well. Among species of pines many of the losses are due to rust diseases such as white pine blister rust caused by *Cronartium ribicola*. This fungus has several spore types, one of which (aeciospores) is produced on the pine and can infect only currants and gooseberries (genus *Ribes*). The infected *Ribes* in late summer produce another spore type (basidiospores), which can infect only pine needles. These basidiospores are delicate, which limits the effective windborne range in which pines can be infected to <1 km under most situations.

Blister rust kills white pines by forming cankers that can girdle the branches and the trunk. The annual losses due to blister rust in North America have been estimated at more than 5.66×10^6 m^3 of lumber (Davidson and Prentice 1967). If this disease is not controlled, the white pine industry will cease to be economically profitable except in a few areas where local climate prevents infection and spread.

In the examples above, we have seen estimates of losses due to airborne plant pathogens. Eventually the estimate of loss should be coupled to a concept commonly called *economic threshold*. Simply put, the concept is supposed to indicate the maximum disease loss that can be tolerated before active control measures need to be initiated. Some of the components that should be included in a study of economic threshold (ET) can be seen in the equation:

$$\mathrm{ET} = \frac{Y \times V}{C} - I \qquad (6.1)$$

where Y is the predicted increased yield due to a disease control program, V is the cash value of the increased yield, C is the cost of the disease control program, and I is the incentive factor, i.e., the minimum net return on investment required by the farmer or other decision maker. The incentive factor is included because if a control measure costs, for example, \$10, one must decide what is the minimum value of yield increase that one wishes to obtain: \$10, \$20, \$30, or more? If Equation (6.1) is equal to zero or greater than zero, the ET has been reached or passed and it is time to initiate control. The concept of ET for plant diseases is still in its elementary stages and awaits further development into a practical tool.

We have seen that aerobiological processes have a significant impact on plant productivity over great distances as well as locally. The more we learn about these processes the better able we will be to anticipate the consequent results on plant communities. Many of the future developments in plant pathology can be expected to be closely linked to those in aerobiology.

Impact of Air Pollutants

Atmospheric pollutants are known to cause abiotic plant diseases and subsequent losses to agricultural and forest ecosystems. A national air pollution problem exists today, particularly near large urban centers. Although this may be a localized situation in some cases, it is a widespread problem on the East Coast from North Carolina to Maine and in California.

The air pollutants known to affect agricultural and forest productivity are the gases ozone (O_3), sulfur dioxide (SO_2), nitrogen dioxide (NO_2), peroxyacetyl nitrate (PAN), fluorides, and ethylene (Heck 1973, Wood 1968). In recent years such materials as trace elements, pesticides, ammonia (NH_3), and chlorides have been added, but little is known about their impact. Relatively little research has been done on the impacts of particulates and most work has focused on specific types of dusts (e.g., cement dust) rather than the mixture found in the atmosphere. Crusts from cement dust can damage plants seriously (Lerman and Darley 1975), but considerable controversy surrounds any general conclusion about the impact of particulates.

The impact of air pollutants can be measured in several ways: biochemical effects, acute and chronic damage including impact on growth and productivity, the impact on plant community structure, the impact of air pollutants on biological disease-causing agents, and the economic impact.

No attempt will be made here to cover in detail the impact of air pollutants on vegetation. For this the reader is referred to Mudd and Kozlowski (1975) and Naegale (1973). A rather brief summary of air pollution impacts is presented here.

Biochemical impacts

If the effects of air pollutants on plants are to be fully understood, it is important that we understand the biochemical mechanisms of their action. Many of the chemical reactions and pathways may be affected, particularly enzyme reaction. Some of the effects of SO_2 are attributable to its acidifying effects. Possible toxic reactions have also been identified for nitrogen oxides, O_3, and PAN. Mudd (1973), however, indicates that despite the amount of published research on biochemical reactions with air pollutants the sequence of events leading to characteristic symptoms cannot yet be described.

Acute and chronic impacts

The impact of air pollutants on plants has been demonstrated through visual symptoms of acute injury. Most of this acute injury is expressed on plant leaves since leaves are the most active part in exchanging gases with the surrounding atmosphere. Table 6.1 indicates levels of important pollutants that cause acute injuries.

Pollutant gases diffuse through stomata or cracks in the cuticle into the water-saturated intercellular spaces. Those gases with high solubility in water are important plant toxicants but those with low solubility, such as carbon monoxide (CO) and nitric oxide (NO), are unimportant.

Acute injuries from exposure to pollutant gases generally result in the death of large areas of leaf tissue. The dead tissue turns various colors ranging from white to ivory to red or dark brown (Taylor 1973). Chlorotic symptoms may represent a transition from healthy tissue to necrotic tissue or the accumulation of a nonlethal dose of pollutant. Premature defoliation may result even when no external symptoms are visible if doses are high enough over a short period of time.

There are many excellent detailed descriptions of acute injury produced by individual air pollutants on both agricultural and forest crops (Hindawi 1970, Davis 1973). A brief description of typical injuries follows. It should be remembered that susceptibility to pollutant damage varies with the plant species. Intraspecific variability also occurs.

Sulfur dioxide

Acute SO_2 injury on broad-leaved plants is represented by white to ivory necrotic patches between the veins. Tissue along each side of the major veins is not affected, making the veins stand out clearly.

Injury to grain crops such as barley, rye, oats, and wheat develops as necrotic streaks between the veins. Injury to conifers, particularly pines, starts at the tip of the needles and extends to the base and is reddish brown in color. Bands of injury typically occur.

Fluoride

Hydrogen flouride (HF) is the most abundant fluoride pollutant although other forms do occur. The tips and margins of leaves are affected first because fluorides are readily transported. Injured tissue turns reddish brown in general.

Nitrogen dioxide

At high concentrations, NO_2 resembles SO_2 injury on broad-leaved plants, with irregularly shaped areas between the veins dying and bleaching to white or light tan. At lower concentrations small, dark, pigmented lesions are produced that may resemble ozone injury.

Ozone

The most susceptible tissue to O_3 is that which is recently expanded. Very young and older leaves are usually resistant.

Ozone may cause "stipplelike" lesions of varying colors from red to black to brown on broad-leaved plants. Ozone-injured pine needles usually have orange red bands that spread to the tips with subsequent defoliation of the tree leaving a tufted appearance, since the very young foliage remains. Interveinal streaks of white necrotic tissue develop on grains, grasses, and corn.

Peroxyacetyl nitrate

A photochemical pollutant, PAN seldom injures woody shrubs and trees but severely injures other plants. Injury tends to develop on the lower leaf surface of broad-leaved plants where a bronzing symptom is apparent. Distinct white or brown bands occur on grains, grasses, and corn.

Minor pollutants

Minor pollutants are ethylene and hydrogen chloride, which are byproducts of combustion, and chloride and ammonia, which are generally accidentally released into the atmosphere. Ethylene is the most important and damage is usually expressed as a growth distortion, chlorosis, and defoliation.

Pollutant combinations

Pollutants rarely exist alone in the atmosphere. Research on this topic has generally involved only two pollutants, however, e.g., SO_2 and O_3, SO_2 and NO_2, and SO_2 and HF (Reinert et al. 1975).

With SO_2 and O_3 greater than additive "synergistic" effects have been noted. Mixtures of these two gases at concentrations below the injury threshold for SO_2 and at or below the threshold for O_3 produce O_3-type symptoms. The SO_2-type symptoms are not observed unless SO_2 concentrations are well above threshold.

Our knowledge in this area is, however, fragmentary and many questions have yet to be answered. For example, does the ratio of one pollutant to another influence response of plants?

Chronic impacts

Effects of low levels of pollutants, i.e., chronic dosages, may be just as important as acute dosages but are not as visually obvious. The exact difference between acute and chronic effects is, however, sometimes difficult to define, but it is related generally to the exposure time and concentration of the pollutant. How is a plant altered when it is exposed for long periods of time below the level at which visible injury occurs?

Plants may exhibit repressed growth and yield without expression of visible injury (Feder 1973). A change in growth rate, a reduction in leaf area, or both have been demonstrated after prolonged exposure to SO_2 (McCune et al. 1967). Both chronic and acute symptoms must be taken into account in explaining plant damage and yield losses.

Impact on plant communities

Natural grassland, desert, and forest communities have been impacted by air pollutants, but the most dramatic damage has occurred in forest ecosystems. Forest ecosystems are dynamic units that change species composition and numbers over time, a process known as *succession*.

Air pollution can influence the natural succession pattern because of the relative susceptibility of each species and its importance (Miller and McBride 1975, Miller 1973). Sulfur dioxide has changed forest communities in the vicinity of smelters at Redding, California; Anaconda, Montana; and Copper Basin, Tennessee. Oxidant damage has occurred in coniferous forest ecosystems in southern California. Ponderosa pine, a favored species, is very susceptible to oxidant damage. As ponderosa pine continues to decline in numbers, conditions favor the regeneration of shrub species (Miller 1973).

Woodwell (1970) has summarized the expected effects of our pollutants on ecosystems: (1) elimination of sensitive species and reduction of diversity in the number of species, (2) selective removal of larger overstory plants and a favoring of plants small in stature, (3) reduction of the standing crop of organic matter leading to a reduction of nutrient elements held within the system, and (4) enhancement of the activity of insect pests and some diseases that hasten mortality. Root diseases may increase while rusts and leaf diseases may decrease (Grzywacz 1971).

The hydrology of forests may also be altered by changing species composition, with perhaps increased snowmelt and runoff, and changes in stream productivity.

Sulfur dioxide and NO_2 are transformed in the atmosphere by hydrolysis to sulfuric and nitric acids. Acid rainfall results, and Likens et al. (1972) have suggested several effects on forest ecosystems including changes in the rates of leaching of nutrients from soil and foliage. The full significance of acid precipitation to forest ecosystems remains to be defined.

Changes in lichen and bryophyte populations have also been noted, especially the decline of epiphytic species with continued exposure to air pollutants (Ferry et al. 1973). The sensitivity of these plants to air pollutants is relatively higher than that of other plant groups, and thus they have been used as biological pollution indicators.

Air pollutants and plant diseases

Air pollutants influence the development of plant diseases by modifying disease progress, the host, and the virulence of pathogens (fungi, bacteria, and viruses) through inhibition of spore germination and reduction of mycelial growth. This topic has been reviewed by Heagle (1973) and Treshow (1975), who concluded that in certain situations pollutant concentrations are sufficiently high to impair fungal development, yet are below the threshold for host injury. An example of this situation occurs with foliage diseases where black spot of roses (*Diplocarpon rosae*) was eliminated in areas where SO_2 concentrations exceeded 100 $\mu g/m^3$ (Saunders 1966). Such high levels are rarely reached with current emission controls.

Fluoride emissions are also no longer sufficiently high to expect much impact on disease progress. Ozone, although more difficult to control at the source than SO_2, is not as toxic to fungi as SO_2 (Treshow 1975) and only certain infection stages are sensitive to O_3.

The interaction of air pollutants with disease is likely to be greatest where accumulations of pollutants occur in high concentrations in the host, or where a major amount of the pathogen is exposed to the atmosphere, e.g., with powdery mildews.

Sulfur dioxide may also affect soil flors. Some investigators found increased diversity in polluted areas whereas others found fewer numbers. Mycorrhizal associations are also affected (Treshow 1975).

Manning (1976) has shown that O_3 can stimulate mycoflora on leaves and roots. No work has been conducted on the response of fungi, bacteria, and viruses to combinations of pollutants.

Microorganisms may also be exposed to air pollutants while airborne. Lighthart et al. (1971) found that SO_2 killed airborne *Serratia marcescens* at low relative humidity but viability was protected at higher relative humidity.

In general foliage diseases and rusts are more likely to be reduced by air pollutants. Heart rots do not seem to be affected by pollutants, but root rots such as *Armillaria mellea* are increased.

The first survey of photochemical air pollution effects on crops was conducted in California in the mid-1950s (Middleton and Paulus 1956). These superficial estimates fixed an annual loss of approximately $8 million on the West Coast and $18 million on the East Coast by the early 1960s for all pollutant types. Yearly losses for the whole country in the late 1960s were estimated to be $500 million (Heggestad 1968).

Heck (1973) reports a different approach to estimating losses based on the thesis that hydrocarbon emissions are related to oxidant production. Reductions in crop yields were empirically related to hydrocarbon emissions from over 100 metropolitan areas in the United States. Estimates of losses for two years show annual losses of $100 to $125 million.

When visible crop injury is not found, there is currently no way of surveying the impact on growth and yield. In addition, there are no estimates of economic damage on ornamentals and natural ecosystems, but losses could amount to billions of dollars. The ultimate concern over air pollution impacts should really emphasize the effects on both simple and complex plant communities.

IMPACTS OF AIRBORNE INSECTS*

S. A. Forbes (1915) stated that "the struggle between man and insects began long before the dawn of civilization, has continued without cessation to the present time, and will continue, no doubt, as long as the human race endures. It is due to the fact that both men and certain insect species constantly want the same things at the same time."

Insects destroy or damage all kinds of growing crops and other valuable plants and disseminate disease organisms to plants, animals, and humans. Pest populations (insects, pathogens, birds, mammals, weeds) are consuming, destroying, or both, nearly one-half the world's food supply (Pimentel 1976). In Asia, one-half the world's human population depends on rice for 80% of its food. Eight years ago Cramer (1967) estimated that more of the rice yield is lost to pests than is left for human populations—57%, of which 36% is due to insects. Pests, especially insects such as leafhoppers and planthoppers, as well as their transmitted diseases, are the main limitation to increased rice production (Way 1976).

---

*M. L. McManus

Direct damage to cultivated crops and forest trees

Insects damage living plants in every conceivable way: by chewing leaves, buds, stems, bark, or fruit; by sucking sap from leaves, buds, stems, or fruits; by boring or tunneling in the bark, stem, twigs, fruit, nuts, and seeds; by producing galls on all of the above parts; by attacking roots and underground stems; by ovipositing on aerial parts of plants; and by using plant parts for nests or shelter (Metcalf et al. 1951). It has been estimated that in 1974 annual food losses due to insects in the United States was ca $7.2 billion (Pimentel 1976). This figure is staggering when one considers that these losses occurred despite extensive chemical control activities.

Most of the economically important arthropods that are of interest in aerobiology are either sapsuckers or defoliators. The list is too extensive to discuss in detail and therefore I have selected a few representative examples; they are discussed in subsequent subsections.

Dispersal is a major component in the life history of at least three major forest insect pests, all of which are defoliators: the spruce budworm, the Douglas-fir tussock moth, and the gypsy moth. The spruce budworm is the major pest of the spruce-fir forests of North America. The outbreak of 1910–1919 destroyed 98 million m^3 of spruce and fir in Maine alone. In 1974 the budworm defoliated 2.1 million ha of spruce-fir in northern Maine; in addition, there is a 36.4-million-ha infestation in the adjacent provinces of Quebec and New Brunswick. In 1975 the state of Maine and industry applied insecticides on more than 1.4 million ha of spruce-fir timber in order to maintain a standing inventory of live trees. (USDA Forest Service 1976, final environmental statement for cooperative spruce budworm suppression project in Maine, 1976. 137 p. Unpublished.)

The Douglas-fir tussock moth is a more sporadic pest but has caused spectacular outbreaks at more-or-less 10-yr intervals in the fir forests of California and the Pacific Northwest (Wickman et al. 1973). The most recent infestation occurred on nearly 1 million ha in Washington, Oregon, and Idaho. The USDA Forest Service estimates that 1.4 million m^3 of merchantable timber was killed in 1973 and that potential losses could reach $70 million (NAS 1975).

The recent gypsy moth outbreak covered ca 7252 km^2 of oak forest in the Northeast and over 0.7 million ha were defoliated in 1973. Losses are not so spectacular in terms of tree mortality because hardwoods are able to refoliate during the same growing season; however, cumulative tree mortality may exceed 50% over a 5-yr period, mainly because of secondary organisms that attack the weakened trees. Immediate impacts on the heavily populated Northeast are severe and include losses in residential property value, recreational use, aesthetics, and cost of control. Both the tussock moth and the gypsy moth have similar life histories and behavior, and both species disperse as newly hatched larvae in the spring of the year.

Transmission of disease-causing organisms

Cultivated crops

Insect vectors of plant viruses are known to occur in at least six of the major orders of insects—Homoptera (aphids, leafhoppers, whiteflies, scales, planthoppers), Thysanoptera (thrips), Hemiptera (plant bugs, lace bugs), Coleoptera (beetles), Orthoptera (grasshoppers), and Dermaptera (earwigs). In addition, one family of mites has been implicated in the transmission of plant viruses. Approximately 90% of all the plant diseases vectored by insects, however, can be attributed to the Homoptera, and specifically to the aphids and leafhoppers. These insects, along with members of the Hemiptera and Thysanoptera, have piercing-sucking mouthparts and directly transfer and inject disease organisms from diseased to healthy plants while feeding on plant fluids.

Both the aphids and leafhoppers share certain similarities: multiple generations each year, high reproductive potentials, piercing-sucking mouthparts, polyphagous feeding habits, small size and flight capability, and migrating potential.

Aphids. Although many new vectors of plant viruses have been discovered in the past 15 yr, including nematodes, fungi, and mites, the aphids still transmit greater numbers of viruses and mycoplasmas than any other group of organisms. The relationship between the viruses and aphids seems to have evolved around three kinds of aphid behavior (Watson and Plumb 1972): (1) host selection (aphids make brief, shallow probes into a plant after migration or their movement between plants); (2) colony establishment (invading winged forms of some species deposit a few individuals on several plants rather than establishing a single colony on one plant); (3) dispersal of the population (aphids characteristically produce alternate generations of winged forms, which then disperse and redistribute the population and the viruses). There are three categories of plant viruses associated with aphids and they are determined by the period during which feeding aphids remain infective after acquiring the virus. (1) *Nonpersistent* viruses are acquired after short probes by aphids and are thus stylet-borne or mechanically acquired. (2) *Semipersistent* viruses can be acquired within about 10 min of feeding and are frequently transmitted by polyphagous species of aphids that change hosts frequently. (3) *Persistent* viruses are acquired by aphids that have fed for an extended period of time on the same plant and usually must spend a latent period within the aphid before they can be successfully transmitted.

The actual number of aphid-transmitted viruses is unknown; however, at least 38 persistent viruses have been identified (Gibbs 1969). The green peach aphid is known to transmit more than 50 different viruses, mostly of the non-persistent type. Some of the more important aphid-transmitted viruses include the complex of *mosaic* diseases of cucurbits and other vegetables and the *yellows* diseases of cereal and grain crops.

Leafhoppers. The leafhoppers are second only to aphids in both the number and severity of diseases that they transmit to plants. Some of the more important diseases include phony peach disease, aster yellows, curly top virus, and Pierce's disease of grapevines. Results seem to indicate that the vectors of plant viruses are all phloem feeders while the remaining species feed on the mesophyll layer of leaves. In addition to transmitting viruses, many leafhoppers cause a condition called *hopperburn,* which apparently results from toxins introduced into plant tissue along with salivary secretions of the insect.

Mites. The only proven mite vectors of plant viruses are certain members of the family Eriophyidae. Because of their small size (150–250 μm) they have often been overlooked as vector candidates, although they commonly cause a wide array of characteristic galls and also cause curling and twisting of foliage. They are readily dispersed by wind at certain stages in their life cycle, and are essentially parasites of perennial plants; however, many species of mites do cross over to annuals if adjacent crops exist. Examples of mite-transmitted diseases include wheat streak mosaic, currant reversion, peach mosaic, and fig mosaic.

Forest and ornamental trees

Actually, very little is known about the relationship between forest tree diseases and the insects that may transmit them. Insects have been implicated in some of the most serious diseases of forest trees, specifically oak wilt, Dutch elm disease, elm phloem necrosis, and beech bark disease. Mycoplasmas or PPLO's have recently been found in association with diseased phloem tissues of a number of the yellows type diseases such as elm phloem necrosis, which were formerly presumed to be caused by viruses. The white-banded elm leafhopper has been identified as the vector of elm phloem necrosis.

It has been estimated that Dutch elm disease, which is transmitted by two species of bark beetles, and phloem necrosis are responsible for the loss of over 400,000 elm trees annually in the United States at a cost of $100 million. About 40% of the 25 million elms that have been planted over the years have now been lost to these diseases (NAS 1975).

Leafhoppers and aphids are now suspect as carriers of other virus diseases such as ash ringspot, black locust witches'-broom, and walnut bunch virus.

Extension of Range vs. Long-Range Dispersal

It has been well established that a considerable population of windblown insects exists at higher altitudes. The literature on this subject was reviewed in chapter 2 of this volume. The economic significance of insects, especially those species that may be influenced by aerobiological processes, was discussed in the preceding section. Basically, one can consider that there are two categories of impacts caused by the atmospheric transfer of economic species of insects—those impacts that result from what I prefer to call *extension of range* and those that result from *long-range* migration or dispersal. Examples of insects and of the distributive processes involved are discussed briefly.

Extension of range

Pest managers are mainly interested in maintaining destructive insect populations below an established economic threshold and in restricting the spread of current infestations. Both objectives are difficult to attain, and assessing the dispersal of insect populations is probably one of the most difficult problems confronting the pest manager.

Meteorological considerations

Air currents are important in the distribution of insects. They have been classified as vertical or horizontal based on their direction of flow, and short or long range based on their powers of distribution (Wellington 1945). Horizontal wind flow is important to insect distribution only within a zone of temperature favorable for insect flight. Therefore deliberate flight plus horizontal wind may greatly extend the flight range of an insect. That is the typical way that most winged species extend their distribution.

Apterous or otherwise passively blown arthropods, however, require vertical eddies for their support and transport, and therefore horizontal winds cannot account for the long-range dispersal of these individuals. Purely vertical currents such as those caused by convection may carry insects to very high altitudes, but the short duration of these processes, their limited areas, and their random surface distribution bar them from long-range transport (Wellington 1945). He further stated:

> The most significant distributions of insects are the result of atmospheric currents restricted to a few square miles in surface area. These short-range processes may be high or low altitude processes. So many factors are required for successful distribution that long-range distribution by winds may ordinarily be neglected by the economic entomologist.

There are exceptions to this statement and they will be discussed later; however, Wellington's comments seem to explain accurately most dispersal processes in arthropods. Examples of local processes that may affect insect distributions are those caused by topography (slope or valley winds) or diurnal phenomena (seashore breezes).

Examples

The dispersibility of newly hatched larvae of the gypsy moth (*Lymantria dispar* L.) was recognized by early workers (Burgess 1913, Collins 1917, Collins and Baker 1934), who estimated that larvae were passively carried up to 40 km by wind currents and were recovered at an altitude of 610 m. Nichols (1961) noted the occurrence of a new outbreak 56 km from the nearest known infestation, apparently from wind dispersal. To this day, the literature on the gypsy moth refers to larval dispersal as basically a long-range phenomenon. McManus (1973), however, considered the dispersal of this species within a systems concept, and further determined that the larvae have large settling velocities (75 cm/s). Mason and Edmonds (1974 and chap. 7, this volume) then developed an atmospheric dispersion model for gypsy moth larvae that incorporated the settling velocity. The model predicts that most larvae would be deposited within 1.6 km of a source area in the absence of extreme turbulence or convection. Preliminary results from field trials verify the predictions.

I suggest that the long-distance dispersal reported in the literature was probably attributable to a buildup of incipient infestations that went undetected. This does not imply that individual larvae cannot be blown for considerable distances—rather it suggests that most windblown larvae are deposited within a short distance of their origin and that those few larvae that are carried for long distances probably do not survive and establish new populations. The history of the spread of this insect supports the notion—the spread has been gradual and not necessarily directional since the insect was introduced into the United States over 100 yr ago. Every new infestation that has been detected outside the generally infested area has been caused by an artificial introduction and not by natural spread. Furthermore, if the larvae are passively blown for long distances, one would expect that the gypsy moth would now be generally distributed throughout the oak forests of North America. That is not the case.

The decision on whether or not to suppress gypsy moth populations is based on egg mass counts that are made in the winter months. These counts are not reliable predictions of summer defoliation, however—the best explanation for this shortcoming is that larval dispersal in the spring may cause major redistribution of the population, which results in defoliation in areas where few egg masses were found, and no defoliation in areas where blow-out exceeded blow-in. State records of defoliation in consecutive years suggest that this occurs. A regional systems approach to management of this insect is needed that also includes some predictions of the extent of annual dispersal.

Many scale insects of economic importance are sessile for most of their life history and rely upon a short period of dispersal in the "crawler" stage. These insects are passively transported by wind and most dispersal is limited to a distance of ca 1 km from the source. Consequently a gradual but continual extension of their range will occur dependent upon the presence of suitable hosts in the deposition area. Beech bark disease, which is caused by an association between a scale insect and a fungus, is spreading in this manner. Willard (1973) found that the wandering time of crawlers of the California red scale increased at high densities and increasing temperatures, and that most wandering occurred in the late morning when most turbulence occurs. This pattern of behavior seems to prevail for all species of insects and other arthropods that are passively transported by wind.

Many winged insect pests may extend their range through periodic migrations (Poston and Pedigo 1975, Rummel et al. 1975). Marked cotton boll weevils were recaptured at distances of 1.6–53 km from their release point (Johnson et al. 1975). Marked sterile male tobacco budworms were recovered after 24 h at a distance of 16.1 km from their release point in northeastern Mexico (Hendricks et al. 1973).

Dobzhansky (1973) repeatedly sampled three populations of *Drosophila* sp. mainly in California since 1945, and concluded that passive transport by wind had resulted in an appreciable expansion of range of two of the species. Howell and Clift (1974) released marked sterile male codling moths in order to monitor their inter- and intraorchard dispersal. They recovered moths frequently at distances of 2.4 km and occasionally at 8.7 km, and concluded that most populations tend to remain localized. These examples point out the variation in dispersibility that exists among species—probably related more to the flight capacity of the species than to the influence of horizontal winds.

The dispersal of economic insects usually results in a gradual extension of range and in major redistribution of populations. This redistribution creates a major obstacle for pest management specialists who are attempting to control damaging populations by maximizing their assessment and control techniques, while restricting their area of treatment to a minimum.

Long-range dispersal

Meteorological considerations

Although Wellington (1945) de-emphasized the importance of long-range dispersal in the successful distribution of insect species, he did not discount the occurrence of such events. A fairly steady supply of the required stage, preferably adult, must be deposited in a suitable habitat in order to effect a successful establishment in a distant area, thus calling for a fixed circulation over an extended period of time. Wellington mentioned that attention has been directed to oceanic islands populated by means of wind. He suggests that population of such islands by long-range wind dispersal is easier to attain than are shorter distributions over land because there exist very steady and durable circulation regions such as the trade-wind belts over ocean surfaces.

High-pressure areas sometimes exist over land and may possess the qualities such as steadiness of wind direction and length of duration to affect the long-range distribution of insects. These high-pressure areas sometimes become relatively stationary over a region for many days at a time and thus result in a fairly regular circulation pattern.

Low-level jet winds that occur from the southern plains into the north central states have been identified as the vehicle for long-distance dispersal of many aphids and leafhoppers (Wallin et al. 1967). A low-level jet wind is defined as a stream of fast-moving air, best developed at night in the layer from the ground surface upward to about 914 m. The winds at 914 m are from 40 to 80 km/h but are 16–40 km/h at the surface. The following meteorological conditions are associated with the development of the low-level jet (Wallin et al. 1967): (1) steady southerly wind flow at the surface from the southern plain to the northern states, (2) cloudless nights, (3) a low-pressure area located over or just east of the central or southern Rockies, (4) a large high-pressure area located over the eastern or southeastern states, (5) strong afternoon insolation of air flowing into the jet region, and (6) a stationary front extending to the northeast for 800–1600 km from the low that is located over the Rockies. These conditions occurred in April/May for a 5-yr period between 1965 and 1969 and appear to be related to mass transport of aphids.

Examples

Spruce budworm. An association between mass flights of the spruce budworm and passages of cold fronts was first reported by Henson (1951) and later confirmed on several occasions in New Brunswick by Greenbank (1957). The budworm disperses in two different larval stages, but this dispersal is not as dramatic as the mass dispersal of moths. Greenbank (1957) recognized two types of moth dispersal: (1) convectional transport, in which large numbers of moths are borne aloft and may be carried many kilometers by the updrafts associated with the passage of cold fronts (high-pressure areas); and (2) turbulent wind transport, which is a local phenomenon similar to larval transport. In both cases, the transported moths are predominantly females retaining a portion of their egg complement.

Convectional transport resulted in tremendous flights of moths that have descended on Canadian cities at times when the nearest budworm infestation was 80 km away (Morris et al. 1958). Moth dispersal can play a major role in the population dynamics of the budworm and, along with larval dispersal, has a tremendous impact on management decisions and the ultimate success of suppression projects against the insect.

Aphids. Some of the best examples of long-range dispersal occur in the aphids and there are two excellent review articles on the subject (Taylor 1965, Jensen and Wallin 1965). Several aphid species including the greenbug are vectors of barley yellow dwarf virus, which attacks barley, oats, and wheat (Jensen and Wallin 1965). The greenbug winters in wheat fields in Texas and Oklahoma and is apparently wind disseminated to the northern plains, where outbreaks of virus occur in the spring. Wallin and Loonan (1971) found that both the greenbug and English grain aphids appeared in trap plots in Iowa several weeks before the local population developed to the winged stage. Aphids appeared coincident with low-level jet winds from the southern plains and yellow dwarf symptoms were noted three weeks later. The same phenomenon was reported by Kieckhefer et al. (1974) in South Dakota. Wallin and Loonan (1971) proposed a system for forecasting the incidence and severity of the disease in Iowa based upon knowledge of the aphid population in the source area, low-level jet wind advisories, early detection in the target area, and weather conditions after aphid deposition.

Leafhoppers. Three or four species of leafhoppers are known to migrate periodically at the same time each season. Temperature is important as a stimulus although overcrowding, deteriorating plant condition, and the formation of grain heads have also been suggested as stimuli (DeLong 1971). Leafhoppers have been recovered at all altitudes up to 4267 m and the duration of migratory flights may require several days or weeks. The sugar beet leafhopper is known to have migrated 320 km in two days on prevailing winds (Dorst and Davis 1937).

Two patterns of leafhopper migration have been observed: (1) the sugar beet leafhopper has two migrations, one in autumn and one in spring, which are in opposite directions; (2) the potato leafhopper migrates once a year in the spring from a permanent breeding area. One of the best examples of long-range dispersal is the spring migration of the six-spotted leafhopper, which transmits aster yellows virus (Chiykowski and Chapman 1965, Nickiparick 1965). Leafhoppers may accumulate in a source area prior to migration; convection currents from plowed soil in southern regions can lift leafhoppers to higher altitudes, where they are transported on warm southerly winds and deposited when the warm air meets a high-pressure front (Huff 1963, Pienkowski and Medler 1964).

Long-range dispersal of aphids, leafhoppers, and the spruce budworm occurs under conditions that were delineated by Wellington (1945): a large population in the source area that can provide a steady supply of pests for dispersal (greenbug populations in Texas were estimated at 6560–16,400/m), and weather processes that are of fixed direction and frequent occurrence so that sufficient numbers can be transported over a period of days. The low-level jet certainly satisfies this requirement.

There is obviously a need for a systems approach to the management of pest species that is based on our capability to predict changes in insect numbers; this approach in turn depends on our ability to predict or assess the dispersal of the species involved. Therefore knowledge of the dispersal behavior of each species is needed along with an understanding of the aerobiological processes that may be involved—this applies to both short- and long-range dispersal processes.

IMPACT OF AIRBORNE POLLEN ON PLANT DISTRIBUTION*

Sexual reproduction among plants requires an extraneous agent to transport pollen from its source to its receptor. Most seed plants effect pollen transfer by passive use of either animals (zoophily) or wind (anemophily).[1] These two transfer phenomena reflect contrasting adaptive strategies for accomplishing reproduction with genetic recombination. They are one portion of an interrelated set of adaptive characteristics that involve differing plant distribution patterns, habitat requirements, and speciation modes. This complex is the subject of the discussion below.

Pollination

Almost all animals used for pollination have two characteristics in common. First, most animal pollinators fly. While birds and bats are the more spectacular pollinators, insects, particularly of the Hymenoptera, Diptera, and Lepidoptera, are the important pollinators. Most authors agree that the earliest flowering plants (angiosperms) of the Cretaceous period were insect pollinated (entomophilous), probably by crawling beetles in search of edible pollen and floral parts (Robertson 1904, Grant 1950, Eames 1959, Cronquist 1968). Subsequently, parallel evolution during the Tertiary period of diverse entomophilous plant groups and of the flying insects (Diptera, Hymenoptera, Lepidoptera) occurred (Cronquist 1968) and may indeed be interdependent. Flying pollinators tend to decrease speciation by reducing geographic isolation.

Second, most insect and bird pollinators possess color vision. Not surprisingly, the plants that attract the animals do so with bright, showy flowers of various colors. The colorful sight attractants are supplemented by the nectar (simple sugars) that serves as a scent attractant and as food for the pollinating agent. Evolutionary change in zoophilous flowers frequently involves greater specialization of animal attractants to ensure more consistent pollination by specific animal pollinators. Even zoophilous flowers that are adapted to attracting a wide variety of pollinator species do so through larger, more colorful, and more nectariferous flowers. Few pollen grains need to be produced because the transfer vector moves pollen directly from one flower to another with little pollen loss.

---

*A. M. Solomon

In sharp contrast, anemophilous flowers require little morphological specialization for their "shotgun" approach to cross-fertilization. Anemophilous flowers are generally much smaller than zoophilous flowers and generally possess no nectaries. If corollas are present, they consist of small, green, inconspicuous petals and sepals. Primary production of anemophilous flowers is funneled into many small flowers per plant, clustered in dense inflorescences, and into large, numerous anthers, which produce great quantities of pollen. Morphological specialization occurs in anemophilous flowers, including expanded stigmatic (pollen-receptive) flowers on different parts of the same plant (monoecious plants), or on separate plants (dioecious plants). These represent few morphological adaptations when compared with the level of specialization achieved by most zoophilous flowers. One explanation of these two strategies follows.

Consider the basic logistic equation developed to describe population density over time by Verhulst (1838), modified by Pearl (1930). If age distribution is stable (i.e., birth and death rates are constant) in a space-constant population, and if crowding affects all individuals and life stages equally, then the density of the population can be described by the following sigmoid growth curve:

$$N_t = \frac{K}{1 + \exp a - rt} \tag{6.2}$$

where N is the number of individuals in the population, a is the y-intercept (ln $[K - N/N]$ at time $t = 0$), r is the intrinsic rate of population increase, t is the time increment from $t = 0$, and K is the maximum population size possible. Thus, as initial population growth occurs, the maximum or intrinsic rate of increase r is realized in the relationship

$$N_t = N_0 \exp rt \tag{6.3}$$

The intrinsic rate of increase, or *reproductive potential,* is independent of population density and can produce the maximum increment of individuals in an unlimited environment. Because the population is space (and therefore resource) finite, however, growth begins to slow thereafter. As time continues, the value of rt increases, exp $(a - rt)$ approaches zero, $1 + \exp(a - rt)$ approaches unity, and the equation approaches the value for K. The finite resources thus allow no greater population density than K, which describes the *carrying capacity* of the environmental resources for the population, and is thus a strictly density-dependent constant.

A currently popular hypothesis among population ecologists describes natural selection as two separate selection systems that follow from the Verhulst-Pearl logistic equation (MacArthur 1972, Gadgil and Solbrig 1972, Pianka 1972). One selection mode is effected through population mortality factors that are density-dependent (DD factors). These include increased mortality and decreased natality due to biological competition for limited light, mineral, space, and water resources, and to similar biologically induced stresses. Because the DD factors operate only near the upper limit of population density (K), the selection they induce is termed K-selection.

The organisms responding primarily to K-selection forces (K-strategists) take on specific characteristics. They have slow growth rates because they must be very efficient in their use of limited resources to avoid resource depletion. They have low reproductive rates, for individuals merely need to replace themselves within their stable environment. In addition, the seeds and seedlings must also be adapted to withstand competition, so they are normally large and few in number. The K-strategists allocate a great proportion of their energy to nonreproductive adaptations that allow them to compete more successfully. Because an adaptive response to a competitor eventually elicits a similar response from the competitor, greater and greater nonreproductive energy expenditures for adaptive responses are required.

The other mode of selection is effected through population mortality factors that are independent of population density (density-independent or DI factors). These factors include increased mortality due to severe weather or substrate parameters, to landslides, fires, and to similar stresses caused by the physical environment. Because DI factors operate at all levels of population density, and so control the rate of population increase (r), the selection they induce is termed r-selection.

The organisms responding to r-selection (r-strategists) take on characteristics that contrast with those of K-strategists. Growth rates among r-strategists are rapid because resources are not limited, but the time period for growth may be very limited. Reproduction rates among r-strategists are also rapid to compensate for the risk that the local population may become extinct. Their seeds germinate quickly and are normally small and plentiful. In addition, r-strategists allocate a great amount of energy to reproduction to make up for the high population losses. The DI mortality factors, of course, do not produce a counter response when adaptations to them by the organisms occur, so adaptive responses require only enough energy for success.

The concept of K- and r-selection is not absolute. A continuum exists between K- and r-selection forces and between K- and r-strategies. Any biological entity must therefore display characteristics of both K- and r-strategies. It follows that the concept is useful only when comparisons are made of the relative positions of similar biological entities on the continuum. The eastern deciduous forest biome, for example, is K-selected when compared with the tundra biome, but is r-selected when compared with the tropical rain-forest biome. Most shrubs are K-strategists when compared with grasses, but are r-strategists when compared with trees. In fact, K- and r-selection and strategy contrasts can be discerned among biomes, ecosystems, communities, species, local populations of the same species, or, indeed, among life stages of the same individuals.

If the concept is applied in isolation to flowering (the gametophyte stage in the life cycle of seed plants), zoophily clearly emerges as a K-strategy, and anemophily as an *r*-strategy. Successful zoophily depends upon successful competition for the animal transfer vector. It results in positive feedback systems whereby more and more energy is allocated to nonreproductive adaptations (showy petals and sepals, nectar production, and so on) with concomitant reduction in pollen production.

If data were available on the topic, they would probably indicate that overall growth rates of zoophilous plants are slower, and the seeds of zoophilous plants are fewer, as well as larger to allow better response to immediate competition (through larger food storage supplies and the like) than are the growth rates and seeds of anemophilous plants. The latter allocate much less material to nonreproductive flower parts, being adapted instead to withstand parameters of a physical environment in which the day's pollen emission can be destroyed by particle-scavenging rain, or a season's pollen production can be lost to a late frost (see chap. 2 under *Pollen*).

Without the positive feedback of biological stress that continuously enhances adaptations for competition among *K*-strategists, amenophilous flowers of *r*-strategists retain only enough adaptive apparatus to survive the more frequent disasters (Tables 2.6 and 2.7 in chap. 2). They emit large amounts of pollen that is notably free of the elaborate geometrically sculptured surfaces found in zoophilous pollen grains. The distributions of anemophilous plants and the timing of their pollination adapt them to the DI factors they must overcome.

Plant Distribution, Habitat Requirement, and Speciation in Anemophilous Plants

The major problem in cross-fertilization that must be solved by each anemophilous species involves the relation of distance between plants versus numbers of pollen grains produced. In essence, the problem is that for successful pollination among plants spaced at arithmetically increasing distances, logarithmically increasing pollen quantities are required. Because there is a practical limit to the amount of pollen plants can produce, spacing of plants becomes a critical variable.

Proctor and Yeo (1972) note that "If effective pollination requires no more than one pollen grain to reach a stigma with an area of one square millimeter (about the area of an oak stigma), every square meter of the plant's habitat must receive around a million pollen grains to make pollination reasonably certain." If pollination were the only variable in reproduction, then the event described above would need to occur only once in the 70- to 600-yr life span (Fowells 1965) of the oak. Subsequent to pollination, however, a whole host of mortality factors operate on all stages of the plant's development from seed production through maturity. Thus the minimum fertilization suggested may be a considerably conservative value.

Distances between individual anemophilous plants of the same species are normally short, providing their populations with very clumped spatial distributions. To maintain clumped distributions, these plants are frequently limited to specific edaphic or climatic situations that allow high plant density while maintaining relatively low species diversity. At first glance, this seems to involve DD mortality, characteristic of *K*-selection. Close inspection reveals that the plant densities involved are not necessarily at the maximum potential density, i.e., the carrying capacity of the population growth sites. In fact, the growth sites are those where low interspecies competition occurs so that relatively low species diversity can be maintained. These unstable and high physical stress environments induce the *r*-strategy of high reproduction rates, rapid colonization, and rapid growth.

Plants adapted to environments characterized by excess physical stress are usually wind pollinated. The short-lived weed communities created by storms or by man's agricultural activities are composed in large part of wind-pollinated members of the Gramineae (grass), Chenopodiaceae (goosefoot), or Compositae (sunflower) families (Wodehouse 1939). The deciduous forest vegetation of fire-prone temperate areas is similarly conducive to wind pollination, and is dominated by fire-tolerant anemophilous Fagaceae (beech), Betulaceae (birch), and Juglandaceae (walnut) families. The drier temperate prairies of North America, the "grasslands," were also fire dominated before they were farmed by Europeans (Wells 1970, McAndrews 1968).

Saline soils, salt marshes, and lakesides are not species rich, but are inhabited by dense stands of anemophilous salt bushes (Chenopodiaceae), salt grasses (Gramineae), and cattails (Typhaceae), respectively, commonly with sedges (Cyperaceae) in the last two areas. Even in the desert, where normal rainfall is too low to support dense growth of any plant species, the arroyos in which precipitation is concentrated support dense, almost pure stands of annual anemophilous weeds (Chenopodiaceae, Gramineae).

Both disturbance situations and those created by edaphic pecularities support and induce the growth of wind-pollinated and *r*-selected vegetation. Their geographic distribution in isolated strips or other shapes containing continuous separated areas presents the necessary patchiness or "clumps" of substrate that will allow anemophilous populations to reproduce. Certainly not all plants of disturbance or edaphic situations are wind pollinated or *r*-strategists, nor are distributions of wind-pollinated plants limited to these areas. Anemophilous plants, however, do dominate such areas of dense, low-diversity vegetation, and they do not dominate areas that contain either low plant density potential or high species diversity (Whitehead 1969).

The geographic separation of subpopulations, because of habitat requirements such as those given above, increases the potential in anemophilous species for development of new races and subspecies. This in turn can lead to speciation. Stebbins (1966:106) points out that "two conditions must be fulfilled before populations can diverge with respect to any characteristic which is determined by many separate genes. First, divergent pressures of natural selection must be acting upon them. Second, they must be well enough isolated from each other so that the initial genetic divergence will not be 'swamped' by gene interchange." The processes that produce new species capable of sympatry (adjacent existence without hybridization) are qualitatively the same as those that give rise to divergent races and subspecies (Stebbins 1966). The unstable or extreme habitats that anemophilous plants prefer contain the selective pressures that promote divergence. At the same time, the isolated distributions of such habitats provide the geographic separation required to prevent gene exchange.

Most species in the vast, anemophilous grass family (4500 species) fit this description. Although the family is of recent geological origin (mid Eocene, $<5 \times 10^7$ yr), speciation among grasses occurred rapidly and into diverse habitats. Grass species are now found in almost every conceivable environment from tropics to tundra, from sea level to treeless alpine conditions above 4880 m elevation, and from arid deserts to water-saturated estuarine soils. Grass species characteristically (1) grow in scattered populations that occur in pure stands of multiple individuals, (2) emit pollen at very specific times of the day for very short periods (Hyde and Williams 1945), and (3) possess pollen grains that are uniformly smooth, lose little water to the atmosphere, and thus are rapidly deposited.

The cumulative effect of these three adaptations is to produce momentarily high atmospheric pollen concentrations in the immediate vicinity of the local population, and to allow little pollen transport from the stand. As a result, local grass species populations are geographically isolated from others only 10^2–10^3 m away (Raynor et al. 1970, 1972a). Ecotypic variations between local populations are common among all grass species investigated (Quinn 1969, Quinn and Ward 1969, Rotsettis et al. 1972, McMillan 1967, Robertson and Ward 1970). The potential for speciation is therefore always available.

Anemophily also seems to intensify the potential for divergence when increases in *r*-selective pressures occur. Despite the occasional records of long-distance transport involving distances beyond 10^3 km (Ritchie and Lichti-Federovich 1967, Maher 1963, Solomon and Harrington, chap. 7, this volume), anemophily is essentially a short-transport-distance phenomenon. While the short atmospheric life span of pollen apparently neutralizes genetic exchange through long-distance transport, changing plant distributions under greater *r*-selective pressures further decrease genetic interchange.

Recent work by Handel (1976) demonstrated pollen dispersal distances of under 5 m from two *Carex* species. Systematic studies by Raynor et al. (1970, 1972a,b, 1973) revealed that concentrations of pollen escaping most small-area or line-source populations have declined by 90% well within 10^2 m of the sources. The decline is more rapid, of course, with smaller sources and with more dense pollen grains.

Smaller, more scattered area and line sources occur on the edges of species ranges where *K*-selective pressure from better adapted species and *r*-

selective pressures from environmental stresses are both more intense; however, *r*-selection is logically the most important pressure. Pielou (1975) concludes that competition produces sharp species boundaries, while environmental stress induces diffuse ones. In situations where environmental stress rather than biological competition controls plant distributions (e.g., the northern limits of *Picea mariana* or *Abies balsamea* in the Canadian tundra, the upper altitudinal limits of *Picea engelmannii,* or *Abies lasiocarpa* in alpine zones of the Rocky Mountains), the populations are reduced to the characteristic form of scattered, clumped, small-area, or line populations.

Similarly, pollen grain density gradients can be found that conform to gradients of *r*-selective pressure. From the data presented by Solomon (1975), *Pinus* sp. pollen grain density and dispersal distance is inversely related to the size of *Pinus* sp. pollen. The smallest pine grains travel the shortest distance. Cain and Cain (1944) found that the smallest *Pinus palustris* pollen occurs on the western and eastern edges of its range along the Atlantic and Gulf Coast plain. Pollen sizes from the more wide-ranging *Pinus echinata* seem more variable, although the largest grains were to the north and east of its southeastern United States range (Cain and Cain 1948a). They found that pollen size increased among three different *Pinus* species (*P. banksiana, P. resinosa, P. strobus*) whose respective ranges were increasingly southerly from the Arctic Circle (Cain and Cain 1948b).

Pinus sp. pollen grain size gradients can also be observed among closely related *Pinus* species that are distributed along altitudinal gradients of the Rocky Mountains in Colorado or the Sierra Nevada Mountains in California. *Pinus* species with small pollen grains are found at the lower (*P. edulis, P. monophylla, P. cembroides*) and the upper (*P. balfouriana, P. aristata, P. longaeva*) altitudinal limits of trees, while those with larger pollen grains occur in the intermediate altitudinal zones (*P. ponderosa, P. lambertiana, P. jeffreyi, P. monticola,* etc.). The intermediate zones are not under the stress of cool summer temperatures that curb growth and reproduction of all *Pinus* species at the upper tree limit (LaMarche 1973), nor the stress of minimal summer precipitation that curbs growth and reproduction of all *Pinus* species at lower tree limits (Fritts 1966).

Thus, where *r*-selective pressures are greatest, genetic isolation increases because of more scattered plant distributions and more rapid attenuation of pollen concentration with distance. The effect is generally to increase ecotypic variation and potential speciation as maximum environmental stress is approached. This effect, in turn, requires greater fitness for species' survival in the harsh or temporary environments that the anemophilous *r*-strategists encounter.

MacArthur (1972) points out that extinction, which all species must avoid, is much less probable among species that consist of many genetically segregated, small subpopulations than among species that consist of few large populations. This is true even when the former are under high local extinction pressure and the latter are under mild extinction pressure. The anemophilous plant species are adapted to avoid extinction through their reliance upon a system of factors that include population divergence and speciation through *r*-strategy responses to selective pressures, and spatial isolation in extreme or temporary habitats.

IMPACT ON HUMAN AND ANIMAL SYSTEMS

IMPACT OF AIRBORNE PATHOGENS IN OUTDOOR SYSTEMS: HISTOPLASMOSIS*

Airborne Diseases

An infection acquired by the inhalation of infectious particles, whose source is found in soil or organic debris and where the etiologic agent exists as a free-living form, is defined in this chapter as *airborne*. Excluded, therefore, are those diseases that are transmissible from human to human or animal to human by droplet nuclei in aerosols, or from contaminated animal products. They are covered in this chapter under "Airborne pathogens in the indoor environment. . . ." Although some pathogens within the order Actinomycetales could be included because of their existence as soil saprobes, they are not considered here, and those that are discussed are fungi.

The major fungal diseases of North America acquired from exposure to foci where the respective causative agent resides in nature are aspergillosis, blastomycosis, coccidioidomycosis, cryptococcosis, and histoplasmosis. The mortality due to four of these diseases for 1961–1970 is shown in Table 6.2. Since these are nonnotifiable diseases, the values shown may be less than the actual death rates for these infections.

Aspergillosis, although airborne, is a secondary disease, and very little is known about the natural existence of the fungus causing blastomycosis, therefore considerations of these two infections are excluded. Cryptococcosis is apparently acquired by inhalation and the causative agent, *Cryptococcus neoformans,* is found in free-living form. Its association with pigeon (*Columba livia*) excreta or soil enriched with excreta from this bird is well known. There is very little information, however, on the epidemiology of cryptococcosis and the presumed aerobiological transport of *C. neoformans*. On the other hand, the ecology of the fungi causing coccidioidomycosis and histoplasmosis and the epidemiology of both diseases have been studied extensively. In their respective endemic regions both infections are significant public health problems. Because of the limitations of space, however, and the author's greater experience with *Histoplasma capsulatum,* our knowledge relating to the free-living existence of this fungus in soil, its transmission to susceptible humans, and its impact upon inhabitants residing in the endemic area will be presented.

*H. F. Hasenclever

Table 6.2. Deaths from four systemic airborne fungal diseases, United States, 1961–1970.[a]

	1970	1969	1968	1967	1966	1965	1964	1963	1962	1961	Av
Blastomycosis	3	2		17	12	29	21	24	17	16	= 14
Coccidioidomycosis	42	60	58	49	45	52	46	71	55	62	= 54
Cryptococcosis	112	117	96	65	90	62	74	73	90	86	= 87
Histoplasmosis	56	61	58	67	60	74	77	70	82	90	= 70

[a]From U.S. Dep. HEW, PHS (1974).

Histoplasmosis

It has been estimated that as many as 500,000 people acquire histoplasmosis annually (Furcolow 1965). Approximately one-third of them develop a clinical illness that varies from a very mild respiratory disorder to a widely disseminated disease. Over 90% of the persons that have some clinical manifestations recover spontaneously without any chemotherapy or hospitalization. The severity of symptoms among those infected varies greatly and, as a result, so does the loss of time at work or school. Most of the clinically apparent but milder forms of this mycosis are undiagnosed or misdiagnosed as being of viral etiology. Laboratory or radiological evidence, or preferably both, are necessary to confirm a diagnosis. Serologic tests, one of the most reliable methods for proving diagnosis in milder forms of histoplasmosis, are performed at a limited number of institutions, and the use of X ray for all people suffering upper respiratory infections is ill-advised. Because of these reasons most of the cases with milder forms of disease are not detected. Diagnosis is made more frequently for those patients suffering from severe pulmonary or disseminated infections because of the need for medical treatment or hospitalization. Histoplasmosis is not a notifiable disease so reliable statistics concerning its incidence are difficult to obtain.

Geographic distribution

Histoplasma capsulatum has a worldwide distribution in temperate, subtropical, and tropical regions. With the exception of Antarctica, autochthonous cases have been reported from every continent. Areas of endemicity exist in Southeast Asia, and a few cases have been recorded from Australia. In Europe the fungus appears to be restricted to the central portion and northern Italy. *Histoplasma capsulatum* is found almost anywhere in the eastern half of the United States, with areas of greatest endemicity occurring in the Ohio, Mississippi, and lower Missouri river valleys and through Kentucky and Tennessee. There are regions east of the Appalachian Mountains, however, where the residents have a high incidence of infection. There is some basis for belief that the fungus is more widespread than now reported and the prediction of Ajello (1971) that it will be found in all states where a suitable habitat exists is entirely reasonable. Locally acquired cases of histoplasmosis also have been reported from southern Canada.

Ecology of *H. capsulatum*

Much has been done to define the ecological conditions related to presence of *H. capsulatum* in soil. The fungus often exists in sites or areas that have been enriched with avian or chiropteran feces. It is found in association with old or unused chicken houses; under blackbird/starling roosts; and in soil, at the base of buildings inhabited by bats, which has been fertilized by their droppings. Caves sheltering bats often provide environmental conditions suitable for the existence and propagation of the fungus. While there is little or no evidence to indicate that avian species are infected, bats, however, may be and are known to excrete viable *H. capsulatum* cells in their feces.

Available information indicates that the fungus resides primarily in these limited foci but is not found in all areas where these conditions exist. It apparently is not dispersed uniformly in soils, and a suitable area presumably must be seeded or inoculated by natural dissemination before establishment can occur. Once established, heavy colonization occurs as long as fecal enrichment continues. Removal of the avian or chiropteran roost appears to have little immediate effect on the presence of *H. capsulatum* in the soil, but after 10 yr the degree of colonization diminishes (Hasenclever and Piggott 1974). Detection of the fungus, however, is still possible.

The mineral clay composition of soil has been proposed as a possible factor controlling the growth or presence of *H. capsulatum* in soil (Stotzky and Post 1967). They tested soils positive for the fungus collected from all parts of the world for their content of clay minerals. Montmorillonite was found to be absent in all samples but two. It has been postulated that this clay alters the microbial ecology to make conditions inimical for the existence of *H. capsulatum*.

Histoplasma capsulatum is a true saprobe of the soil. It can compete with other soil microorganisms in its ecological niche; however, studies attempting to define these microbial interactions have produced only fragmentary evidence. Pathogenicity apparently reflects the remarkable physiological and morphological versatility of this fungus and is not a mechanism to satisfy a complex life cycle. With our present knowledge it is hard to visualize how the disease-producing capacity of this microorganism can result in an evolutionary advantage enhancing its chances of survival. This teleological concept is, of course, based upon our present incomplete information, and as more knowledge is acquired our ideas may have to be revised.

Epidemics

The point source or focal origin of the infective spores or particles has been recognized for many years in the epidemiology of histoplasmosis. Mechanical disturbance of soil at the point source, when conditions are such that an airborne cloud of particulates can be created, is prominently associated with epidemics of this disease. The extent of the outbreak is dependent upon the location of the point source, whether rural or urban, on a farm where only a few members of a family may be exposed, or in a city or on a school yard where numerous people may be involved. Sarosi et al. (1971) have described most of the recorded epidemics of histoplasmosis in the United States. The number of people that were ill, according to their clinical criteria, as a result of these outbreaks ranged from 2 to 87.

One point source that caused two of the most extensive recorded epidemics of histoplasmosis occurred at Mason City, Iowa, in 1962 and 1964 (D'Alessio et al. 1965, Tosh et al. 1966a). The first outbreak, in August of 1962, resulted in two deaths and a total of 28 people suffering from clinically diagnosed histoplasmosis. During the second epidemic, in February and March of 1964, which was studied more extensively, 270 sought medical treatment and 87 were ill enough to be hospitalized, but there were no deaths. Estimates based upon skin testing surveys conducted during this study indicated that about 6000 people were infected. The source of infection for these outbreaks was a blackbird roost located in a wooded area along a creek in the central part of town. In August of 1962 the site was being cleared with heavy motorized and power equipment, when environmental conditions were dry, and much dust was raised. The epidemic developed and, when the disease was diagnosed and the source was identified, work was halted. The clearing operation was resumed in February 1964 at a time when soil and vegetative debris in the site were wet and partially frozen, and with the assumption that airborne infectious particles would not be raised. The resultant epidemic of extensive proportions proved the unpredictability of these aerobiological phenomena.

A more recent epidemic occurred at a school in Delaware, Ohio, in May 1970 (Foss and Saslaw 1971). The airborne cloud of infective particles was created, paradoxically, as a response to Earth Day, an ecological program to clean the school yard. The secluded area from which *H. capsulatum* was isolated subsequently was frequented by blackbirds and pigeons and was near air intakes for the school ventilating system. This point source was cleared of vegetative debris and raked manually. Of 949 students and faculty, 294 (31%) were ill enough to be absent from school. Eighty-nine more had symptoms but attended in spite of their illness. All recovered without any specific therapy. The school was closed for an undisclosed period of time because of the outbreak.

Exposure in a cave colonized by fungus represents another point source of infection. Since fewer people are involved, outbreaks resulting from this type of exposure usually are smaller. Illustrating this form of epidemic is a report from the Panama Canal Zone in 1967 (Hasenclever et al. 1967). From a total of 15 people known to have visited the cave, in groups from a single individual to seven at a time, eight experienced illness. Symptoms lasted 3–4 days in those with the mildest infections, but for the most severely affected patient hospitalization for 6 wk was required. Attention was focused upon this site when five of seven soldiers who had entered became sick. Subsequently, two of four scientists who entered, investigating ecological conditions, also developed disease, and at this visit proved by air sampling that inefective particles of *H. capsulatum* were present in the atmosphere of the cave. A large colony of *Pteronotus rubiginosa* (greater mustached bat) roosted intermittently in the cave, and many of these bats were shown later to be infected with *H. capsulatum*. Extensive studies were conducted on the ecology and aerobiology of the fungus at this site (Shacklette et al. 1967, Shacklette and Hasenclever 1968).

Aerobiology

The infective spore or particle of *H. capsulatum* is presumed to be its conidiospore. These microconidia, also known as *aleuriospores,* average about 2 μm in diameter. In soil cultures they are produced in myriad numbers and when conditions are favorable they become airborne. There is no fungal mechanism for forcible discharge, and in culture the spores appear to be rather tightly attached. A consistent observation for all reports thus far is that mechanical disturbance of the soil at the point source always preceded the epidemic. In Mason City, Iowa, the blackbird roost identified as the point source existed for years before the first epidemic occurred (D'Alessio et al. 1965). The fungus, obviously present in vast quantities in the soil and vegetative debris after the first outbreak, was apparently not a public health menace until further mechanical alterations at the focal point resulted in the second epidemic. *Histoplasma capsulatum* existed in the point source on the school yard at Delaware, Ohio, for an indeterminate period of time before the soil in the secluded area was disturbed, producing dissemination of fungal spores through the ventilating system. Human activity appears to be a prime factor in the creation of a spore cloud and its subsequent dissemination by wind currents. Aerobiological studies by Hasenclever and Piggott (1973) at two point sources where the surface was undisturbed confirmed that *H. capsulatum* was very seldom found to be airborne. At one site only one of 257 samples, obtained by two sampling techniques taken over a 4-yr period under most meteorological conditions and during all seasons, was positive for the fungus. At the other site none of 51 samples taken over 2 yr and under similar conditions was positive.

Air sampling in areas where air movement is restricted, however, has been more productive. Ibach et al. (1954) reported the isolation of *H. capsulatum* from the air in the interior of two different chicken houses. Care was taken before and during the sampling procedure not to disturb the organic debris in the houses. At the cave in the Panama Canal Zone (Shacklette et al. 1967, Shacklette and Hasenclever 1968) the fungus could at times be isolated from the air, and the presence or absence of the colony of bats during sampling was not important. These studies suggest that once infective particles are suspended and since dispersion cannot freely occur, they remain in the air of these restricted areas for an indeterminate period. Studies on external point sources indicate that without mechanical disturbance few infective particles become airborne. When disturbance of a point source does occur, vast numbers of infectious particles are raised and dispersed creating a health menace to the susceptible population in the vicinity. Although the studies of Hasenclever and Piggott (1973), using air sampling techniques at two point sources, showed that few infective particles could be detected, Tosh et al. (1970) have evidence from skin testing surveys near large starling/blackbird roosts that dissemination of infective fungal particles by natural air currents does occur. Children living or attending school near a roost had a higher percentage of skin test reactions than those at a greater distance. During these studies no outbreaks of histoplasmosis were reported to have occurred.

The second epidemic at Mason City, Iowa, provides the best information as to how far infective particles of *H. capsulatum* can be carried by aerobiological transport (Tosh et al. 1966a). Skin testing of children after the disease had subsided indicated that those who attended school within 0.8 km and on the downwind side of the point source had a higher percentage of positive reactions than did those on the windward side or at a greater distance. Although the actual distance was not stated, the fungus was isolated from rooftops of buildings within several blocks of the point source 2 mo, but not 7 mo, after disturbance of the soil occurred. It was assumed that the fungal particles had been deposited on the roofs by air currents.

Impact

It is impossible to determine the annual economic and social impact of histoplasmosis in the United States. The disease is very widespread, diagnosis of the most common clinical forms is difficult, and, since even diagnosed infections are not reportable, the statistics available represent a small fraction of the total. During an epidemic, though the actual loss of time at work and school would still be hard to determine, a more meaningful overview of the impact on a community could be made. This may be accomplished by examining the data from the second outbreak at Mason City (Tosh et al. 1966a). According to epidemiological evidence, they estimated that 6000 residents were infected. Approximately one-third of those infected would be expected to have symptoms of the disease (Furcolow 1965). Therefore about 2000 people had some, although in most instances very mild, illness. It is known that 270 were sick enough to seek treatment and that about 87 required hospitalization. Five received chemotherapy and, in three, surgical resection to remove pulmonary cavities was performed. Mason City had a population of 30,000 and an event of this magnitude would without question exert profound economic, social, and psychological effects upon a community.

The epidemic at Delaware, Ohio (Foss and Saslaw 1971), resulted in the loss of thousands of student and faculty hours in educational time. Not only did 30% of the 949 students and teachers miss school, but the institution was closed for an undisclosed period in an attempt to control the spread of the disease. Because of the unique dissemination of the fungal particles, i.e., through the ventilating system, only those within the building or in the cleaning process at the point source were exposed, and the surrounding community was not involved. Histoplasmosis was not suspected until clinical studies on a patient from the outbreak indicated this disease, and subsequent ecological and epidemiological data obtained by investigations at the point source confirmed it.

Even small outbreaks such as the one that occurred in the Panama Canal Zone (Hasenclever et al. 1967) can result in a surprising amount of occupational loss. Sixty workdays were lost by seven of the individuals that were ill. Two-thirds of this time was accounted to one person, who was hospitalized for 41 days. Excluding this case, however, the other six with mild infections lost a total of about 20 workdays.

The result of an epidemic on a community is usually measured in terms of economic loss or the inability of those affected to work or to engage in their normal activity. Also to be considered is the cost of medical treatment and hospitalization. The psychological and emotional impact on the families of victims that contract serious or fatal disease cannot be measured. The outbreaks cited illustrate how these events may involve and interfere with normal community functions. The financial losses and the public health significance of these occurrences are of such magnitude that preventive measures are necessary.

Control

Some infectious diseases may be controlled by immunization of a susceptible population, disinfection or pasteurization of water or food, eradication of a vector, or the administration of chemotherapy to a victim with an infection. Numerous studies in laboratory animals have shown that a low level of immunity to histoplasmosis can be induced. Because a rather small percentage of people infected, even under epidemic conditions, require treatment, it is questionable that vaccine development would be warranted especially if other means of control are available.

The clinical nature of histoplasmosis contraindicates transmission by food, water, or any arthropod or animal vector. The fungus, however, is known to remain viable in water for long periods of time (Ritter and Culp 1956, Cooke and Kabler 1956). This condition may result in the dispersal of viable particles by rivers and streams to new foci where conditions are favorable for colonization by the fungus.

When effective drugs are available, one of the most common means of controlling infectious diseases is with the use of chemotherapeutic agents. Amphotericin B, an antifungal antibiotic, is highly effective in treatment of fungal disease, but because of frequent serious side effects it is used only when the life of the patient is in jeopardy. Deaths due to histoplasmosis, however, have been greatly reduced because of this antibiotic, but it is never administered if spontaneous recovery appears likely. There is a great need for another drug with equal effectiveness to amphotericin B, but with less toxicity. Surgical resection of pulmonary cavities in an advanced and chronic form of histoplasmosis often refractile to amphotericin therapy may be a successful form of treatment.

The unique environmental existence of *H. capsulatum* permits a control not applicable to most other infectious agents, which is to restrict the dissemination of contagion from the point source. Emmons and Piggott (1963) were able to eradicate the fungus in a site by the application of pentachlorophenol in fuel oil. Similar attempts by lowering the pH and altering the microbial flora of the soil were unsuccessful. They also noted that an overlay of 15–20 cm of soil on an area colonized by the fungus acted as a physical barrier. One year later *H. capsulatum* could be isolated from depth samples taken under but not on the surface of the overlay. They recognized that this might be a temporary control, and studies by Hasenclever and Piggott (1974) 7 yr later showed that recolonization at the surface had occurred.

At Milan, Michigan (Dodge et al. 1965), a spread of infectious particles from a point source was controlled by covering a school yard with asphalt. The success of this method was measured by a drastic reduction in positive skin test reactors among children attending the school after the asphalt was laid (Dodge et al. 1970). The restricted use of the point source, a part of the area not covered with asphalt but overlaid with sod, as a playground and parking lot for school buses was also cited as a possible contributing control factor.

Laboratory studies (Tosh et al. 1966b) have shown that 3%–5% formalin or cresol solutions applied to soil colonized by *H. capsulatum* were effective in destroying viability of the fungus. The results obtained by the application of 3% formalin to the surface under field conditions at two point sources have been reported (Tosh et al. 1967). At one site in Mexico, Missouri, 1260 m^3 of the disinfectant was sprinkled on 3 ha. Three years after treatment all surface and depth samples were negative for *H. capsulatum*. Eradication of the fungus at the site in Mason City, Iowa, required more applications (Weeks and Tosh 1971). After the first application of 882 m^3 of 3% formalin, the percentage of positive surface soil samples was reduced from 62 to 0.5. One year later the surface was still negative but in some areas depth samples were still positive. Two years after the first treatment a second was made in areas where depth samples were positive. A year after the second application (3 yr after the first) a small percentage of depth samples were still positive and a third treatment, only to positive areas, was made. The final subsurface sampling taken from soil excavated for a bridge foundation 1 yr after the third treatment, resulted in 2 of 27 samples being positive. Surface samples taken from around the foundation after the bridge construction was completed were negative. It appears that the treatment of two point sources with single or multiple applications of 3% formalin has successfully eradicated *H. capsulatum*.

The ideal agent for point source control according to Smith et al. (1964): (1) controls the pathogen in the soil; (2) is relatively nontoxic for animals, humans, and vegetation; and (3) is inexpensive and easy to apply. Of the agents tested, 3% formalin was the most ideal (Weeks and Tosh 1971). It permits the fastest return of grass and plants in the area treated and during application is relatively nontoxic for those handling it. Its cost when purchased in large quantities does not preclude its use. The disadvantage of this disinfectant is its volatility and it therefore has little residual action. While cresol and pentachlorophenol have long-lasting effects, they also restrict the growth of grass and plants in the areas treated.

Control of infectious particles by chemical treatment of soil at the point source has been accomplished. The process, while somewhat laborious, appears to be effective and long lasting. Concomitant with soil treatment, dispersal of the flocks of birds roosting at the point source is also necessary. The overlaying of a point source with asphalt, sod, or soil at least temporarily controls dissemination of fungal particles. Which of these methods of control is used depends upon the size, location, and future use of the site.

Future

Very little is known about the microbial relationships that exist in soil and affect the presence of *H. capsulatum*. Our knowledge that enrichment of soil with avian or chiropteran excreta favors colonization by the fungus is but a casual observation, and we have no basic understanding of this association. Although aerobiological studies at external point sources have been disappointing, more at larger and more extensive sites should be conducted. It is certain that testing done at selected locations would yield significant aerobiological data.

With present techniques, a factor that restricts more extensive investigation into ecological relationships is the laborious and time-consuming procedure required for the isolation of *H. capsulatum* from its natural sources. Direct isolation in culture from soil samples has been accomplished but the reliability of the method is quite poor. Mouse inoculation is the most reliable means of detecting the fungus in soil or air samples, but it takes 5–6 wk. The expense, space, and personnel required for large-scale studies therefore are limiting. The development of a simple and reliable technique for the detection of *H. capsulatum* in its natural environment is necessary. In spite of many attempts such a technique has not been developed, and continued efforts should be directed toward the problem.

Methodology does exist that permits the control of *H. capsulatum* in its natural environment. Control of epidemics, however, is dependent upon discovery of point sources that, by their locations, are public health hazards.

Blackbird/starling roosts and buildings housing bats should be located and soil under the roosts and near the buildings should be tested for the presence of *H. capsulatum*. If the fungus is present, location of the site determines if eradication procedures need to be initiated. If the fungus is absent and roosting in the area continues, periodic testing should be done. Because of the variety of capabilities required for the successful use of methods available, the cooperation of local, state, and federal agencies is necessary. Future epidemics of histoplasmosis can be prevented if concerted efforts and the rational use of our present knowledge are applied.

AIRBORNE PATHOGENS IN THE INDOOR ENVIRONMENT WITH SPECIAL REFERENCE TO NOSOCOMIAL (HOSPITAL) INFECTIONS*

The hospital is an indoor environment especially conducive to the spread of airborne pathogens (Riley 1972, 1974). It contains a reservoir of individuals infected with pathogenic microorganisms that are potentially transmissible to other persons including patients, hospital personnel, and patients' visitors. Furthermore, the hospital serves as a geographic site of collection for persons whose underlying disease processes render them unusually susceptible to infection with organisms not usually pathogenic, as well as to standard pathogens. Through interaction with the community, the hospital is a potential source for disease that may be transmitted later within the community outside the hospital environment. Thus, with respect to airborne pathogens, the hospital can be expected to reflect, in heightened and exaggerated form, the general situation when airborne microorganisms and susceptible hosts are brought together in an indoor environment.

It is believed that most nosocomial (hospital) infections are transmitted by direct contact, by routes such as contamination of hands of hospital personnel by infective discharge from a patient with subsequent carry-over of that organism to other patients. Such infections are also spread by large wet droplets that occur during coughing or sneezing (particularly true of viral respiratory infections), but some hospital infections are spread by droplet nuclei with equivalent diameters in the 1- to 3-μm range and are thus true airborne infections. These true airborne nosocomial pathogens include organisms that are pathogenic for normal individuals such as the rubeola virus, *Mycobacterium tuberculosis,* the variola virus, the varicella virus, and the influenza viruses, among others. Organisms not pathogenic for the normal host can also be airborne within the hospital environment; the most notable example documented in recent years is the increased incidence of infection with *Aspergillus* sp. among immunosuppressed hosts.

Here we review selected examples of airborne spread, or potential airborne spread, of microorganisms within the hospital, with selective discussion

*J. M. Leedom and C. G. Loosli

of other indoor environments. As Riley (1972) has so aptly stated, "The enclosed atmosphere of the hospital building and its human occupants constitute an ecological unit." Airborne microorganisms may originate from the human occupants of the hospital or reside and replicate in the inanimate hospital environment; but regardless of origin, airborne microbes have the potentiality for becoming dispersed throughout the entire exchangeable atmosphere within the hospital building (Riley 1972, 1974).

Perusal of literature relevant to the problem of airborne infections reveals confusion attributable to failure to define terminology. Langmuir (1961) has pointed out that four mechanisms that spread microorganisms are usually discussed when contact vs. airborne infection is scrutinized. These four mechanisms are contact, droplets, droplet nuclei, and dust.

The term *contact* usually does not lead to confusion. Contact, unless otherwise specified, means actual touching of the contaminated person or object either directly as in kissing or touching, or indirectly as in the use of contaminated surgical instruments, contaminated surgical dressings, or the passing of a contaminated toy from a sick or infected child to a well or uninfected child (Langmuir 1961). The term *droplets* leads to the most confusion. Droplets originate primarily from the mouth, and to a lesser extent from the nose, during the activities of talking, coughing, and sneezing. These droplets are generally large, moist particles, greater than 10 μm in diameter, and usually do not remain airborne for a distance of more than about 1 m from the producing individual (Langmuir 1961). As Langmuir (1961) has so aptly stated, "Droplets actually pass through the air and in a literal sense are airborne, but at the same time they exist only in the immediate vicinity of their source." Furthermore, Langmuir (1961) emphasized that control of the spread of these large droplets can be accomplished by such simple actions as covering a cough or a sneeze with a handkerchief or wearing a mask. Thus the approach to the control of infections that spread by droplets is immediate and related directly to the persons involved, exactly as is the control of infections spread by direct or indirect contact.

Droplet nuclei are small residues 1–3 μm in equivalent diameter that remain suspended from dried droplets (Langmuir 1961). These droplet nuclei are capable of remaining suspended in the air indefinitely and may be transmitted on air currents throughout a room or enclosed vehicle or may pass through ventilating ducts, stair wells, elevator shafts, laundry chutes, and so on (Langmuir 1961).

Dust consists of the usually large particles 12–18 μm in diameter found on floors, clothing, bedding, and other surfaces in a given room. This dust is capable of periodic suspension and resuspension in the atmosphere by all sorts of activities, especially dressing, bedmaking, sweeping of floors with dry mops, and vacuuming of floors with unfiltered vacuum cleaners (Langmuir 1961).

In contrast to the control of infections transmitted by large droplets, which, as has been noted, is essentially the same as control of infections transmitted by direct contact, the control of infections transmitted by dust and droplet nuclei is within the partial province of the engineering and architectural experts who design and maintain a given hospital facility. Theoretically, measures capable of extracting dust from surfaces and filtering dust and droplet nuclei from the air should prevent spread of airborne infectious aerosols. Largely because of this difference in approach to control, Langmuir (1961) has recommended that droplet infection be classified as one form of contact infection and that the term *airborne infection* be limited to spread by droplet nuclei and dust. Throughout the remainder of this paper that convention will be followed, and the term *airborne infection* refers to organisms transmitted by droplet nuclei or dust. This definition of airborne transmission has also been adopted by the United States Public Health Service (USPHS). In its manual on isolation techniques for use in hospitals, the USPHS has defined airborne transmission as "the dissemination of either droplet nuclei (residue of evaporated droplets that may remain suspended in the air for long periods of time) or dust particles containing the infectious agent. Organisms carried in this manner are subsequently inhaled by, or deposited upon, the susceptible host" (Brachman 1970).

Importance of Nosocomial Airborne Infection

Brachman (1971, 1974) has summarized some of the difficulties of assessing the true frequency of airborne nosocomial infection. Occasionally, sharp epidemics of airborne nosocomial infection occur and provide indisputable examples of nosocomial airborne disease. Assessing the importance of airborne spread in the routine endemic situation is very difficult, but it is probable that some fraction of endemic nosocomial infection is transmitted by the airborne route. Surveillance data and reports classify infections as to anatomic site affected rather than method of transmission, and it is difficult to obtain precise data as to mode of transmission for endemic situations (Brachman 1970, 1974).

In a review of nine reports of nosocomial infection rates published during the period 1950 through 1967, Sanford and Pierce (1971) found the overall occurrence rate of nosocomial infections varied from 3.5 to 16.8 per 100 patients studied. The proportions of these infections involving the respiratory tract ranged from 15% to 35%, with a prevalence of nosocomial bacterial respiratory infections ranging from 0.5 to 5.0 per 100 hospital admissions (Sanford and Pierce 1971). These data were later extended and refined by the National Nosocomial Infection Study (NNIS) of the Center for Disease Control (CDC) for the calendar years 1970 through 1973 (Brachman 1974). The NNIS data yielded a nationwide picture gathered from monthly surveillance of 117 hospitals located throughout the United States. These data came from approximately 1% of American hospitals, represent almost 4 million discharged patients (approximately 1% of all patients discharged from United States hospitals during the study period), and are very useful in ascertaining relative risks and trends of nosocomial infections.

The NNIS data showed an overall nosocomial infection rate of about 5% and an overall nosocomial respiratory infection rate of about 0.6% (or 17% of the total nosocomial infections; Brachman 1974). Most (54%) nosocomial respiratory infections occurred on surgical services (0.8 per 100 surgical patients); medical services had 37% (0.8 per 100); the remainder were divided among gynecology 3% (0.3 per 100), pediatrics 3% (0.3 per 100), and the nursery 2% (0.1 per 100); (Brachman 1974). Brachman also reported that 27,409 organisms were isolated from the NNIS patients with nosocomial respiratory infections. Klebsiellas were isolated from 11%, pseudomonades from 10%, *Staphylococcus aureus* from 9%, *Escherichia coli* from 8%, enterobacter from 7%, protei from 5%, *Streptococcus pneumoniae* from 5%, and *Candida* from 3% (Brachman 1974). As will be apparent later, these infections probably were not all airborne and, similarly, respiratory infections do not account for all of the nosocomial infections transmitted via the airborne route. Respiratory infections, however, do represent an important sample of airborne nosocomial infections, particularly of those occurring on a day-to-day endemic basis.

Specific Airborne Diseases

General

Most indisputably documented episodes of airborne nosocomial infections are of the epidemic variety, because they are easier to detect (Brachman 1971). Although airborne spread from droplet nuclei and dust undoubtedly contributes to the usual endemic level of hospital-acquired infections, it has proved impossible to separate this component in most instances. Indeed, there is considerable controversy about whether the airborne route of infection is important enough in the causation of the usual nosocomial Gram-positive or Gram-negative pneumonias, urinary tract infections, or wound infections to warrant any attempts to control airborne organisms within hospitals. For these reasons any discussion of documented airborne nosocomial infections perforce addresses itself mainly to epidemic episodes.

Viral infections

It is probable that the respiratory viruses that cause the majority of endemic and epidemic upper and lower respiratory diseases are ordinarily transmitted by person-to-person contact via large droplets. There is no doubt, however, that under certain conditions some of these viruses can be true airborne infections. Couch et al. (1969) and Knight (1973) have produced convincing experimental and epidemiological evidence for airborne transmission of coxsackievirus A type 21 and adenovirus type 4.

Knight (1973) and Couch et al. (1970) were able to show natural airborne transmission of respiratory infection with coxsackievirus A type 21 upon exposure of antibody-negative adult volunteers to other volunteers previously infected by administration of known viral aerosol. The uninfected volunteers were separated from the infected men by a double-walled wire screen 1.2 m in width. The wire screen prevented direct contact and passage of large airborne droplets but allowed smaller particles to circulate freely. Clinical illness was noted 4 days after administration of aerosol to volunteers. Some 2–4 days after that, on day 6, infection was detected in sentinel volunteers on the opposite side of the screen who had not received viral aerosol. These illnesses in the sentinel volunteers began some 3 days following the recovery of coxsackievirus A type 21 from room air. The investigators (Knight 1973, Couch et al. 1970) stated that there was sufficient agitation of the air by the fans to lead to continued airborne suspension of particles 20 μm in diameter or larger, which might otherwise have sedimented to the floor in a very short time. It was noted that, when inhaled, particles of this size would almost exclusively sediment on the nasopharynx, the site of greatest susceptibility to infection with coxsackievirus A type 21.

In experiments with adenovirus type 4, Knight (1973) reported that volunteers became infected with a smaller number of viable virus particles inoculated by aerosol than by deposit with nasal drops. Aerosolized virus is deposited to some extent throughout the respiratory tract. If the nose rather than the bronchi or alveolar ducts were the site of preferred virus deposition and replication, however, one would expect the infectious dose of virus to be smaller when given by nasal drops than when given by aerosol. In fact, for adenovirus type 4 the infectious dose was smaller when aerosolized, leading to the conclusion that the lower respiratory tract is more susceptible to infection with this virus than are the nasopharyngeal mucosa.

Adenovirus type 4 is a known cause of epidemic respiratory disease among military recruits under conditions that strongly suggest airborne transmission (Couch et al. 1969, Artenstein et al. 1968). Artenstein et al. (1968) have detected adenovirus type 4 in the air of a room occupied by military recruits ill with adenovirus respiratory disease. This finding, coupled with a low dose of aerosolized adenovirus type 4 required to produce infections and illness experimentally, has been interpreted as strong evidence for a significant role of airborne transmission in the spread of adenovirus respiratory disease among military recruits (Couch et al. 1969, Artenstein et al. 1968).

The data just reviewed (Couch et al. 1969, 1970, Artenstein et al. 1968) strongly suggest that adenoviruses, group A coxsackieviruses, and similar agents could be transmitted via the airborne route in the hospital. Unfortunately, the opportunities for contact transmission via droplets are manifold in the hospital environment, and no good studies that dissect out the airborne component of illness due to the common viral respiratory agents have been done on hospitalized civilian patients under natural conditions.

Portnoy et al. (1966) compared the frequency of double simultaneous, or nearly simultaneous, respiratory virus infections in children hospitalized for viral respiratory syndromes versus "control" children hospitalized for trauma or elective surgery. They found serologic evidence of infection with two viral respiratory agents in 36 of 258 (14%) children hospitalized with acute lower respiratory disease syndromes (laryngotracheobronchitis, bronchiolitis, and pneumonia) as compared with 17 of 131 (13%) controls. These differences were not significant. The high frequency of double respiratory virus infection in the study of Portnoy et al. (1966) certainly illustrates the facility with which the common viral respiratory agents spread among children. It was expected, however, that children in wards devoted to the care of those ill with viral respiratory syndromes would share their viruses and that the frequency of double virus infection would be higher than among the controls. The control children were hospitalized on a different floor of the hospital than the children with viral respiratory disease, and no child with overt symptomatology of viral respiratory infection was admitted to the control ward.

Portnoy et al. (1966) did not interpret their data in reference to possible airborne spread of respiratory viruses in the hospital. They believed that asymptomatic infections with the common viral agents were so ubiquitous in the households of the children prior to the hospitalization that any nosocomial spread, airborne or otherwise, was not detectable against the high background rate of infection brought into the hospital by the children from their homes. Nevertheless, it is highly probable that there is an indeterminate, but real, endemic level of nosocomial respiratory virus transmission and an occasional identifiable epidemic outbreak. Certainly, the medical literature contains many reports of nosocomial spread of respiratory virus infections, including influenza viruses, parainfluenza viruses, adenoviruses, respiratory syncytial virus, and psittacosis. Although most such episodes probably represent contact infection via droplets, some are truly airborne. For example, Mufson et al. (1973) noted that: (1) the wintertime air in their hospital was warm with a low relative humidity, conditions favoring survival of parainfluenza type 3 virus in aerosol; and (2) infants were usually provided with humidified air in "croupettes" during the first few days of their hospitalizations. Secondary cases of parainfluenza type 3 virus infection usually occurred after infants and children were removed from croupettes, out of the humidified atmosphere and into the dry air of the ward, which would be expected to facilitate survival and airborne transmission of the parainfluenza type 3 virus.

Epidemic influenza

Epidemic influenza is usually caused by variants of the type A influenza virus and less frequently by type B strains. The type A influenza strains prevalent since 1957–1958 have been variants of influenza type A2. There is reasonably good evidence that influenza viruses can be transmitted by the airborne route.

It is not known with certainty which anatomic site is initially invaded by the influenza virus, i.e., the nasal, pharyngeal, or lower respiratory tract mucosa, although all these sites can support virus replication. It is known, however, that influenza can be transmitted experimentally by either intranasal instillation of a viral suspension or by the inhalation of droplet nucleus–sized viral aerosol (Knight 1973, Knight and Kasel 1973). In fact, in the studies cited by Knight (1973), it was shown that the infectious dose of influenza A/21 Bethesda/10/63 virus for volunteers via the aerosol route was lower than the infectious dose of the same virus strain given by intranasal instillation. This would certainly support the thesis that influenza virus infection is at least sometimes initiated in the lower respiratory tract by small-particle aerosols containing inhaled viral particles (Knight 1973). The fraction of these inhaled aerosols retained in the upper respiratory tract contained a dose of virus considerably smaller than a minimum dose necessary for production of infection by intranasal instillation (Knight 1973, Knight and Kasal 1973).

During the 1957–1958 epidemic of A2 influenza, a study of transmission and prevention of influenza was performed at the Veterans Administration Hospital at Livermore, California (McLean 1961). During this study the air throughout the main building of the hospital was irradiated with ultraviolet (UV) light. Neighboring buildings did not have irradiated air. No mixing of patients from the "test" (irradiated air) and "control" (unirradiated air) buildings was permitted. Among the unprotected patients there was a sharp outbreak of A2 influenza in January 1958, which resembled airborne spread of disease, during the second wave of A2 influenza that swept the country. No comparable outbreak occurred among the patients protected by ultraviolet air disinfection. The frequency of serologically diagnosed influenza among 209 patients housed in the irradiated building was 2% in contrast to a frequency of 19% among 396 patients living in the other buildings (McLean 1961). This nosocomial outbreak suggested airborne spread of influenza in two ways: (1) The epidemic among patients in the unprotected buildings was explosive in character, and (2) the virus would have had to be transmitted by the airborne route for UV irradiation of air to have protected patients housed in the main hospital building (McLean 1961).

Airborne influenza has been studied in mice (Schulman and Kilbourne 1968, Schulman 1967, 1968). Schulman observed that if infected mice were separated from contact with uninfected controls by two wire screens 1.9 cm apart, the separated uninfected mice acquired influenza at the same rate as uninfected contacts housed in the same cages with experimentally infected animals. Thus airborne infection was deemed to be as efficient a method of spread as contact in this experimental situation. In other experiments Schulman (Schulman and Kilbourne 1968, Schulman 1967, 1968) found that if infectors and uninfected mice, housed in separate cages, were placed in a closed chamber with regulated airflow, the rate of transmission decreased as the rate of airflow increased, as would be expected if transmission were adversely affected by dilution of airborne virus by airflow. These investigators also found that the relative humidity of the air had a profound effect on transmissibility of influenza virus infection. Rates of transmission are significantly greater at lower relative humidities than at higher relative humidities. This finding is consistent with the known relationship of relative humidity to the viability of the influenza virus in air and also correlates with the usual time of influenza epidemics in temperate climates—the winter, a season when buildings are artificially heated and relative humidity tends to be low. All those observations are consistent with the hypothesis that natural transmission of influenza viruses can occur by means of airborne droplet nuclei.

Rubeola

Rubeola is a virus disease that provides a classic example of airborne transmission when exposure occurs within indoor environments. Epidemic rubeola has been documented as airborne in schools by Wells et al. (1942) and others (Riley and O'Grady 1961). A single exposure to rubeola infects about 80% of susceptibles.

Varicella

Varicella, like rubeola, is capable of causing explosive airborne outbreaks upon exposure of susceptibles in an indoor environment, particularly hospitals (Riley 1969, Gordon 1962). Gordon (1962) estimated an 80% infection rate of susceptibles following a single exposure to varicella. It should also be remembered that herpes zoster is caused by the varicella virus. Persons with herpes zoster are not as infectious as those with varicella because they generally shed fewer virus particles into the air. There are instances, however, in which single patients with varicella, hospitalized among susceptibles, have caused subsequent explosive epidemics of chicken pox (Riley 1969, Gordon 1962). About 15% of susceptibles are infected upon a single exposure to a zoster patient.

Variola

An ancient, and we hope almost extinct disease, variola, or smallpox, has recently been shown to be capable of transmission via the airborne route within a hospital (Wehrle et al. 1970). Wehrle studied an outbreak of smallpox in a hospital in the town of Meschede, West Germany. A 20-yr-old German youth, not successfully vaccinated against smallpox, had returned to Meschede from Karachi, West Pakistan. He became ill with fever some 10 days after returning to Germany and was hospitalized the next day in an infectious disease ward on the ground floor of a three-story building. The patient was placed in a private room because physicians suspected typhoid fever, and he was also convalescent from hepatitis. He did not leave his room during the first 4 days of hospitalization. On the 4th hospital day a rash was noted, and on the 5th day smallpox was suspected. The diagnosis was confirmed on the 6th day by electron microscopy. Immediately subsequent to the diagnosis of smallpox, the patient was transferred to a recently constructed smallpox isolation hospital. At the time of transfer he had extensive oral and cutaneous lesions and coughed frequently.

In all, 19 additional cases of smallpox were attributed to this importation. Seventeen cases were judged to be secondary and two tertiary. Cases occurred in 13 other patients in the original hospital, two contacts of other patients, one visitor, and three nurses. There were four deaths. Wehrle et al. (1970) ruled out direct personal contact and transmission by fomites because no direct contact between the index patient and the others had occurred, and none of his fomites were interchanged with the remainder of the hospital. Nursing personnel who had been assigned to the isolation ward did not work on other floors of the hospital.

Airborne dissemination seemed the most probable explanation. Several pieces of data supported this route of spread. The patient visitor who contracted smallpox had visited the hospital on only a single occasion, during the 3rd day subsequent to the patient's admission, and had remained in the building for only about 15 min. The visitor had entered the hospital through the front door on the ground floor and his only contact with a member of the hospital staff was with a physician. The visitor did not enter the isolation room corridor. His smallpox lesions made their appearance 11 days later.

Another episode involved a patient who had been confined to her room on the third floor of the hospital and had not left the room for an entire month. No personnel from the isolation ward had entered her ward or her room. This patient could have been infected only through the airborne route. Finally, attack rates by floor in the hospital did not differ significantly. They were 20%, 16%, and 20% on the first, second, and third floors, respectively (Wehrle et al. 1970).

Wehrle et al. (1970) studied airflow patterns within the building with smoke bombs. A smoke bomb placed within the index patient's room sent dense smoke into the corridor and into adjacent rooms, passed down the corridor through a door that was normally kept ajar and into an entrance hall where the visitor who had contracted smallpox had talked with the physician. Smoke then passed to the central stairwell, which acted as a chimney carrying smoke up into the second and third floors. Furthermore, smoke from the index patient's room flowed out the partially opened window, up the outside wall of the building, and readily entered open windows of rooms on the floors above. The spatial distribution of cases on the three floors was consistent with the distribution of the smoke, and thus airflow, from the index patient's room.

It is unusual for a patient with smallpox to infect more than a few persons, and most of the patients he infects are usually close personal contacts.

The Meschede outbreak (Wehrle et al. 1970) was very unusual in the large number of second-generation cases and in the fact that the index patient had no face-to-face contact with any of the secondary cases. This outbreak was probably related to the patient's being a particularly effective disseminator of smallpox virus because he had a confluent rash that involved mucous membranes, and a severe bronchitis and cough that would tend to promote formation of virus-laden aerosols. Furthermore, the relative humidity was low—a condition associated with better survival of smallpox virus in aerosol (Wehrle et al. 1970, Harper 1961).

Aerosol transmission of arboviruses in the laboratory

Some viruses not ordinarily transmitted by the airborne route or into the respiratory tract are of significant importance as airborne pathogens only in the indoor environment of the laboratory. Hanson et al. (1967) reviewed laboratory-acquired infections with arboviruses. They surveyed 91 laboratories in 38 countries and were able to find 428 overt laboratory-acquired infections with the arboviruses, with 16 fatalities. Most of the arboviruses known to have caused laboratory infection belong to group B, although the largest single number of infections due to a single virus found in this survey were 118 due to a group A arbovirus, the Venezuelan equine encephalitis virus. In many cases the specific mode of transmission of the virus to the laboratory worker was not known. It was only known that an individual was working with a specific virus and subsequently became ill. Many instances of disease acquisition, however, occurred under circumstances that pointed to the inhalation of an infectious aerosol. For example, contaminated dust from mouse cages has apparently been responsible for several infections with Venezuelan equine encephalitis virus and for a fatal infection with Machupo virus (Hanson et al. 1967).

Rickettsial infections

Coxiella burnetii, the etiologic agent of Q fever, is the only rickettsial organism regularly transmitted to man without the mediation of an arthropod vector. *Coxiella burnetii* is shed in large numbers in the parturient fluids of animals, particularly cattle and sheep. The organism is highly resistant to extremes of environmental temperature and humidity and remains viable in dust indefinitely. The usual route of infection for man is the inhalation of viable *C. burnetii* in particles of suspended dust contaminated with dried birth fluids from infected animals.

The most important epidemiologic reservoir of *C. burnetii,* so far as man is concerned, is domestic ungulates. Thus the major impact of *C. burnetii* as an airborne pathogen is in the outdoor, or at least "outbuilding," environment. The organism is highly infectious in the laboratory environment, and transmission is especially frequent among laboratory personnel working with infected chicken embryos (Johnson and Kadull 1966); however Q fever can be extensively airborne throughout a laboratory building. In a review of laboratory-acquired Q fever, Johnson and Kadull (1966) found that 29 (58%) of 50 persons with laboratory-acquired infections did not work with *C. burnetii* themselves, but did work in the laboratory building in which experiments with the organism were carried out. In their discussion, Johnson and Kadull (1966) pointed out that such transmission of *C. burnetii* to janitors, secretaries, and other persons who have simply entered buildings containing Q fever laboratories has occurred many times in spite of extensive precautions to avoid the spread of the organisms.

Recently there has been a report that sheep with unsuspected *C. burnetii* infection have transmitted the organism to laboratory workers and could conceivably be the source of infectious aerosols, which might cause other cases within buildings (Schachter et al. 1971). One confirmed and another probable case of clinical Q fever, as well as a high prevalence of complement-fixing antibodies to *C. burnetii,* were detected among employees of the University of California Medical Center, San Francisco, who had been exposed to sheep used in experiments in anesthesia and surgery (Schachter et al. 1971). The asymptomatically infected sheep had been transported through corridors to which patients had access. No analysis of airflow was included in the report, but Schachter et al. (1971) correctly pointed out that such asymptomatically infected laboratory animals could constitute a significant nosocomial hazard.

Person-to-person transmission of Q fever is unusual but does occur, and the relevant literature has been reviewed recently by Leedom (1974). It is worth noting that several epidemics among hospital personnel have been initiated by attendance at necropsies of Q fever patients. Presumably, *C. burnetii* was aerosolized in an infectious form during the necropsy procedures. In one report, 16 of 17 persons who attended such a necropsy contracted Q fever (Babudieri 1953).

Tuberculosis

Perhaps the archetypical disease whose principal mode of transmission is via the true airborne route (i.e., via droplet nuclei) is tuberculosis. The elegant studies of Riley and co-workers (1962) at the Veterans Administration Hospital in Baltimore, Maryland, established beyond doubt that human patients with tuberculosis generated viable tubercle bacilli in aerosols and that those bacilli were capable of infecting guinea pigs. Riley (1974, Riley et al. 1962) designed a study in which guinea pigs held in a penthouse on the roof of the hospital breathed air vented from rooms containing patients with active pulmonary tuberculosis. The guinea pigs thus served as samplers and sentinels for the detection of any infectious droplet nuclei present in the exhausted air. Approximately 120 guinea pigs were kept in the colony at one time. The colony's census was kept roughly constant by the replacement of sick or infected animals. All animals were tuberculin tested at monthly intervals to identify those that had been infected with tubercle bacilli. During a 4-yr period 134 guinea pigs were infected with *Mycobacterium tuberculosis* (Riley 1974). This study provided indisputable proof that tubercle bacilli are regularly disseminated from human sources on particles small enough to be exhausted through ventilating ducts and can be inhaled into the lower respiratory tract of guinea pigs. The only particle capable of carrying infection and traversing the ventilating duct network between the tuberculous patients and the animal chamber and being randomly distributed with the air would be a particle with the aerodynamic characteristics of droplet nuclei (Riley et al. 1962, Riley 1974).

Autopsy examination of the guinea pigs revealed that each infected guinea pig characteristically showed a single tubercle randomly distributed in the lungs. Thus the minimum infective dose of *M. tuberculosis* for guinea pigs is one viable bacillus (Riley et al. 1962). Calculations by Riley (1961) involving the rate of tuberculin conversion of nurses working on tuberculosis wards would be in agreement with the hypothesis that on the average a human being, like a guinea pig, becomes infected after having inhaled one tuberculous quantum of air—a single viable bacillus.

Riley (1961, Riley et al. 1962) has emphasized not only the airborne nature of tuberculous infections but the fact that some sputum-positive patients are more efficient producers of infectious aerosols than others, are better infectors of the air, and are thus more dangerous to persons breathing that air. For example, during the first 2 yr of the studies of Riley et al. (1962), only 8 of 61 patients who were untreated and infected with organisms susceptible to antituberculous drugs infected any guinea pigs at all. Patients infected with resistant strains of *M. tuberculosis* did not seem to produce infectious aerosols as readily as those with susceptible organisms, although the comparison was based upon a small number of disseminators in each group (Riley et al. 1962).

An example of the havoc causable by a single efficient disseminator in the hospital environment can be seen in a recent report by Ehrenkrantz and Kicklighter (1972) from the Jackson Memorial Hospital in Miami, Florida. A 65-year-old man was seen and treated for 3 h in the hospital emergency room with a diagnosis of acute pulmonary edema. The patient was then transferred to a medical floor where he stayed for 57 h. A cardiac arrythmia supervened, and he was transferred to an intensive care unit. He died after 67 h in the intensive care unit. This patient produced large amounts of sputum throughout his hospital stay and therefore had frequent and vigorous nasotracheal suctioning. Active cavitary tuberculosis was diagnosed at autopsy. Ehrenkrantz and Kicklighter (1972) reported that routine tuberculin skin testing of personnel at Jackson Memorial Hospital in preceding years had disclosed a 2%–3% tuberculin skin test conversion rate during any given 6-mo period. Subsequent to the diagnosis of tuberculosis in this 65-year-old man, personnel were again skin tested. Among the personnel employed in the units where the patient had been hospitalized, 50% of those with repeated and close contact with the patient had converted to skin test positivity, and 25% of those with little or no personal contact had converted. In all, 25 skin test conversions and two active cases of pulmonary tuberculosis eventuated from exposure to this patient. The highest rate of skin test conversion occurred in personnel on the medical floor. That unit is a ward with 45 beds in two- and three-bed rooms, cooled by a 20-t air conditioner that recycles 70% of the air without filtration. This mix of an active, efficient disseminator, susceptible persons, and high concentrations of infectious organisms and recycled unfiltered air led to the high rates of tuberculosis infection and skin test conversion.

Another explosive epidemic of tuberculosis in a different kind of indoor environment is further illustrative of the infectious nature of *M. tuberculosis* in air. This outbreak occurred aboard the naval vessel U.S.S. *Richard E. Byrd* (Houk et al. 1968). The air within this ship was entirely recirculated, resulting in minimal dilution of any infectious particles added to the air. A person with an unrecognized case of tuberculosis remained aboard the ship for several months. Subsequent tuberculin testing of the entire ship's company showed that nearly half the men had become infected with tubercle bacilli during this period. Some of the infected men had had direct contact with the index case, but many had no direct contact but shared a common air supply. In fact, there was no significant difference in the likelihood of infection among men with direct personal contact with the index case in comparison with men sleeping in a compartment through which air was circulated after passing through the compartment housing the disseminator. It was concluded that direct personal contact played no part in transmission of infection during this outbreak, and in fact this study helped to establish the view that primary pulmonary tuberculosis is exclusively droplet-nucleus borne (Riley 1974, Houk et al. 1968).

Streptococcus pneumoniae

The pneumococcus is currently an important cause of nosocomial pneumonia. Graybill et al. (1973) estimated that 10% of nosocomial pneumonias in the Johns Hopkins Hospital were caused by the pneumococcus. The pneumonias reported as pneumococcal by Graybill et al. (1973), however, occurred mainly in postoperative patients, a situation in which aspiration, of at least minor degree, of pharyngeal organisms could be expected. It is highly probable that most of these pneumococcal pneumonias were not true nosocomial infections, airborne or otherwise. Most probably resulted from aspiration of pneumococci from pharynges colonized prior to admission. The trauma of intubation and the action of anesthetic agents combine to facilitate entry of pharyngeal organisms into the lower respiratory tract and interfere with normal "mucociliary" clearance mechanisms.

Graybill et al. (1973) did not report any instances of apparent airborne epidemic pneumococcal pneumonia in their hospital population, and such reports are a rarity in modern times among either hospitalized patients or other individuals in indoor environments. Outbreaks of pneumococcal pneumonia have been observed, however, in hospitalized patients, in schools, in dormitories, in prisons, and among military populations. Many of these epidemics, which occurred during the preantibiotic and early chemotherapeutic eras, have been reviewed in detail by Finland (1942).

In most of these epidemics the precise mechanisms of transmission of *S. pneumoniae* from person to person is not clear. It is believed that the most common mode of transmission is by inhalation of large droplets containing pneumococci originating in the nasopharynx of a carrier. These droplets are retained in the nasopharynx of an uninfected person, and the pneumococcus subsequently replicates upon his nasopharyngeal mucosa.

There is evidence, however, that some airborne spread of *S. pneumoniae* can occur. Finland (1942) recorded several instances in which pneumococci were found in the dust of rooms occupied by patients or carriers, suggesting, at least, that airborne spread from resuspended droplet nuclei would have been possible from those rooms.

Hodges (Hodges and MacLeod 1946a,b,c,d, Hodges et al. 1946) presented classic studies of epidemic pneumococcal pneumonia in a military training school during World War II. Hodges and MacLeod (1946b) found that circumstances were highly suggestive of airborne transmission of pneumococci within the classroom environment. There were no barracks-related clusters of cases. Further, in relation to sleeping quarters and military units (squadrons), the disease did not spread centrifugally. There was a sharp and simultaneous rise among all the squadrons of trainees. The classrooms were the only sites where prolonged mixing of troops, trainees, and cadre alike occurred.

In examining pneumonia rates among the cadre of classroom instructors at the training base, Hodges and MacLeod (1946b) found the rates for all respiratory diseases were higher for classroom instructors than for any of the other permanent personnel except for members of the medical detachment. (Members of the latter group of personnel were in constant close contact with hospitalized patients with pneumonia and other respiratory diseases.) In fact, the pneumococcal pneumonia rate for classroom instructors was approximately four times that of the remainder of the permanent personnel except for the medical detachment. Instructors and students used classrooms in shifts. Hodges and MacLeod (1946b) stated that if classroom transmission had been based solely on man-to-man or large-droplet contacts, one would have expected considerable differences in the rates of disease among instructors in various teaching shifts because contact between the personnel in different shifts was almost nil. In fact, the different teaching shifts all reacted in a uniform manner during the early stages of the epidemic of pneumococcal pneumonia. Thus pneumococcal pneumonia spread uniformly and quickly throughout the entire population of the school.

Carrier studies by Hodges et al. (1946) also showed uniform distribution of pneumococcal serotypes throughout cadre and trainees. With the limited man-to-man contact between personnel in different teaching shifts, the authors (Hodges et al. 1946) believed that some mechanism other than contact or droplet spread was operative in the classrooms, which permitted the passage of infectious agents from one shift to another because the same rooms were used in rotation by different shifts. They believed that airborne dissemination was the likely mechanism. Using rather crude techniques, Hodges and MacLeod (1946b) were able to find *S. pneumoniae* by swabbing the floors in the classrooms in 22 of 83 (26.6%) swabs studied.

Thus it seems likely that *S. pneumoniae* infections can be airborne under proper conditions and that airborne transmission was important in the epidemic studied by Hodges and MacLeod (1946a,b,c,d) and Hodges et al. (1946). The lack of reporting of epidemics of pneumococcal pneumonia among closed populations in indoor environments since the end of World War II probably reflects changes in transmission and ecology of these organisms effected by the widespread use of antibiotics.

Streptococcus pyogenes

The group A, beta-hemolytic streptococcus is capable of causing pharyngitis, tonsillitis, minor skin infections, scarlet fever, major infections of operative wounds with massive cellulitis, abscesses, lymphadenitis, lymphangitis, bacteremia, and occasionally primary pneumonia. The most important mechanism of spread of *S. pyogenes* from person to person is by contact via large droplets (Hamburger et al. 1945). These large droplets fall to the floor, dry into dust, and are ordinarily no longer part of the chain of transmission. Under certain conditions, however, these dustborne organisms can become important in the transmission of group A, beta-hemolytic streptococci to open wounds of postoperative patients, to weeping skin of patients with eczema or burns, and to pharynges in susceptibles, causing streptococcal pharyngitis.

This ability of *S. pyogenes* to survive in dust and subsequently to become airborne when activities supervene that reraise the dust was well studied during World War II in military personnel, as well as in numerous hospital outbreaks after that (Hamburger et al. 1945, Loosli et al. 1952). Studies of outbreaks of streptococcal disease in hospitals and army barracks have disclosed that where secondary group A streptococcal infections are occurring the inanimate environment becomes highly contaminated with the infecting organisms. Contaminated materials in the inanimate environment include air, dust, bedclothes, and clothing of the occupants, among others.

Loosli et al. (1950), in a review of this problem, pointed out that many possibilities always exist for the transmission of infections among occupants of barracks or hospitals, but the exact mechanisms of transmission vary in importance in relation to the environment, the age of the population, underlying conditions, and so on. In studies of the *S. pyogenes* problem among the military during World War II, Hamburger et al. (1944, 1945) established the concept of the "dangerous carrier" of streptococci. "Dangerous carriers" are able to disperse the organism more efficiently than other carriers (Hamburger et al. 1945). They found that nasal carriers of streptococci were more likely to disperse the bacteria than were throat carriers, and nose blowing was the most efficient method of distributing streptococci. The "dangerous" nasal carriers heavily contaminated their bedding and subsequently the air of their rooms when dust was raised by disturbing their bedding (Hamburger et al. 1944, 1945, Hamburger and Green 1946). The patient who was a nasal carrier and also *symptomatically* infected was found to produce more organisms and to be a more efficient disseminator of *S. pyogenes* (Hamburger et al. 1945, Hamburger and Green 1946). Under hospital conditions, Loosli et al. (1950) demonstrated probable airborne transmission with group A, type 33, beta-hemolytic streptococci traceable to admission to the ward of an infant with infected eczema who was a "skin disperser."

Numerous studies in hospitals and in army barracks demonstrated the effectiveness of oiling floors and blankets in the prevention of dustborne dispersal of group A, beta-hemolytic streptococci. Many of these studies were reviewed by Loosli et al. (1952). The studies showed a generally favorable impact on streptococcal disease transmission among military recruits when floors and blankets were oiled; however, in that same report the authors (Loosli et al. 1952) emphasized that oiling was only partially effective at best and that dustborne and airborne spread of group A, beta-hemolytic streptococci represented only one mode of spread of infection. They reemphasized that direct contact with a heavily contaminated environment or with a "dangerous disperser" is probably equally as important as the inhalation of dust or airborne organisms.

Staphylococcus aureus

Staphylococcus aureus is no exception to the general rule that whenever there is the possibility of airborne transmission of the microorganism, it is always impossible to rule out entirely other modes of transmission. While the relative frequency of transmission of *S. aureus* via the airborne route in hospitals and in other indoor environments is not known, a great deal is known about the ability of the organism to become airborne and to be available for transmission. As is the case with *Streptococcus pyogenes,* persons colonized with staphylococci may or may not be efficient disseminators (White 1961, Varga and White 1961). Persons with staphylococcal infections causing a purulent discharge are more likely to disseminate the organism into the environment than asymptomatic carriers. Furthermore, the studies of White and others (White 1961, Varga and White 1961, White et al. 1964) disclosed that some persons were more heavy nasal carriers of *S. aureus* than others. These heavy nasal carriers contaminated their skins with greater frequency and were more efficient dispersers of organisms than were light carriers. When nasal organisms were suppressed, there was a reduction in both skin carriage and dissemination.

Solberg (1965) confirmed these data and further showed that perineal carriers of *S. aureus* were more efficient disseminators than were nasal carriers and that such dispersal could be lessened by hexachlorophene washing. Hare (1964) reported that very few staphylococci were disseminated directly from the nose but rather dissemination occurred after contamination of the skin from the nose.

Noble (1975) has recently reviewed the problem of dispersal of microorganisms, particularly staphylococcal, from skin. Noble pointed out that human skin has a surface composed of scales that are about $30 \times 30 \times 3$–$5\ \mu m$, with approximately 10^8 scales needed to complete the skin surface. Females have larger scales on the average than males, and a complete layer of cells is lost and replaced about every four days. Scales are released during all natural movements or even when standing naked. These scales can form "rafts" of airborne particles that range in size from less than 4 μm to more than 25 μm equivalent particle diameter. The particles behave, in essence, as if they were spheres of approximately 14 μm in diameter. These larger particles are involved in the contact spread of airborne infection over short distances to the respiratory tract or by the settling of contaminated rafts on susceptible skin or in a wound. Many experts do not regard these large particles as true airborne microorganisms because they are large enough to have a tendency to settle into the dust on the floor of the room. They do travel through the air, however, and can also be resuspended if dust is disturbed by bedmaking, floor sweeping, and other activites.

In his review, Noble (1975) quotes sources that show that males are more heavily colonized with aerobes than are females, and that the microorganisms are not distributed equally over the skin but live in discrete microcolonies containing some 10^2–10^5 viable cells. As the colonies may be relatively far apart on the skin, this accounts for the phenomenon that only about 10% of squames (a squame is a flake of desquamated skin) carry viable microorganisms (Noble 1975). In disease states such as eczema, psoriasis, and pityriasis rosea, the skin is frequently heavily colonized with *S. aureus*. These organisms are then dispersed on skin scales, which contaminate the environment.

These desquamated skin scales increase in number in the air of the hospital ward during activities such as bedmaking (Davies and Noble 1962, 1963). During bedmaking, counts of more than 1766 *S. aureus* particles/m^3 in a 28-m^3 room have been recorded (Noble 1971). Normally clothed individuals, performing gentle exercise, may liberate up to 10,000 bacterial particles/min (Hare 1964, Sciple et al. 1967, Riemensnider 1967).

As stated previously, persons with diseases of the skin disseminate *S. aureus* most widely into the air and into the environment (Noble 1971, 1975). A few persons with normal skin are also efficient dispersers (Noble 1962, Schaffner et al. 1969).

Some instances of direct dispersion of airborne particles of *S. aureus* from the nose have been reported. One of the most interesting of these was the so-called cloud-baby phenomenon reported by Eichenwald et al. (1961). They found babies without symptoms of skin infections who dispersed *S. aureus,* bacteriophage type 80-81, into the air. When a baby with skin lesions was enclosed in a plastic bag so that only his head remained exposed, a marked drop in aerial contamination occurred, whereas if the "cloud babies" were so enclosed, the rate of bacterial dissemination remained unchanged. The authors (Eichenwald et al. 1961) concluded that in the cloud baby the body skin, the extremities, and the umbilical stump played little role in the aerial dissemination of *S. aureus*. Most bacteria spread by these infants originated from about the head, the logical source being the respiratory organs. The cloud babies had concomitant asymptomatic nasal infection with both *S. aureus* and a cytopathogenic virus.

Eichenwald et al. (1961) have reported another kind of bacterial-viral interaction promoting spread of staphylococci, which is probably closely related to the cloud-baby phenomenon. This was the epidemic "stuffy nose syndrome" of premature infants. Infants with the stuffy nose syndrome were symptomatic only if they were simultaneously infected with adenovirus type 1 and *S. aureus*. Nasal infection with either agent alone did not yield symptoms. Thus bacterial-viral interaction was shown to act synergistically to promote infection by *S. aureus*.

The data cited above provide ample evidence that *S. aureus* is disseminated widely in indoor environments, particularly in the hospital. These potentially pathogenic organisms are thus available to be inhaled by, and colonize, susceptible hosts, perhaps resulting in spread to skin with later wound infection, or aspiration into the lungs with consequent staphylococcal pneumonia. Transmission may not occur immediately after *S. aureus* is shed, and the organisms may remain suspended in dust in the environment. They may fall to the floor and be resuspended during ward activities and then fall out upon a susceptible area of a compromised patient, such as an operative wound or a large burn, thus causing an infection. These airborne staphylococci have been shown to cause many epidemics of infections in hospitals.

Cockroft and Johnstone (1964a,b) reported several probable airborne staphylococcal infections occurring after open-heart surgery. Organisms had been found throughout the operating theatre, and a perineal carrier of *S. aureus* was considered to be responsible.

Contemporary studies of air hygiene in operating rooms stem largely from the work of Bourdillion and Colebrook (1946a,b) in a burns unit treatment room. These authors reported reduction of infection rates in major burn wounds subsequent to improvement of air hygiene and considered this reduction to prove the importance of airborne transmission of *S. aureus*.

Shooter et al. (1956) used the concepts of Bourdillion and Colebrook (1946a,b) in a study of the incidence of operative wound infections. During an 8-mo period Shooter et al. (1956) found the incidence of operative wound infections was 9% of 427 wounds. Some 2.5 particles/m^3 containing *S. aureus* were found in samples from air in the operating theatre during operations. Air currents were followed with titanium chloride smoke and proved to flow into the operating theatre from surgical wards. A simple alteration of ventilation, which generated a positive pressure in the operating room and thus excluded staphylococcus-contaminated air from the adjacent wards, was followed by a substantial reduction in the general air bacterial count (the number of *S. aureus* in the air was not reported) and a reduction to less than 1% in the incidence of wound infections of theatre origin among 532 wounds. Shooter et al. (1956) assumed that the reduction in operative sepsis rate could be attributed to a reduction in the number of staphylococci circulating in the ambient air and subsequently settling into wounds and onto sterile instruments and equipment.

Many studies have demonstrated that both newborn infants and adult patients often become nasal carriers of the *S. aureus* bacteriophage types prevalent in a particular hospital ward. It is not certain in all instances whether the nose is colonized by inhalation of staphylococci in the air. The most convincing demonstrations of the sequence of colonization with *S. aureus* are in newborn infants. The umbilicus and abdominal skin have been shown to be generally colonized before the nose (Hurst 1960, Simpson et al. 1960).

In studies by Wolinsky et al. (1960) it was shown that a nurse carrier transmitted her staphylococci to infants only if she handled them. In further studies, the same research group (Mortimer et al. 1962, 1966) found that it was possible to delay colonization of infants by increasing the aseptic precautions directed against contact infection. With very stringent precautions the staphylococcal acquisition rate was reduced from 43% to 14%. It was assumed that the 14% acquisition rate represented aerial transfer of *S. aureus* (Mortimer et al. 1962, 1966).

The relative unimportance of the inhalation route for *S. aureus* colonization of newborn infants is understandable if one considers that the infant has a minute volume of only about 500 ml and that he is handled frequently, usually by nurses who handle many other infants (Williams 1966). Williams (1966) has pointed out that the reported airborne acquisition rate of 14% (Mortimer et al. 1966) during an average hospital stay of 4 days is equivalent to an acquisition rate of some 3%–4%/day, which is of the same order of magnitude as that observed during studies of hospitalized adults.

In summary, it has been amply demonstrated that symptomatic and asymptomatic individuals can disseminate *S. aureus* into the atmosphere. Furthermore, the organisms can be recovered from the air in hospital rooms and on occasion from dust on the floors and surfaces of hospital rooms. Some outbreaks have been discussed in which the airborne route seemed to be the major route of transmission of *S. aureus;* however, the exact proportion of staphylococcal infections transmitted by the airborne route is still problematical. Airborne transfer in hospitals, as noted by Williams (1966), gains special significance from the fact that when this route actually operates, a single disperser is potentially able to infect a large number of other individuals, often situated at some distance from him. Williams (1966) has emphasized the multiple sources of infection and routes of spread of staphylococci in hospitals. He stated that "we have insufficient precise evidence" on the relative importance of airborne versus other routes of spread. He has further noted that the difficulty of separating out airborne staphylococcal sepsis from staphylococcal sepsis transmitted by other means is predicated not only upon the multiple opportunities for transmission of infection that always exist in hospitals, but also upon the fact that the overall incidence of infection is only 1% or 2%. The precise analysis of specific routes of spread becomes exceedingly difficult when the overall incidence of staphylococcal infection is not more than 1% or 2%, and this small proportion must be factored into infection derived by different routes from different sources. Practical justification for attempting the analysis, as noted by Williams (1966), is that it can provide one of the bases for judging how best to construct and ventilate hospitals. If the airborne route of transmission is important, future hospital construction should take that fact into account. That is. hospitals should be built with air-handling systems capable of filtering *S. aureus* from the air and preventing recirculation of aerosols containing *S. aureus* or other potentially dangerous microorganisms from one patient to another.

Gram-negative aerobic bacteria

Gram-negative aerobic bacteria are generally derived from the environment in hospitals. When they are airborne, aerosols are more often from contaminated equipment such as inhalation therapy equipment than from animate reservoirs (Bate and James 1958, Anderson 1959, Rogers 1951). In addition, Gram-negative aerobic bacteria do not survive as well in aerosols at the relative humidity of indoor environments, such as hospitals, as do the Gram-positive cocci discussed previously.

Airborne cross infections with pseudomonas have been reported (Lowbury and Fox 1954, Williams et al. 1960). Lowbury (1954) showed in a controlled trial that a significant decrease in the frequency of pseudomonas infections of burns could be achieved by the use of an air-conditioned, filtered-air, dressing station. Smaller, but also significant, reductions of *S. aureus* and *Proteus* sp. infections were also encountered (Lowbury 1954).

Bate and James (1958) reported airborne transmission of *Salmonella typhimurium* on a pediatric ward. The mode of transmission was dust raised by a vacuum dusting device. Some 21 infections, thought to be transmitted in this manner, were recorded during an 11-mo period.

Other bacteria

Sevitt (1949, 1953) has observed in hospitals clostridial infections that were associated with airborne transmission of the etiologic organism through the vehicle of contaminated dust. Both tetanus and gas gangrene thought to be transmitted by this route were reported by Sevitt (1949, 1953).

Aspergillus species

Aspergillus species have become a major cause of nosocomial infections in patients with altered host defenses. Aspergillus infections are especially important as causes of mortality of patients with leukemia (Rose and Varkey 1975). The epidemiology of nosocomial aspergillosis is not completely understood. Infection usually begins in the lung, and aspergillus is not part of the normal flora; therefore it is usually assumed that the infection is airborne. The two most common species causing disease, *Aspergillus fumigatus* and *Aspergillus flavus,* are frequent airborne saprophytes in the environment (Bennett 1974).

In the study of Rose and Varkey (1975) 10 patients with hospital-acquired aspergillosis were seen. Eight of these 10 patients acquired their aspergillus pneumonia during the period "1964–66" in an old hospital building that was naturally ventilated (Rose and Varkey 1975). Only two patients with aspergillus pneumonia were seen during the period "1966–1973," following a move to a new hospital building that was mechanically ventilated, and both these were community-acquired infections (Rose and Varkey 1975). The air in both hospital buildings had been sampled for the presence of fungal spores using identical settling plate methods. Airborne spores of *Aspergillus fumigatus* were present in the old hospital building but were not detected in the new one. Rose and Varkey (1975) suggested that hospital-acquired aspergillus infection of the lung might be eliminated if all incoming hospital air were filtered, properly ventilated, and not recirculated.

A reverse sort of correlation between aspergillosis and moving to a new hospital building was reported by Aisner et al. (1976). They observed a cluster of eight cases of pulmonary aspergillosis occurring in leukemia patients. These patients acquired their disease shortly after the Baltimore Cancer Research Center was relocated into a new hospital facility. Epidemiologic investigation implicated a dry, brittle, cellulose-based, fireproofing material that had been sprayed above a false ceiling on pipes and on ceiling panels. This fireproofing had been sprayed into place while wet during construction and allowed to air dry. The dried fireproofing material from seven rooms and two hallway areas all yielded aspergillus species, primarily *Aspergillus niger, Aspergillus flavus,* and *Aspergillus fumigatus*. Aisner et al. (1976) postulated that aspergillus spores settled into the fireproofing while it was wet, and vegetated. They believed that subsequently the dried fireproofing material served as a source of aerosolized aspergillus spores especially when the false ceiling was disturbed during renovation or maintenance. Aisner et al. (1976) recommended that attention be given to hospital construction materials to avoid using materials that might support the growth and sporulation of aspergillus species.

Bennett (1974) has emphasized the importance of proper maintenance of air conditioner filters as well as housekeeping practices that reduce dust in the reduction of the airborne aspergillus spore count. Obviously, specially designed rooms with laminar airflow and high-efficiency air filtration would theoretically help protect susceptible patients from the airborne spores of aspergillus.

Candida species

The incidence of all opportunistic deep fungal infections including those caused by *Candida* sp. has increased during the past 20 yr. Infections with *Candida* sp. consistently lead the list of mycotic infections in the compromised host (Rose and Varkey 1975). For example, in a 10-yr study by Rose and Varkey (1975), carried out between 15 November 1963 and 15 November 1973, the total number of fungal infections had almost doubled during the second 5-yr period. The greatest proportion of the increase was caused by deep-seated *Candida* infections.

In another study of deep fungal infection among inpatient populations within two different hospitals, Hart and Associates (1969) reported a fourfold increase in opportunistic infections in compromised hosts during the period 1964 through 1967 compared with the period 1960 through 1963. As in Rose and Varkey's study (1975), *Candida* infections showed the greatest increase in frequency and became the most common fungal infections in their hospitals.

Unlike aspergillus, *Candida albicans* is part of the normal flora of the mouth and gastrointestinal tract in humans. Clayton and Noble (1966), however, found *C. albicans* in the mouths of 39.2% of 376 adult hospitalized patients as compared with 8.6% of the mouths of 23 undergraduate medical students with no patient contact. They (1966) also observed increased rates of carriage among children when they were patients in the hospital, with 16.4% of 73 child patients sampled carrying *C. albicans* in the mouth as compared with 5.4% of 503 healthy children sampled.

The studies of Clayton and Noble (1966) also showed frequent contamination of the air of hospital wards by *C. albicans*. The airborne particles were of a size similar to those observed in studies of airborne staphylococci. The median equivalent diameter of the particles containing *C. albicans* was between 10 and 15 μm, thus closely resembling the size of particles carrying bacteria and other fungi of human origin.

Candida albicans was found on the bedding of 8.8% of 635 adult patients and 9.6% of 73 child patients as compared with 0 of 279 healthy children and 0.4% of healthy police cadets (Clayton and Noble 1966). These data indicate that *C. albicans* colonization increases in hospitalized patients, and that this colonization is reflected by an increased frequency of contamination of bedding by *C. albicans*. By analogy with streptococci and staphylococci, bedding contamination often yields airborne particles of microorganisms capable of dissemination to other persons.

There is undoubtedly a multifactorial explanation for the spread of *C. albicans* in hospitals, with direct contact and fingers to appliances such as plastic intravenous catheters accounting for most of the spread. The presence of *C. albicans* in the air of hospital wards, however, argues that air may at least be a minor mechanism for spread and may account for some of the increased colonization of hospitalized patients.

Coccidioidomycosis

Respiratory acquisition of coccidioidomycosis is a recognized phenomenon. Natural transmission occurs from the inhalation of the arthrospores generated by the mycelial phase of the fungus, which has proliferated in soil.

It is generally believed that transmission from patients in the indoor environment of the hospital does not occur. The yeast phase found in tissue is not transmissible. Eckman et al. (1964), however, described an epidemic of coccidioidomycosis constituting six cases among persons caring for a patient with acute disseminated disease. *Coccidioides immitis* was recovered from cultures of surgical dressings and a plaster cast of a patient who had a draining coccidioidomycosis lesion of a lower extremity. Presumably, drainage from the patient contaminated the cast and the dressings with yeast-phase organisms. The yeast-phase organisms proliferated into the mycelial phase and formed arthrospores, which became airborne and infected the hospital personnel.

It should be remembered that *C. immitis* has the ability to germinate in dressings, plaster casts, or bedclothes contaminated with purulent discharge containing the yeast phase of the organism. The germinating organism produces a mycelial form with infectious arthrospores. Patients with discharging coccidioidomycosis lesions should therefore have dressings and soiled casts changed frequently. Likewise, bed linen should be changed frequently, and there should be environmental monitoring of the patient room to rule out dissemination of infectious arthrospores.

Prevention of Spread of Airborne Infection

Ultraviolet light

Riley (1972, 1974) has recently reviewed the effects of UV light on the viability of microorganisms suspended in aerosols. He also has reviewed the evidence for actual effectiveness of UV light installations in preventing the spread of infectious diseases.

In Riley's experimental setup (1972), upper air irradiation with UV disinfected the air of a room, and irradiation of air in corridors minimized airborne spread of orgnisms within a building. Quantification of results in Riley's studies (1972) was based on the rate of disappearance of airborne test organisms (*Serratia marcescens*) with UV on as compared with the rate with UV off. Riley (1974) found that the test organisms disappeared from the room with UV light on at a rate that would have approximated 50 air changes per hour if the only factor operating to remove the organisms had been a dilutional effect. (One air change washes out 63% of airborne organisms. By analogy, when 63% of airborne organisms are killed by UV irradiation, the effect is called one equivalent air change [Riley 1974]). In contrast, forced-air ventilation of the same room yielded only about the equivalent of five air changes per hour reduction in airborne test organism content.

In other experiments, Riley (1974) reported that the specific susceptibility of *Mycobacterium tuberculosis* to UV irradiation was approximately one-half that of *S. marcescens*. That is, it was 60 equivalent air changes per hour for *S. marcescens* and 30 equivalent air changes for *M. tuberculosis*.

As stated by Riley (1972), the ultimate test that would uphold the benefit of UV light installations would be epidemiologic evidence of disease prevention. By Riley's (1972) own account, "Such evidence in relation to air disinfection in hospital is meager at this time." At the Veterans Administration Hospital at Livermore, California, McLean (1961) reported that there was no outbreak of Asian influenza in the irradiated main hospital building, whereas during the same study period an outbreak did occur in unirradiated adjacent hospital buildings. Wells et al. (1942) reported successful control of measles in irradiated schools. Perkins et al. (1947) reported similar studies of efficacy of UV irradiation of air in a large rural central school with suggestive evidence of control of measles. Other successes attributed to UV irradiation have been reported from hospital ward studies (Sommer and Stokes 1942, Rosenstern 1948).

On the other hand, a classic report of failure of UV irradiation of air to affect measles transmission in a school came from the British Medical Research Council Air Hygiene Committee based on studies in the Southall District of London (MRC 1954). Riley (1972) postulated that the failure of UV air disinfection to ameliorate the spread of measles in the British study was predicated on the fact that school children remain in the irradiated room only 5–6 h/day, and air disinfection in the schools cannot protect the child from infection during the time spent outside the protected environment. As these children live in the congested area of a big city, Riley (1972) argued that they had ample opportunity to acquire measles virus outside the irradiated classrooms.

Riley and O'Grady (1961) have provided other interesting evidence of the killing power of UV irradiation against *M. tuberculosis* when the UV irradiation was directed against the organisms in a ventilating duct. Air from a tuberculosis ward was divided in half so that equal amounts went to each of two animal exposure chambers. The air being conducted to one of the chambers was irradiated in the duct so that only disinfected air reached this chamber. Guinea pigs were housed in each exposure chamber. During two years of study, infections occurred in the colony receiving unirradiated air at an average of about three infected guinea pigs per month, but not one of 150 guinea pigs breathing the same air, disinfected by passage through the UV-irradiated duct, contracted tuberculosis. This provided a convincing demonstration of the ability of UV irradiation to sterilize *M. tuberculosis*–containing air.

Ultraviolet irradiation has also been used in an attempt to rid air in operating rooms of other potentially airborne pathogens, most notably *S. aureus*. Overall, the results of such studies have been disappointing.

In a discussion of the general issue of the importance of supplying pathogen-free air for operating rooms, Dixon (1973) has pointed out that the mode of transmission for even epidemic surgical wound infections is seldom adequately defined, and endemic transmission is even more difficult to study. Many reports in the literature omit adequate investigations of either the operating theatre or the postoperative environment, have failed to differentiate between airborne and contact infection, or have instituted control measures against both these modes of spread (Dixon 1973, Ayliffe and Collins 1967, Browne et al. 1959, Walter et al. 1963).

It is true that it is very difficult to design studies that will properly differentiate between contact spread and airborne exogenous contamination of operative wounds. Within the constraints of difficult experimental designs, however, some conclusions are possible from a large-scale cooperative study (NAS/NRC 1964). This study was sponsored by the National Academy of Sciences/National Research Council (NAS/NRC). It compared rates of postoperative wound infection in various categories of surgical operations in operating rooms irradiated by UV light versus other rooms in the same institutions that were sham irradiated.

No benefit was noted overall in patients operated upon in rooms with air exposed to UV irradiation. A subsegment of the study population, however, patients undergoing "refined-clean" surgical procedures, had significant reductions in wound infection rates associated with operations being performed in UV-irradiated rooms. (Wounds unlikely to be contaminated from endogenous sources were termed *clean* wounds. They were further subdivided into *refined-clean* and *other clean*. The former were elective, primarily closed, and undrained wounds, and the latter were not elective, were not primarily closed, or were drained mechanically through the incision or through a separate stab wound.) Patients with refined-clean wounds constituted a group not exposed to prior traumatic contamination and unlikely to be exposed to endogenous contamination from their own flora, and would theoretically represent the best group to see a decrease in postoperative infection rate by lowering chances for airborne contamination during surgery. In this group of patients, essentially all contamination of wounds should be exogenous.

In the NAS/NRC study (1964) 3277 refined-clean wounds were made in operating rooms with air irradiated by UV light. Some 3379 refined-clean wounds were made in control, unirradiated operating rooms. There were 94 definite infections, for a rate of 2.9%, among patients operated upon in UV-light-irradiated rooms. There were 18 more possible infections for a total of 112 definite plus possible infections, for a rate of 3.4%, among the UV-light group. In the control group there were 128 definite infections, for a rate of 3.8%, and 154 definite plus possible infections, for a rate of 4.6%. The difference between the UV-irradiated versus control groups were statistically significant ($p < 0.5$) if the definite infections alone were considered or if the definite plus possible infections were compared.

Goldner and Allen (1973) reported on 35 yr of experience with UV irradiation of air in orthopedic operating rooms at Duke University Hospital. The overall infection rate among refined-clean cases of less than 1% during the 35-yr was believed to stem from the presence of ultraviolet irradiation. Unfortunately, no direct control population was studied at Duke. During the period 1958–1962, the Veterans Administration Hospital in Durham, North Carolina, associated with Duke University, did not have ultraviolet lights. The infection rate for refined-clean wounds at that institution was 1.7%, whereas the overall rate at Duke for 23,000 refined-clean operations during the same 5-yr period was 0.34% —a fivefold difference.

The present authors would make the following comments about UV irradiation. Despite enthusiasm and advocacy for the use of UV air-sterilizing procedures, there is a paucity of evidence to support its actual day-to-day clinical use within hospitals generally or within operating rooms after more than 30 yr of experience. There is good evidence to indicate that properly applied UV light can kill airborne microorganisms, but this evidence has not been convincingly translated into reduced attack rates of clinical infection among patients at risk in hospitals.

The American Hospital Association's Committee on Infections in Hospitals (AHA 1974) has made several comments about UV air sterilization. First of all, the committee reiterated that properly applied UV irradiation can kill nearly all disease-producing organisms "provided the light rays impinge upon the organisms with sufficient intensity for the minimum time required." The committee listed four difficulties with the UV air-sterilization process in actual practice:

> 1. UV rays have very limited penetration, and organisms are easily protected within dust particles, dry mucus, etc. 2. UV rays lose strength rapidly as they pass through air or water surrounding the source tube. In water, the intensity is reduced to 38% at a distance of 2 inches [5 cm] from the tube. In air, the intensity is reduced to 25% at a distance of 8 to 9 inches [20–23 cm]. 3. UV ray emissions from tubes are markedly reduced by fine films of dust, moisture, or oily substances on the glass surface. 4. The emissions of UV rays from the source tube in the lethal range (2,500–2,600 Angstroms) gradually weakens without any discernible change in visible light emissions.

The committee (AHA 1974) commented that the difficulties associated with UV irradiation make it imperative that the conditions of use be carefully controlled if they are to be effective at all. The equipment must be very regularly serviced and monitored with UV meters by an expert. Installations must also be planned with care so that the UV irradiation does not cause conjunctivitis or burns of the skin of personnel or patients.

The committee concluded by stating, "Extensive studies have not encouraged general usage of UV radiation in hospitals."

Unidirectional airflow (UAF) and high-efficiency particulate air (HEPA) filters

As indicated previously, the control of surgical operative wound infections is a complex matter as there are many different ways in which bacteria might reach the wound. The relative importance of airborne bacterial contamination in surgical wound infection is a matter of controversy. Airborne bacteria in the operating room may arise from desquamation from the skin of the patient or members of the operating team, from the respiratory tracts of operating room personnel or the patient, or even from fomites and from horizontal surfaces such as the floor.

It has been difficult to establish the relative importance of airborne contamination in operative wound infection, but most authorities believe that anything that would decrease the potential hazard from this and other sources would be beneficial. For this reason there has been great interest in laminar flow, or more precisely speaking, unidirectional airflow clean rooms with the air entering and leaving the room being passed through HEPA filters.

The HEPA filters developed for the U.S. Atomic Energy Commission are of glass-asbestos construction and are 99.9% efficient for the removal of particles of 0.3 μm diameter (Zeterberg 1973). According to Zeterberg, filters can remove 24 billion particles/h at airflow rates of 30 m/min through a 1858-cm^2 surface. If efficient prefiltration is accomplished for the removal of large particles, an HEPA filter may last as long as 5 yr and absorb as much as 1.6 kg of submicron contaminants before it is reduced in efficiency.

The HEPA filters are used in conjunction with ventilating and air-conditioning systems set up to produce laminar airflow, a concept of linear, or unidirectional, low-turbulence airflow in parallel lines. Actually, the imposition of the operating table, the operating team, and other fixtures within the surgical theatre interferes with strict columnar laminar airflow as defined by physicists. Therefore the term *unidirectional airflow system* (UAF) is preferred for installations in hospitals.

The UAF systems work because the introduction of sufficient air uniformly in a single direction in a room causes the lines of airflow to be nearly parallel, and there is negligible interchange of air and airborne particles across the flow lines. The direction of airflow may be horizontal or vertical. The exact arrangements of a room are variable within considerable limits without disturbing uniform airflow in the greater part of the room's volume. The minimum rate of airflow for stability depends upon the magnitude of thermal differences within a room, as well as other disturbing factors, but 0.2 m/s is generally sufficient, and lower air velocities may be satisfactory. This type of ventilation not only excludes external airborne contamination in a highly effective manner, but (combined with an HEPA filter) it also rapidly removes contamination generated within a room, and there is no transfer to the patient from areas downstream or to his side. Also basic to the understanding of the proper functioning of laminar-flow operating rooms is the concept that the unturbulent flow of columns of air creates a barrier of clean air between the operative wound and the surgeon.

That UAF-HEPA filter installations are very efficient in reducing the number of airborne bacteria has been noted in many studies as reviewed by Zeterberg (1973). Typical results in terms of reduction of airborne bacteria were reported very recently by Alexakis et al. (1976). They found that the number of bacteria isolated during 263 surgical procedures were reduced by 95% in a horizontal-laminar-flow (UAF-HEPA) room compared with a conventional room, and a further 72% reduction of airborne bacteria was observed when a suction mask system was used for operative personnel.

The key question about the effectiveness of UAF-HEPA filter systems is, Do these systems result in a lower frequency of operating room or hospital-acquired infection? As summarized recently by Dixon (1973), there have been no convincing controlled studies that establish beyond doubt that UAF-HEPA filter systems or other systems of air filtration or air control are of value in preventing postoperative wound infection.

Charnley and Eftekhar (1969, Charnley 1972) have reported interesting data relative to postoperative wound infections in major orthopedic, refined-clean, surgical cases during a time when progressively more sophisticated means of elimination of possible airborne bacteria were being introduced. During this time they noted a definite trend toward a reduction in surgical wound infections from 7%–9% in 1960, when the program began, to under 1% in 1970.

Recent work published by Alexakis et al. (1976) emphasized the efficacy of a horizontal-laminar-airflow room equipped with HEPA filters in reducing the numbers of airborne bacteria. Alexakis et al. (1976) noted, however, that the species of bacteria isolated from air at the operative site by means of a slit sampler differed from those isolated from settling plates on the instrument table located at the periphery of the room. Thus studies of possible airborne contaminants in operating rooms should include both sites. Alexakis et al. (1976) concluded that the exact importance of airborne bacterial contamination of operative wounds is still unknown. Therefore they regard installation of UAF-HEPA filter systems and the use of suction-hood systems as "experimental at this time."

There has been interest in various UAF and isolator systems for the protective isolation of patients with severe burns. In the report by Lowbury (1971), results of attempts to prevent airborne bacteria from contaminating burned patients appeared encouraging but did not conclusively prove the value of a laminar-flow-type isolator or an open-top, plastic, life-island type system.

There has also been considerable interest in the possibility of reducing morbidity and mortality from severe infections in patients suffering from leukemia, especially acute nonlymphocytic leukemia. In discussing their own studies and those of others, Schimpff et al. (1975) concluded that patients cared for in laminar-airflow rooms and given oral, nonabsorbable antibiotics had more remissions of their leukemia and fewer infections than those who received routine ward care alone. Patients receiving routine hospital ward care and oral nonabsorbable antibiotics, however, did almost as well as those who received oral nonabsorbable antibiotics and were nursed in isolated laminar-flow rooms. Results of other studies discussed by Schimpff et al. (1975) seem to trend in the direction of benefit for laminar-airflow isolation plus nonabsorbable antibiotics, but it is difficult to factor out the role of laminar-airflow isolation alone.

Nonpermeable drapes and garments

Noble (1975), in his review of the dispersal of skin microorganisms, pointed out that many investigators have shown that the skin of personnel present in the operating room can yield squames. These squames can penetrate ordinary operating room clothing or can drop from the perineum down pant legs and thus become suspended in the air and in dust. Squames can contain microorganisms and have the potential to transmit bacteria to patients.

Various methods of controlling such dispersal of microorganisms have been attempted. Charnley (1973) has described an impermeable gown and hood with a face mask and a suction exhaust ventilation system to remove all air and contaminating particles from around the surgeon and other operating room personnel. Dineen (1973) has reported a 90% reduction in airborne organisms in operating room environments when single-use, impervious, nonwoven, disposable drapes and gowns were employed as compared with airborne bacterial counts done when cloth gowns and drapes were used by the same operating room teams at different times. Clothing that occludes the perineum significantly reduces dispersal of organisms (May and Pomeroy 1973, Hill et al. 1974, Mitchell and Gamble 1974).

There is a very large difference in the frequency of dispersers of *S. aureus* between men and women (May and Pomeroy 1973). In tests of 238 men and 37 women, wearing their own clothes in a sampling chamber, Hill et al. (1974) found that 30 (13%) of the men shed *S. aureus,* some very profusely, in contrast to only 5 (1%) of the women. Rates of nasal carriage of *S. aureus* were similar for both sexes—28% in men and 27% in women. Dispersal was increased by wearing of clothes, presumably by squames shed from skin due to friction by clothes. Dispersal could be decreased by the wearing of bacteria-proof underpants.

Close-weave, bacteria-proof clothes and underwear are uncomfortable, however, and Ayliffe et al. (1974) pointed out that impermeable clothing obviously cannot prevent dispersion by nasal carriers and dispersers. They recommend that special clothing not be used except for high-risk operations such as insertion of hip prostheses.

May and Pomeroy (1973) measured the dispersal of bacteria into the air by naked males compared with naked females and found that the colony-forming particulate output from naked men was five times greater than from naked females. In another group of experiments, the colony-forming particulate count was measured first from men while naked, and later from the same men while wearing tightly fitting neoprene sponge rubber drawers so that there was no output of colony-forming particles from the perineal area (May and Pomeroy (1973). These experiments showed that 80% of the male output came from the area covered by drawers; and while wearing neoprene sponge rubber drawers, otherwise naked males put out the same number of colony-forming particles as naked females.

Conclusions

Potential airborne pathogens remain of great interest in indoor environments, particularly hospitals. For some organisms, such as *M. tuberculosis,* the airborne route is the most important if not the exclusive method of transmission under natural conditions. For many of the important pathogens causing endemic nosocomial disease, such as *S. aureus,* the precise importance of airborne transmission in the multifactorial equation of transmission remains undefined.

Potentially pathogenic microbes are regularly shed by humans and can be found in the air in indoor environments. Several different modalities including UV irradiation and UAF-HEPA filter installations can, under the proper conditions, reduce the count of viable microbes in the air in hospital rooms and other indoor environments. Both UAF-HEPA filter installations and UV irradiation of air have their enthusiastic modern advocates. Unfortunately, neither of these methods of air hygiene can be unequivocally recommended as valuable in reducing the transmission of disease within hospitals and other indoor environments.

More research, incorporating sophisticated experimental design and data analysis, is needed to factor out the fraction of endemic nosocomial infections that are airborne. With better definition of the incidence of endemic airborne nosocomial infections, definition of the effectiveness, if any, of various methods of air sterilization will become possible.

IMPACT OF AEROALLERGENS*

Allergic responses to aerobiological agents impose major adverse effects on the physical and economic health of mankind (Davis 1972). The association of epidemic nasal disease (rhinitis) and asthma with sensitivity to inhaled pollens was perceived over a century ago (Feinberg 1946). Since then, natural emanations of fungi, actinomycetes, algae, arthropods, and vertebrates also have been implicated as aeroallergens, and additional allergic syndromes have been recognized. Substantial insight into the properties of specific allergens, their patterns of prevalence, and the tissue mechanisms that underlie human responses has been gained only recently, however.

Impact in Terms of Recognized Mechanisms of Hypersensitivity

Aeroallergen health effects reflect at least two of the four recognized mechanisms of immune injury (Gell and Coombs 1968) and may involve a third as well. The impact of these processes is most widely recognized in hay fever (seasonal nasal allergy) and allergic asthma provoked by airborne agents. A predisposition (Leskowitz et al. 1972, Levine 1973) to synthesize antibodies (of the immunoglobulin E class), reactive with inhaled and ingested materials, characterizes persons with such atopic conditions; however, asymptomatic subjects also may display this trait. Following allergenic exposure (Levine 1973), specific antibodies (termed *atopic reagins*), are synthesized, circulate in extracellular fluids, and fix to surface sites on blood basophils and tissue mast cells. On reintroduction of a specific allergen, such "sensitized" cells release substances that induce dilation and heightened permeability of blood vessels, smooth muscle contraction, and eosinophilic chemotaxis—tissue changes that are characteristic of allergic rhinitis and asthma. The allergen-induced release of such mediators of inflammation has been demonstrated in allergic human nasal (Kaliner et al. 1973) and pulmonary (Parish 1967) tissues, as well as with primate tissues passively sensitized with allergic sera (Patterson and Kelly 1974). Similar mechanisms appear to underlie respiratory responses to ragweed pollen in spontaneously sensitive dogs (Patterson and Kelly 1974).

Examples of allergic eczematous contact dermatitis due to airborne biogenic particles are relatively uncommon and reflect quite different though well-defined mechanisms (Fisher 1973). Skin eruptions, marked by reddening, thickening, and extreme itching, typify this disorder, which often is restricted to exposed areas of epidermis. These tissue changes reflect the activation of lymphoid cells that have acquired sensitivity to complexes of specific contact

*W. R. Solomon, I. L. Bernstein, and R. S. Safferman

allergens, such as pollen oils, with components of host epidermis. Neither a contributory role for humoral antibody nor associated respiratory symptoms is evident in this condition.

A variety of organic dusts have been shown to produce inflammatory changes involving the minute bronchioles and gas-exchanging alveolar membranes of selected, heavily exposed persons (Pepys 1969). The resulting "hypersensitivity pneumonitis" or "extrinsic allergic alveolitis" generally appears hours after implicated exposures; cough, breathlessness, fever, and prostration often are prominent, but audible wheezing is not. Unlike the rapidly reversible airway changes of the asthmatic, the tissue infiltrates of allergic alveolitis require weeks to resolve or may lead to scarring and permanent functional loss. Prolonged exposures of modest intensity to materials such as avian emanations or actinomycetes in air-processing systems may provoke insidious respiratory insufficiency and death without evident bouts of acute illness (Hargreave et al. 1966, Banaszak et al. 1970). Although the basis of this disease pattern is obscure, interest and etiological conjecture have centered about the common finding of specific serum precipitating antibodies (Pepys 1969) reactive with components of responsible dusts. Antigen-antibody reactions in distal air passages, with activation of the complement "cascade" or other tissue amplification systems, or both, have been postulated. Specific antibody also may be detectable in many persons who remain well despite repetitive exposure (Roberts et al. 1975), and seronegative persons rarely may manifest overt illness (Edwards et al. 1974). Animal models have provided analogous findings (Wilkie et al. 1973) and have suggested recently that reactivity to aerosol exposure may be transferrable by intact lymphoid cells (Bice et al. 1974). These findings imply that cell-mediated processes (delayed-type hypersensitivity) also may contribute to human hypersensitivity pneumonitis.

Attributes of Allergens in Relation to Human Responses

Molecular characteristics of airborne allergens

Among allergens implicated in atopic allergy, only components of a few pollens have been characterized chemically in detail. Aqueous extracts of short ragweed pollen have received special attention and, on chromatographic analysis, have yielded a major allergen, antigen E, and additional allergens designated K, Ra3, Ra5, and so on. Antigen E, although composing only 6% of the extractable protein of pollen, appears to contribute 85%–90% of its allergenic activity (King et al. 1964). Among pollens of many temperate zone grasses, allergenic activity also has been associated predominantly with medium-sized linear proteins (viz., the group 1 antigens; Johnson and Marsh 1965, Marsh et al. 1970). Carbohydrate fractions of grass (Johnson and Marsh 1965, Mullan 1973) and ragweed (King et al. 1964) pollens generally have not shown skin reactivity in sensitive subjects. Preliminary studies of major allergens of birch (Belin 1972) and alder (Herbertson et al. 1958) pollens have suggested that these, too, are proteins—with molecular weights close to 20,000. More recently, cat dander allergens have been associated with albumin and alpha-2 globulin moieties (Ohman et al. 1973) in sera and pelts. Attempts to characterize allergens of other epidermals, fungus spores, and house dust (see below), however, are either too preliminary to summarize or have yielded inconclusive data (Berrens 1971).

The isolation of major pollen allergens has provided a basis for judging the clinically relevant similarities among emanations of related plant species (e.g., *Ambrosia* sp. and other composites [Yunginger and Gleich 1972]). Similarly, it has been possible to "rate" the comparative potency of pollen extracts (Baer et al. 1974) employed for clinical testing and immunotherapy with the use of appropriate, specific antisera. Such reagents have been employed also to localize allergen distribution on and within individual pollen grains (Knox et al. 1970), as well as in association with additional plant structures including vegetative portions (Shafiee et al. 1973). The possibility of employing immunochemical markers in the microscopic analysis of airborne particulates (Busse et al. 1972), including plant humus, has scarcely been exploited, however.

Physical properties of airborne allergens

In considering airborne carriers of allergenic activity, principal attention has been directed traditionally to particles of 1–60 μm—the size range of common pollens and spores. For agents such as house dust, animal danders, and various vegetable dusts, however, no units of characteristic form and size are recognized. Furthermore, instances of reaginic sensitivity to simple industrial chemicals (Kammermeyer and Mathews 1973, Pepys et al. 1972) have suggested that smaller particles, including submicronic fractions, might carry allergenic activity. This view is supported by observations of ostensibly allergic asthma occurring on exposure to cooking vapors (Feinberg and Aries 1932), and implies that a variety of biogenic emanations including pheromones could act similarly. Small allergenically active aerosols also might result from reflotation of partially degraded, previously deposited materials—including insect emanations (Perlman 1961).

Recently it has been reported that ragweed pollen allergens may be associated with aerosol species below 5 μm in size (Busse et al. 1972). Inhaled particles in this range are carried readily to terminal, gas-exchanging units and are known determinants of hypersensitivity pneumonitis. Whether pollen asthma also might reflect deposition of such small particles has been considered since remarkably few intact pollen grains appear to pass the larynx and trachea (Wilson et al. 1973); furthermore, asthma has not been produced reflexly by nasal or pharyngeal challenge with pollen (Hoene and Reed 1971). By contrast, the ease with which pollens and large spores impact upon the ocular and nasal membranes suggests that allergic conjunctivitis and rhinitis are direct effects largely of heavy particle loads; the extent and effects of mucosal absorption and later systemic circulation of eluted allergens remain unknown.

Aspects of Human Response to Airborne Allergens

Allergen absorption and dissemination

Although leaching of soluble antigens from mucosally deposited materials may be assumed, neither the determinants nor the kinetics of elution and tissue penetration have been described. Differences in the morphology and exudation patterns of pollen grains incubated in vitro with saliva of normal and allergic donors have been reported (Mayron and Loiselle 1972), and a role for lysozyme in allergen elution has been postulated. Limited mucosal penetration by intact pollen grains (Jorde and Linskens 1974) has been reported recently in man but also requires confirmation.

Mucosally encountered antigens are known to elicit the synthesis of specific reagins in atopic subjects, while such immunoglobulin E responses are minimal or absent in normal persons (Salvaggio et al. 1964). This difference is not observed with injected antigens (Schwartz and Terr 1971), and has suggested that "defective" (distinctive) mucosal processing may be a basic factor in atopic disorders. Increased blood vessel permeability in active allergic rhinitis is known to foster the passage of plasma components (such as albumin) to the mucosal surface (Steinberg et al. 1975). It is reasonable to postulate that (inward) absorption of materials reaching the mucous membrane is also augmented in inflammatory states such as hay fever. Direct comparisons of allergen absorption by normal persons and *asymptomatic* atopic volunteers with unrelated sensitivities, however, have *not* shown evident differences (Kontou-Karakitsos et al. 1975). Both inhaled and ingested soluble allergens do pass mucosal barriers and can be disseminated by circulating blood. Whether such translocated materials regularly elicit reactions in tissues (e.g., the bronchi) inaccessible to inhaled, intact particles remains a priority area for study.

Dose-response relationships

Clinical experience suggests that, among sensitive, atopic subjects, marked variations exist in symptom severity under conditions of natural exposure to specific agents. These differences correlate significantly, though crudely, with skin test reactivity and reagin titers determined in vitro (Lichtenstein et al. 1973). Site-to-site variations in particle prevalence and differences in activity patterns among subjects, however, generally vitiate efforts to estimate or compare allergen exposures. Furthermore, factors such as ambient temperature and humidity, concurrent infection, the presence of additional allergens and irritants, and the time course of exposure may influence allergen-induced respiratory responses. In view of this established complexity, it is not surprising that diurnal patterns of symptom severity and allergen levels have been observed to differ (Goodwin et al. 1957). Progress in this area will require observation, under controlled laboratory conditions, of responses to aerosol levels simulating natural exposure. Although continuous dispersion of biological particles to create "clouds" of constant concentration is difficult, initial successes have been reported (Solomon 1969, Fontana et al. 1974); however, to date, threshold levels have not been established in this manner for any particulate allergen. Exposure to grass pollen levels of 20 grains/m^3 (following a progressive rise over days) has been observed to elicit symptoms in virtually all British pollinosis patients (Davies and Smith 1973). Reports for ragweed pollen suggesting 7.7 grains/m^3 as a "clinical threshold" are based on questionable transformation of gravity slide data, and in practice higher levels have seemed necessary for the onset of symptoms. It is also common, however, to recognize ragweed pollinosis persisting, late in a given season, at levels that seemed inadequate to elicit symptoms at the onset of pollination. This effect is consonant with observations that progressively lower pollen challenges suffice to elicit nasal changes when sequential daily exposures are employed (Connell 1969). Quite heavy allergen doses have been required generally to provoke symptoms during brief laboratory exposures, although exceptions have been reported. In one study, extraseasonal exposure periods exceeding 1 h were needed to elicit symptoms in most sensitive subjects with levels of native ragweed pollen exceeding 800 grains/m^3 (Solomon 1969).

Humoral immune responses

Although reactions of specific E immunoglobulins are basic to atopic conditions, immune responses to airborne agents are not confined to this antibody class. Both normal and atopic persons synthesize and secrete, at mucous surfaces, IgA and, to a lesser extent, IgG antibodies specific for pollen components and probably other aeroallergens (Turk et al. 1970, Von Maur et al. 1975). Although higher levels of allergen-specific IgA and IgG antibodies have been noted in respiratory secretions of pollinosis patients, neither adverse nor protective effects may be attributed to these immunoglobulins at present (Von Maur et al. 1975). Serum IgG responses are not prominent with natural exposure but are elicited readily by injected allergen extracts in both pollinosis patients and normal subjects. The protective role of such IgG "blocking" antibodies remains controversial. In persons receiving injection therapy for pollinosis, however, an inconstant but generally positive correlation between symptomatic improvement and blocking antibody levels has been observed (Norman 1969).

Sampling Techniques and General Concepts of Prevalence

The value of data describing airborne pollen prevalence was appreciated early in the study of human allergy as an aid to etiological diagnosis and a correlate of clinical symptom severity. More recently studies of additional particles have been attempted, complementing extensive field surveys of the density and phenology of source species. In the case of potentially allergenic fungi and algae, emanations often arise from widespread but inapparent foci (e.g., leaf surfaces or soil), and atmospheric recoveries have provided the sole indicator of local prevalence.

Microscopically identifiable units or the capacity to produce typical growth in artificial culture remains essential for particle enumeration by currently available aerometric techniques. As a result, human exposure to materials such as house dust components, mammalian epidermal scales, insect emanations, vegetable dusts, and many fungus spores that are nonviable or derived from parasitic or heterothallic species still must be judged intuitively from observations of source density and distribution. Until recently, studies of aeroallergens have employed exclusively particle collection by fallout on greased microslides or agar surfaces in horizontal culture dishes. A full discussion of the comparative merits (and evident limitations) of such methods is presented in chapter 5 of this volume and will not be repeated; however, their effects upon traditional concepts of particle prevalence are worthy of note. The sparse recoveries obtained by gravity slides and open culture plates have served to focus attention upon the most abundant particle types and obscure the regular occurrence of many less prevalent aerosol species. In the case of micronic particles with low terminal settling velocities, even abundant types have been recovered with difficulty (Solomon 1975). As a result, the local abundance of many asco- and basidiospores as well as emanations of form species of *Sporobolomyces, Cladosporium, Fusarium,* and the like were overlooked prior to the advent of volumetric devices, especially suction traps. Concepts of relative pollen prevalence have been distorted also because of the preferential fallout on gravity slides of more robust forms, while smaller grains (e.g., those of *Urtica* sp.) have gone largely undetected. Similarly, the potent effects, upon aerosol capture and retention by fixed surfaces, of factors such as wind velocity and rainfall have compromised efforts to relate these variables to atmospheric particle prevalence.

The present generation of rotating arm impactors and suction traps has provided a clearer view of air spora components for clinical correlation and study. Resulting volumetric data readily admit spatial and temporal comparisons and have facilitated initial dose-response studies for inhaled allergens. The higher cost and limited availability of mechanical samplers, however, have slowed their acceptance by clinicians schooled in less precise, though simpler, collecting techniques. Clearly, future progress will require a broader awareness of the merits of volumetric methodology, the evolution of standardized procedures for collecting and reporting data, and opportunities to test clinically hypotheses based upon emerging concepts of prevalence.

Status of Specific Particle Types

Pollens

Pollens of anemophilous plants compose the longest recognized and best-studied group of aeroallergens. Humoral antibody responses to most prevalent pollen types have been noted in exposed humans, and symptoms are well documented following natural exposure to emanations of grasses, weedy forbs (principally chenopods, amaranths, ragweeds, and their allies), and certain trees (principally angiosperms of temperate regions). Notable preliminary progress has been made in defining chemically the principal allergens of ragweed (King 1964) and of (temperate zone) grass (Johnson and Marsh 1965) pollens and in modifying these compounds to provide more optimal therapeutic agents (Haddad et al. 1972, Patterson and Suszko 1974). Furthermore, the efficacy of immunotherapy (injections of specific aqueous pollen extracts) in mitigating symptoms of grass and ragweed pollinosis has been demonstrated statistically in formal and meticulously controlled clinical trials (Norman 1969).

The positive aspects must be viewed against the uncertainty that still surrounds much of pollen aerobiology and its relevance to clinical illness. It is noteworthy that for many regions supporting large human populations, especially tropical and subtropical zones, no data describing atmospheric pollen are available. By contrast, even in well-studied temperate areas, the clinical importance of pollens derived from sources including species of *Urtica, Plantago, Rumex, Parietaria,* and the like, as well as from a host of tree species having brief, overlapping periods of anthesis, remains speculative. The significance of most anemophilous species in provoking symptoms is, similarly, contentious, although the pollens of many are acknowledged sensitizers. Resolution of these problems will require determination of threshold levels and dose-response relationships through comprehensive observations during natural exposure, or, more probably, controlled laboratory challenges. Attempts to apply such data clinically must recognize the potential effects of diverse variables including infection (Nadel 1974), as well as preceding and concurrent allergen exposure (Connell 1969), in promoting heightened end organ responses to specific agents. The clinician must also wrestle with the uncertain allergenic relationships of microscopically indistinguishable pollens derived from closely related species (e.g., the several oaks or the grasses) in any locality. Multiple-point sampling and careful studies of floral phenology may be helpful in resolving these common problems and in delineating the effects upon specific pollen-sensitive subjects of small area or point sources.

Fungi

Airborne spores and hyphae of imperfect microfungi commonly provoke allergic asthma and rhinitis; a few types also have been implicated with more or less certainty in occupationally associated hypersensitivity pneumonitis. In addition, reaginic responses to emanations of species of the Zygomycetes, Oömycetes, and Heterobasidiomycetes are well recognized, suggesting possible clinical importance for these agents as well.

Symptoms ascribable to fungus sensitivities commonly accompany exposure to stored or composting plant materials such as ensilage, fallen leaves, or grass clippings. In addition, digging in dry soil or merely walking through dry standing vegetation (McDonald and Solomon 1975) may impose major exposure burdens.

Despite these characteristic associations, it is frequently difficult to define the contribution of fungus sensitivities to the symptom patterns of selected subjects. For specific taxa, predictable "seasons" of atmospheric prevalence have been noted in temperate areas; however these periods often are prolonged; vary annually with temperature, available moisture, and local agriculture practices; and overlap each other broadly. In many frost-free areas, abundant emanations of diverse fungi (especially *Cladosporium* form species) occur continually from sources in nature. Colonization of domestic, industrial, and commercial interiors, however, may ensure perennial exposure (principally involving yeasts, zygomycete taxa, and form species of *Penicillium* and *Aspergillus*) in colder regions (Van der Werff 1958). Contamination of humidifying devices, including poorly maintained, cold-mist vaporizers (Solomon 1974), has received special emphasis recently, and it is clear that fungi in enclosed spaces warrant further intensive study.

In assessing the clinical role of airborne fungi, it is necessary to confront all of the uncertainties mentioned previously with regard to pollens. The resulting difficulties are compounded further by the great diversity of mycotic particles that may occur abundantly at specific stations and a lack of even preliminary characterization of the responsible allergens. In addition, the uncertain affinities of morphologically defined groups such as the yeasts and deuteromycete form genera provide no tractable basis for inferences of allergenic similarity.

Despite several decades of study by allergists and others, data describing accurately the prevalence of fungus particles in free air remain severely limited. Uncertain levels of particle viability, the selectivity of individual media, and a general lack of reference materials providing morphological criteria for identification have posed limiting problems for workers in this field. Reliance on gravitational collection methods, with their decreasing capability to recover progressively smaller micronic particles, however, has probably biased earlier studies most seriously.

Recent volumetric air sampling at scattered points has documented the abundance of spores of soil and leaf saprophytes such as *Cladosporium, Alter-*

naria, Epicoccum, Helminthosporium-Drechslera, Fusarium, Sporobolomyces, Stemphylium, and *Pithomyces* form species. In addition, studies of spore trap deposits have begun to provide comparative prevalence data for many particle types that usually defy cultural identification, such as spores of myxomycetes, rusts, smuts, and downy and powdery mildews as well as those of taxa including *Cercospora, Polythrincium, Helicomyces* spp., and so on. Abundant recoveries of ascospores and basidiospores at several points have excited particular interest, and their possible role as aeroallergens has been suggested (Gregory and Hirst 1952, Frankland and Hay 1951). Skin and mucosal sensitivity to extracts of readily gathered basidiospores now are well documented in selected atopic subjects (Herxheimer et al. 1966, 1969, Bruce 1963). Similar reactivity has been attributed to a few ascospore extracts (Bruce 1963, Herxheimer et al. 1966); however more definitive work must await improved means of harvesting spores from scattered and frequently minute ascocarps.

Algae

Since algae have not been closely associated with human disease in the past, it is not surprising that investigation of hypersensitivity reactions to algae has lagged behind that of other microbial groups. In view of the fact that algae are widely distributed in various ecosystems, their importance as possible sources of human exposure and sensitization can no longer be ignored. Initial investigations in this field emphasized the significance of direct exposure to algae-infested waters (Heise 1949, Cohen and Reif 1953). More recently, however, a number of investigators have clearly shown that aerial transport is an important mechanism in the geographic distribution of these organisms (Gregory et al. 1955, Brown et al. 1964, Schlichting 1964). Such observations suggest that airborne algae may have significant biomedical impact on the atopic segment of the population.

Apart from the obvious importance of wind scavenging as a mechanism of introducing algae into the atmosphere, Woodcock (1955) as well as Blanchard and Syzdek (1970) have suggested the bursting bubble phenomenon at water-surface microlayers as an equally important alternative. Algae would be expected to follow the same aerobiologic integration and modeling scheme as other airborne components, namely, the pathways of production and release, dispersion, deposition, and biomedical impact. Investigation of the last compartment has been facilitated by identification of a population of patients in whom clinical sensitivity to green algae was documented.

Algae from the order Chlorococcales were selected for study on the basis of reports that members of this order have been commonly encountered in the course of sampling air, soil, and water for algae (Figures 6.1 and 6.2). Skin testing with substances extracted from various algae of this order soon revealed that *Chlorella* sp. was a suitable model for investigating the biomedical impact of algae in human hypersensitivity (Bernstein and Safferman 1966).

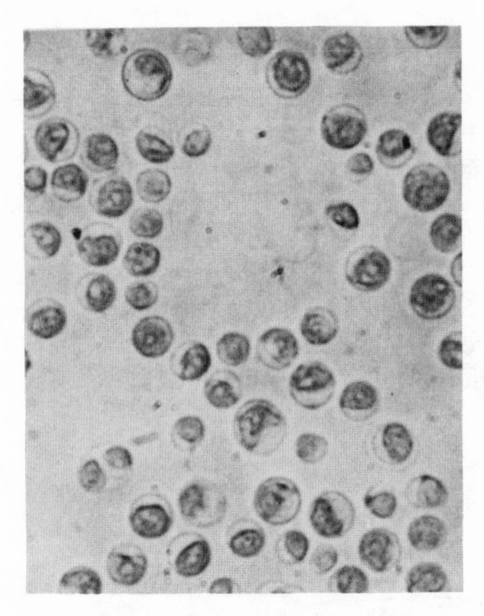

Figure 6.1. *Chlorella* sp.

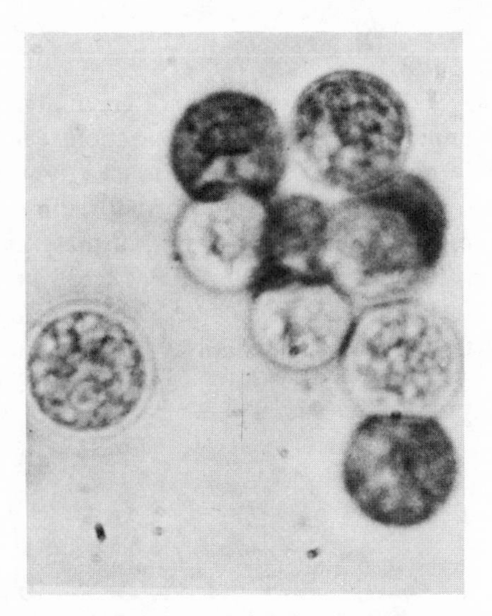

Figure 6.2. *Chlorococcum* sp.

Interestingly enough, members of this order resemble common pollen allergens in size and shape.

In investigating the immunogenicity of several *Chlorella* species, it was noted that closely related species (e.g., *Chlorella vulgaris* vs. *Chlorella pyrenoidosa*) failed to show cross-reactive determinants (Bernstein et al. 1969). This lack of cross-reactivity suggests that geographic differences in algal dominance must be considered as an important variable in the investigation of clinical hypersensitivity to these organisms.

Other studies revealed that there was a high incidence of positive reactions to algal organisms in house-dust-sensitive patients. These observations led to the investigation of viable algae in house dust samples (Bernstein and Safferman 1970a). A large number of viable algae were isolated; the most frequently noted were *Chlorella* (Figure 6.1) and *Chlorococcum* (Figure 6.2) species. Thus it was demonstrated that indoor localities were significant deposition sites in the algal aerobiology model, and as such they could be used to study the biomedical effects of algae in atopic individuals.

One of the *Chlorella* sp. isolates from house dust was selected for subsequent clinical and laboratory use because it contained antigenic determinants common to three different commercial house dust extracts and to a purified house dust fraction, fraction E, isolated by Professor L. Berrens, University of Utrecht (Bernstein and Safferman 1970b). Although skin reactivity to an extract of this isolate was frequent in tests of our local atopic population, we obtained widely varying results in other population centers. Thus, as might be expected from previous experience with pollen and mold allergens, clinical sensitization to algae depends upon the proper exposure conditions to specific organisms.

From a test battery of green algal extracts in atopic patients, a significant number of positive skin reactions (about 60%) occurred (Bernstein and Safferman 1966). This skin reactivity was equated with reaginic-type antibodies by passive transfer to normal recipients using serum of sensitive donors. Direct provocation of several of these atopic individuals with dilute solutions of aerosolized algal allergens caused positive bronchial mucosal responses that included instances of severe wheezing. Subsequently the scope of challenge testing was augmented by using the nasal provocation route (Figures 6.3 and 6.4), and these results were compared with positive skin tests and in vitro leu-

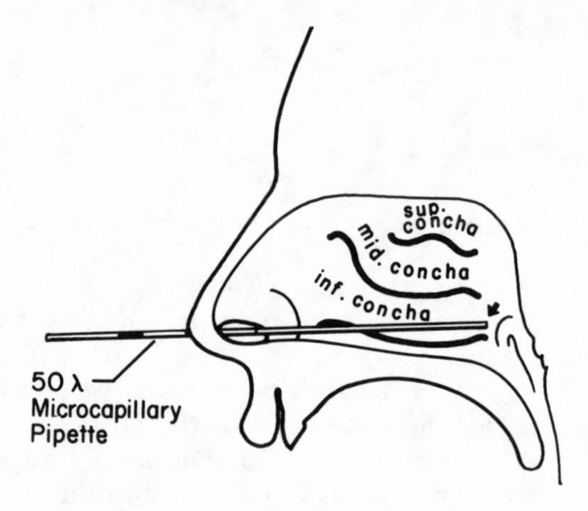

Figure 6.3. Nasal challenge test. Arrow indicates mucosal deposition site of nasal challenge.

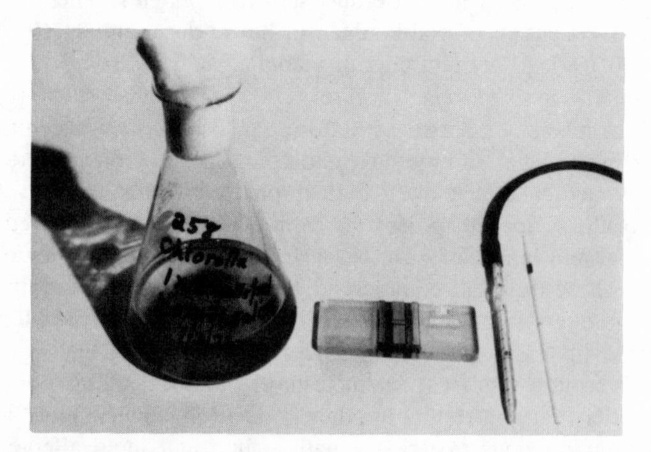

Figure 6.4. Materials (from L to R, algal suspension, cell-counting chamber, standard diluting pipette, microcapillary pipette) for the nasal challenge test.

cocyte histamine release tests (Bernstein and Safferman 1973). A general correlation between nasal responses and the degree of skin sensitization was observed. Consistent with previous bronchial provocation experiences, several patients with strong skin reactivity failed to show positive nasal response. When such patients were studied further, it was demonstrated that their leucocytes also failed to release histamine upon algal allergen challenge in vitro. On the other hand, patients with positive skin and leucocyte histamine release tests almost invariably gave significant nasal responses (Figure 6.5). Despite some of these problems in the interpretation of positive skin tests, they are considered to be useful screening procedures in determining the relationship of allergic symptoms and exposure to various types of algal organisms.

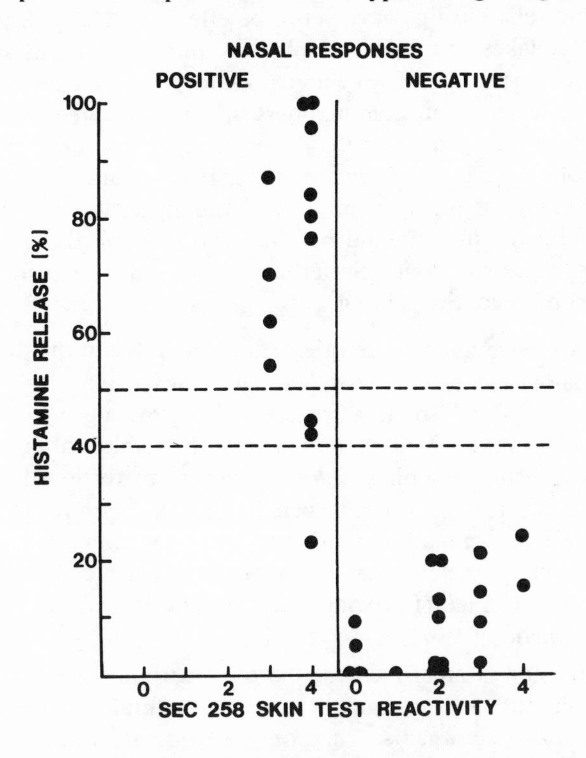

Figure 6.5. Comparison of skin reactivity and leucocyte histamine release induced by *Chlorella* sp. Three quadrants formed by dotted lines (0%–40%, 40%–50%, and 50%–100%) indicate negative, moderate, and high leucocyte histamine release, respectively.

Quantitation of exposure

Useful exposure data may be obtained by challenging mucosal surfaces of suspected allergic individuals with known amounts of allergenic substances. Such tests can be interpreted properly only by using objective measuring devices and randomization of test samples. We found that nasal tests performed with either whole or extracted algal preparations are simple, safe, and inexpensive diagnostic procedures (Figures 6.3 and 6.4). In addition, they enable us to compare the minimal thresholds of nasal allergenicity with airborne concentrations of algae in specific localities. Some patients, for example, have reacted to as few as five algal cells/ml.

Algal emission, dispersion, and deposition

The component elements of the green algal aerobiological model and their functional relationships have yet to be effectively defined. Inland bodies of water such as lakes and ponds are obvious sources of potentially allergenic species of algae, especially if accelerated eutrophication occurs. Not as well appreciated is the fact that large numbers of algae are present in soil (Smith 1944), from which they may be disseminated by wind currents and thermal updrafts. Viable algae in house dust most likely originate from contamination of the indoor environment with soil or airborne algae. We had also suggested that indoor aerated aquaria could be ideal pathways of dispersion for many species of algae (Bernstein and Safferman 1971). This interesting possibility was recently confirmed (Schlichting 1974).

Recently we have undertaken other studies relative to the dispersion and deposition compartments of atmospherically transported algae. From these studies we were able to isolate a variety of airborne algae by using a high-volume cascade impactor in conjunction with viable cultural techniques. Eukaryotic organisms including *Chlorella* spp. were readily detected in this system. Figure 6.6 reveals seasonal peaks in the airborne dispersion pattern of viable algae in Cincinnati, Ohio. When these data were converted to algal impactions per cubic meter of air, numerical estimates of aerial cell density were in the general range of previous air sampling from other locales (Brown et al. 1964, Schlichting 1964).

The dispersion compartment of the algal aerobiological model is more difficult to define in precise terms because the natural sources and emission characteristics of these organisms can only be estimated. Several years ago a small eutrophic lake in the northern sector of Cincinnati was assumed to be a source of algal dissemination. Analysis of skin test surveys in 2000 patients showed an incidence of positive reactions, ranging from 30% to 40% in those patients residing within a radius of 4.8 km downwind from the point source, whereas the range of positive reactions was lower (10%–25%) in other localities upwind from the source. Although these data revealed certain trends, the environmental complexity of possible algal sources does not permit clear-cut interpretation at this time. In the future, however, this approach could be refined to compare sampling data at multiple depot sites downwind from the point source(s) with clinical manifestations in those areas.

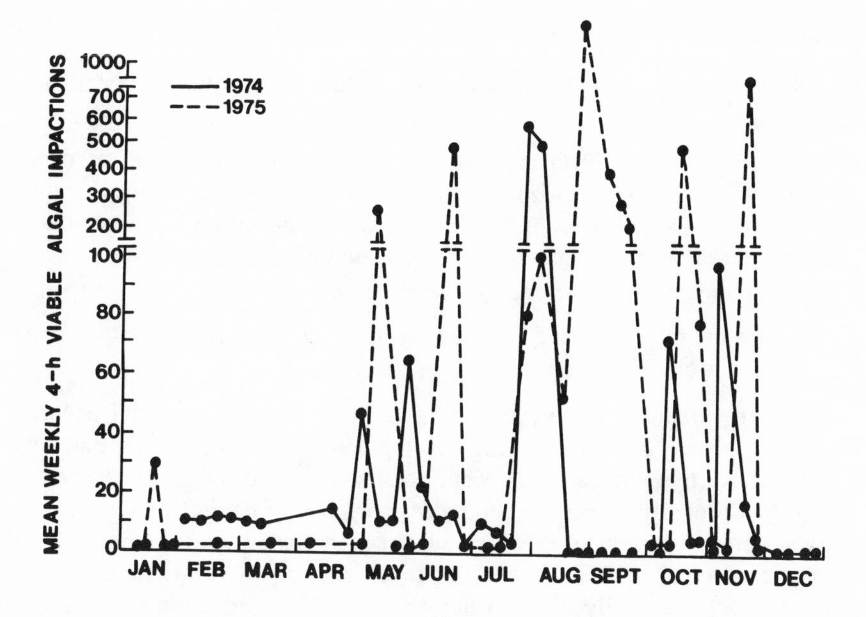

Figure 6.6. Mean weekly viable algal impactions, April through August 1974. Sampling was conducted for 4 h twice weekly, using a cascade impactor and a 0.22-μm Millipore filter with air pressure calibrated to deliver 28 m^3 of air through the inlet during the sampling period.

Summary

We have identified a population of patients who appear to be clinically sensitive to green algae. The techniques described should enable more meaningful correlation between clinical sensitivity and predominating organisms in specific localities. This clinical model could also serve a unique purpose in defining the biomedical impact of algae in geographic areas adjacent to bodies of water that may be undergoing eutrophication. Indeed, combined studies of aerial flora, skin sensitization, nasal challenge responses, and in vitro tests of immediate hypersensitivity in an atopic population should provide a new dimension of defining ecological changes on a sequential basis. The allergenic potential of many forms of green algae may be as significant as some of the commonly accepted aeroallergens.

Bacteria and actinomycetes

Despite an extensive medical literature on the public health impact of respiratory infections, the exact role of airborne bacteria as direct causes of allergy is as yet unresolved. The following mechanistic proposals summarize the current hypotheses of bacterial allergy:

1. Bacterial aerosols are true inhalant allergens.
2. Bacterial agents are pathogenetic by induction of a state of delayed hypersensitivity.
3. Bacteria or their products (exoproteases, endotoxins) are adjuvants, thereby enhancing sensitization by other allergens or creating a state of increased responsiveness of human airways to various chemical mediators. The latter effect has been specifically equated with blockade of beta-adrenergic receptors in respiratory tissues.
4. Bacteria cause nonspecific stimulation of hypersensitive respiratory receptors.

Failure to identify clinical commonality of specific bacterial agents in respiratory allergy has impeded the development of suitable models that could test the above hypotheses. The recent documentation of occupational sensitization by a proteolytic enzyme produced from *Bacillus subtilis,* however, represents a unique system from which such a model could be developed (Flindt 1969, Franz et al. 1971). Individuals sensitized to dust containing this bacterial product eventually develop hay-fever-like and asthmatic symptoms on exposure, which can be corroborated by the presence of specific IgE in the serum as well as positive respiratory challenge and skin tests. Atopic persons are most likely to manifest such immediate hypersensitivity reactions, but under proper conditions of skin exposure to this product any person may develop contact (delayed) sensitization (Ducksbury and Dave 1970). Because the majority of cases occurred in industrial plants incorporating enzymes into household laundry detergents, preventive measures could be instituted by prilling enzyme dust particles and introducing closed transport and mixing systems. Before these measures could be applied on a wide scale nonoccupational sensitization was observed in a small number of people living in close proximity to enzyme detergent plants (Izraylet and Feoktistova 1970), and in some consumers whose only exposure was to subtilisin enzymes contained in packaged detergent products (Belin et al. 1970, Shapiro and Eisenberg 1971, Bernstein 1972). Current occupational and public health hazards, however, have been effectively neutralized by an industry-wide control program that includes serial skin test and pulmonary function surveys in workers and periodic aerobiological monitoring of the plant environment (Weil et al. 1971).

Spores of several species of thermophilic actinomycetes have been demonstrated to induce a unique type of immunological lung damage termed *hypersensitivity pneumonitis*. This association was first recognized in farmers who developed characteristic pulmonary symptoms (farmer's lung) after exposure to moldy hay containing the spores of *Micropolyspora faeni* (Dickie and Rankin 1958). Four to six hours after exposure patients develop cough, dyspnea, chills, fever, and malaise. If there is chronic exposure to the antigen, weight loss and inanition is evident. Physical examination reveals rales at the lung bases while wheezing, usually associated with IgE-mediated asthma, is not present. Serum-precipitating antibodies to actinomycete extracts can be demonstrated in the majority of cases. The histological appearance of the lungs in these diseases, however, is more compatible with delayed hypersensitivity responses than the pathology associated with tissue deposition of antigen-antibody complexes. A similar disease has been reported in sugarcane workers who may be exposed to stored stalks infested with *Micropolyspora faeni* and *Thermoactinomyces vulgaris* as well as *T. sacchari* (Salvaggio et al. 1967). More recently, sources other than vegetable dusts have been discovered to be contaminated with these organisms. Biological aerosols emanating from home and office humidification (Tourville et al. 1972) and air-conditioning units (Banaszak et al. 1970) have been implicated as causes of hypersensitivity pneumonitis in a number of patients. The extent and significance of these non-agricultural sources of thermophilic organisms will depend on increased awareness of this association by members of the medical community.

Arthropods

Insects

Human allergic reactions to arthropods may be induced by: (1) inhalation of "dust" (i.e., exoskeletal scales, hairs, and decayed fragments); (2) injection of venoms of stinging insects; and (3) absorption of salivary secretions of biting insects through the skin. In this discussion only the first type of reaction will be considered relevant to aerobiology.

Seasonal epidemics of allergic respiratory symptoms in susceptible individuals may occur after exposure to large numbers of swarming insects. Such allergic responses have been documented in atopic persons residing near the natural habitats of caddis flies (Trichoptera; Parlato 1929) and mayflies (Ephemeroptera; Figley 1929). The extent and severity of clinical hypersensitivity to allergens derived from these insects can usually be corroborated by positive direct skin tests with whole-body aqueous extracts.

A variety of other insects have been implicated as causes of human respiratory allergy (Feinberg et al. 1956, Wiseman et al. 1959, Perlman 1961). Representative species include housefly, moth, carpet beetle, cockroach, aphid, cricket, mosquito, and weevil. It should be emphasized, however, that evidence for the association of these insects with clinical sensitivity is circumstantial as it is based almost entirely on positive skin test results. A notable exception is the cockroach; positive bronchial responses have been demonstrated in patients with positive skin responses to an extract of this insect (Bernton et al. 1969).

Insect fragments are commonly observed on impaction devices used for pollen sampling. As yet no adequate method has been devised for either quantitatively investigating dispersion and dissemination of insect particles or verifying species by morphologic characteristics of these particles. Obviously, similar constraints are applicable in testing the hypothesis that finely disintegrated insect end products are commingled with soil and indoor dusts. It is therefore understandable that the current status of inhalant allergy to nonswarming insects is highly speculative.

Acarids

In recent years there has been considerable interest in the role of household acarids as possible etiologic agents in human allergy (Voorhorst et al. 1967). The two species studied most extensively are the European house-dust mite, *Dermatophagoides pteronyssinus,* and the North American house-dust mite, *Dermatophagoides farinae* (Wharton 1970). They thrive best in dwellings of man where the milieu of mattresses, pillows, carpets, and overstuffed furniture apparently affords optimum conditions of humidity and food sources —especially human dander. The number of mites in house dust collected from these various sites may range from a few to greater than 200/0.1 g dust (van Bronswijk 1971). Mite counts are generally higher in those articles of household furniture used more frequently (Mitchell et al. 1969). Survival of mites is directly dependent on their ability to absorb water from ambient air (Arlian and Wharton 1974). Thus geographic distribution of these arthropods is primarily in temperate, humid climates.

Workers in Europe and Japan have reported a high degree of correlation between skin sensitivity reactions induced by mite and dust extracts (Voorhorst et al. 1967, Miyamoto et al. 1968). Investigations designed to confirm these results in the United States have resulted in equivocal findings (Kawai et al. 1972). It is possible that these differences reflect climatological factors affecting the prevalence of the household mite. Van Bronswijk and Sinha (1971) indicated that in the Netherlands the seasonal fluctuation in density of *D. pteronyssinus* could be correlated with seasonal temperature and humidity changes. Moreover, there was also a significant relationship between the density of mites in house dust and allergic symptoms in inhabitants of these homes. Obviously, the relative importance of bacteria, molds, algae, and mites as house dust allergens in various parts of the world is a complicated ecological problem, the ultimate resolution of which will require a multidisciplinary approach along the lines fostered by the U.S. Aerobiology Program of the IBP.

Vegetable and wood dusts

Respiratory disease in grain sifters and dressers of flax and hemp was first recognized by Ramazzini (1713). With few exceptions, health hazards due to vegetable and wood dusts are occupationally related. In the United States, byssinosis is the most prevalent problem in this category of diseases (Ayer

1971). It occurs chiefly in workers exposed to dust in the cotton textile, baking, grain, bean and seed, coffee, and wood-processing industries.

Biomedical effects of certain vegetable pollutants have been reported in atopic individuals residing in close proximity to the sources of these contaminants (Figley and Elrod 1928). Similar effects, however, were not substantiated in an investigation of recurrent episodes of asthma in the greater metropolitan area of New Orleans despite the fact that many biogenic dusts are prevalent in that community (Salvaggio et al. 1971).

THE ROLE OF BIOLOGICAL MATERIALS IN ATMOSPHERIC PRECIPITATION PROCESSES*

This section briefly summarizes aspects of atmosphere-biosphere interactions that are perhaps not part of the conventional definitions of aerobiology. Emphasis is laid here on the active roles biological matter may play in atmospheric processes. What is meant by *active role?* In brief, it refers to the now demonstrated possibility that biogenic materials are not only transported passively through the atmosphere, but that various components of this atmospheric biomass actively modify cloud and precipitation processes. By doing so, they indirectly modify the transport processes also. Thus we have an extremely complex system, abundant with feedback mechanisms. Only the very first manifestations of this system have yet been revealed.

In some respects, the topic discussed here falls between the disciplines of atmospheric chemistry and aerobiology, and has a strong tie with cloud physics. No details will be mentioned here of the multitudinous ways in which the biosphere influences and is influenced by the composition of the atmosphere. Treatises on atmospheric chemistry (Junge 1963, 1972a, Rasool 1973, among others) contain ample references on this subject. The point is worth adding, though, that there are indications, described later, for a significance of biogenic materials to cloud and precipitation processes considerably beyond what might be expected from their concentrations as trace atmospheric components. In a very general sense, this importance may be ascribed to the propensities of biogenic materials to interact with water—the dominant constituent of clouds and of precipitation.

The main interactions to be discussed here are the capacities of biogenic substances to be active as (1) condensation nuclei, (2) ice nuclei, (3) coalescence centers, and (4) surface-active agents. All these interactions influence the formation and evolution of clouds. Since cloud formation is almost exclusively a phenomenon of the troposphere, attention will be restricted here to the troposphere, even if the more general term, *atmosphere,* is used.

---

*G. Vali

A distinction is made in cloud physics between particles that catalyze the condensation of water droplets from the vapor at high supersaturations ($>200\%$), and those that do so at very low supersaturations ($<3\%$). The former are referred to as *Aitken nuclei* (AN), the latter are called *cloud-condensation nuclei* (CCN). Normally, only CCN are of direct importance for the formation and development of clouds. Aitken nuclei are indirectly important as sources and modifiers of other aerosols.

Aitken nuclei are fairly numerous: concentrations vary from less than $10^2/cm^3$ in very clean tropospheric air to over $10^5/cm^3$ in polluted urban air. Particle size can be as small as 10^{-2} μm; practically all particles, independent of their chemical composition, are able to act as AN.

The predominant mechanism for the production of small aerosol particles, and hence of AN, is by condensation from the vapor phase. Thus the composition of AN is primarily determined by the reactivity of the vapors and by the presence of sunlight to catalyze the condensation of particles. Organic vapors are thus natural candidates for the formation of AN. That this is not only speculation has been clearly demonstrated by the work of Went (1964, Went et al. 1967) and of Paugam (1975). The experiments of Went have shown that in remote locations the exhalations of vegetative matter are the primary sources of AN. In his own words:

> In conclusion we can say that both in a desert and in a montane forest, where no air pollution through human activities existed, large numbers of solid condensation nuclei are formed every day. This is a photochemical process in which volatile plant products, terpenes, are activated to agglomerate into Aitken particles of less than 10^{-5} cm in diameter. In the course of hours these grow into particles of more than 2×10^{-5} cm diameter. From this same unpolluted air can be filtered a black carbonlike material consisting of submicroscopic particles that cluster together into larger microscopic or even macroscopic masses, which blacken filter paper. The only logical interpretation of these two sets of data is that in the photochemical condensation process of molecularly dispersed terpenes in the atmosphere, carbonlike so-called "combustion nuclei" are formed which cluster together into microscopically visible condensation nuclei.

A somewhat less general but nonetheless significant finding is that of Paugam (1975) who showed that, in coastal areas, organic materials emitted by drying algae are abundant sources of AN. Although there is no definite evidence for it, the possibility clearly exists that open ocean surfaces also produce AN through the abundant organic films and the numerous sea-air transport processes.

Our knowledge of the particle sizes and composition of CCN is fairly meager. It is known from theory and from laboratory experiments that particle size and chemical composition have to combine in much more critical ways to produce CCN than those required for AN. Large particle size and hygroscopicity or wettability are the general requirements for active CCN. While sea salt was long thought to be the most prolific source of CCN, recent evidence does not support that view. The work of Dinger et al. (1970) and of Twomey (1971) shows that sea salt is in fact only a minor contributor to CCN. Volatility tests are the major source of information in this regard. The observations support the likelihood that ammonium sulfate particles constitute the majority of CCN, but the evidence is not very specific. Consequently, no assessment can be made of the contribution that may be made by biogenic materials to the populations of atmospheric CCN, but the possibility that some contribution is made seems to be there.

One specific example of biological materials that could participate in cloud formation through condensation activity was described by Dingle (1966). Drawing upon evidence from rain scavenging studies and referring to the works of Durham (1943) and of Harrington and Metzger (1963), Dingle argued that pollen grains can serve as effective CCN. Confirmation of this assertion from more direct studies has not been forthcoming, but the suggestion nonetheless deserves attention.

Another chain of events may also lead to biogenic CCN. Recent studies by Jiusto and Lee (1975) reemphasize the possibility that CCN are formed by the coagulation of AN, at least in urban atmospheres. Assuming that their point is more generally valid, supporting that assumption by the general observation that large fractions of particles of the size of CCN are produced by coagulation of smaller particles, and combining these facts with the previously mentioned evidence for biogenic sources of AN, one may surmise that a larger fraction of CCN have their origins in biological matter than is possible to demonstrate at present.

ICE NUCLEI

If ice particles and water droplets coexist in a cloud, the ice particles grow rapidly (the vapor pressure over ice being lower than over water). This provides a relatively rapid means of developing precipitation-size particles. This so-called ice-phase process of precipitation development is thought to be responsible for a great part of natural precipitation that occurs over middle and northern latitudes, not only in winter but throughout the year. The actual processes of growth are diffusion from the vapor and the accretion of water droplets onto the ice particles when the ice crystals become large enough to begin to fall.

The importance of the ice-phase or Bergeron—Findeisen process in meteorology has generated keen interest in the nuclei that are capable of initiating the formation of ice crystals. While it is now recognized that secondary ice-generating processes may change things away from an expected one-to-one correlation between nuclei and crystals, the importance of ice nuclei is still held to be very great.

Ice nuclei constitute a very small fraction of the total atmospheric aerosol. Their concentrations range from 1 to 1000/m^3. There is as yet no clear evidence on what distinguishes ice nuclei from other aerosol particles except for the very general knowledge that rather subtle physicochemical surface characteristics govern ice nucleating ability. From experiments with large varieties of substances and from examination of the particulate contents of ice crystals, the consensus gradually emerged over several decades that natural ice nuclei are most likely to be clay particles of 0.1 μm and larger. While this view appears to be the most tenable, the literature abounds with exceptions and contradictions.

Evidence that biological materials may also act as atmospheric ice nuclei was first given by Soulage (1957), who found bacterial cells at the centers of ice crystals. Rosinski and Parungo (1966) showed that natural plant oils can produce moderately active ice nuclei. Vali (1968) demonstrated that soils that contain large fractions of organic materials are better sources of freezing nuclei than the clay constitutents of the soils are by themselves. Pursuing this finding, Schnell and Vali (1972) showed evidence that production of ice nuclei accompanies the decomposition of tree leaves, for example. Fresh (1973) discovered that the development of these nuclei depended on the presence of bacteria, which were subsequently isolated and grown in pure culture. Maki et al. (1974) identified the bacteria involved as *Pseudomonas syringae;* later work by Maki (private communication) showed the existence of a few other species of bacteria capable of ice-nucleating activity. In all cases, the ability to nucleate ice seems to be associated with the bacterial cells themselves.

A review of the research on biogenic ice nuclei is given in recent papers by Schnell and Vali (1976) and by Vali et al. (1976). The highlights of the results are the following: (1) decayed plant leaf litters contain copious numbers of ice nuclei; (2) the abundance of nuclei in litters varies according to the climate of the plant's origin; and (3) active ice nuclei are also found in seawaters rich in phytoplankton. These findings raise the possibility that a considerable portion of natural atmospheric ice nuclei derive from biogenic materials. There are several sets of measurements that support this assertion but it will require further extensive research to demonstrate the real extent of the biogenic contribution to atmospheric ice nuclei.

COALESCENCE CENTERS

The development of rain by the gradual coagulation of cloud droplets into larger raindrops is a well-demonstrated process, often called the *warm rain* process, in distinction to the growth of precipitation elements through the *ice-phase* process. Calculations of coalescence rates (Berry 1967, Cotton 1972, Leighton and Rogers 1974, and others) clearly show that the critical step is the formation of a few droplets that are much larger than the most common cloud droplet size. It is one of the central issues of cloud physics research to establish how these initial large droplets are generated. One of the solutions is to postulate that "supergiant" particles (>50 μm will generally suffice) are present along with the populations of CCN in the cloud-forming air. The concentrations of giant particles necessary to start coalescence, and thus to upset the colloidal stability of a cloud, are generally considered to be in the range of 10–100/m^3. Jaenicke and Junge (1967) and Junge (1972b), among others, reported observations of comparable concentrations. Unfortunately, the abundances of giant particles are difficult to measure, and there is therefore no clear evidence for judging the generality of those observations. The giant particles are generally believed to be composed of soil materials detached from the ground by wind erosion. The sizes and concentrations of pollen, however, are often also sufficient to be of significance in initiating coalescence. Direct empirical verification of this role for pollen is not yet available and will have to come from extensive sampling and examination of the aerosol populations in air in which cloud formation is imminent (convective updrafts, for example). The occurrence of pollen in rain collected at the ground is well known but cannot be taken as evidence for the initiation of the coalescence process, since the contribution of passive scavenging by falling precipitation cannot be readily separated from the active input. Deductions that can be made from the dependence of pollen concentrations on rain rate tend to indicate that passive removal dominates. That does not prove, however, that initiation of coalescence does not also contribute, and there certainly may be situations where the presence of pollen would be crucial to initiating rain.

SURFACE-ACTIVE MATERIALS

There is extensive evidence (Garrett 1965, for example) for the existence of a thin layer of organic substances at the surfaces of oceans and also of lakes and rivers. Through the mechanisms of bubble bursting, and to lesser degrees by other processes, these organic materials produce some aerosols of wholly organic composition and coat with an organic layer all other particulates emerging from the water surface. The processes are well documented and are discussed in detail by Blanchard (1975).

Soil particles lifted from the earth's surface into the atmosphere also tend to carry organic coatings. Evidence is mounting for the ubiquity of this phenomenon (Delany and Zenchelsky 1976).

Aerosol production at ocean and soil surfaces together accounts for well over half (by mass) of the total atmospheric aerosol. When these particles become incorporated into water droplets, through nucleation or through capture, then the soluble fractions of the organic coatings are transferred from the particle surfaces to the drop surfaces. The presence of such surface-active layers has two effects, both of some importance for the evolution of the droplets into precipitation.

First, the kinetics of condensation and evaporation are altered. Retardation as well as enhancement of evaporation rates have been observed (e.g., Bigg et al. 1969, Snead and Zung 1968, Hughes and Stampfer 1971). Alterations of the activities of condensation nuclei have also been reported (e.g., Leonov et al. 1971, Beitel and Heideger 1971). The possible impact of these effects on the evolution of cloud droplet spectra has been studied by Podzimek and Saad (1975). They showed that the effect can be considerable, but concluded that there is not enough background information available to predict in which direction influences might be expected in natural clouds.

Second, the presence of surface-active films on water drops alters the coalescence efficiencies of the drops. Coalescence efficiency is a parameter that describes the probability of fusion between drops upon coming into contact with one another. List and Whelpdale (1969) showed that relatively small changes in surface tension can profoundly affect coalescence efficiency. Thus it is to be expected that the presence of surface-active agents would lead to changes in the rate of coalescence of droplets in clouds and hence to changes in the rate of rain formation.

In summary, organic coatings of aerosol particles, derived from biological activities in oceans and in soil, can alter the activities of CCN or, when transmitted to cloud droplets, can influence the evolution of the droplets into precipitation. Once again, determination of the extent to which these interactions are of importance must await much further research.

CONCLUSIONS

There is, admittedly, an undertone of tentativeness in much of what has been said here. A great deal of detail has been left out, for the sake of brevity. The real reasons of uncertainties are, of course, the relative newness of the ideas involved and the complexities of the topics. The basic elements of the interactions are not in question. What is missing is the demonstration of just how much the interactions modify atmospheric processes. Even if it turns out that the contributions of biogenic condensation and ice nuclei to the formation of atmospheric precipitation and the other influences of biological materials on precipitation processes are only minor, the very existence of such direct feedback mechanisms between biospheres and atmosphere should warrant their further scientific study.

FOOTNOTE

[1]A few specialized seed plants use other pollination agents, notably water (hydrophily). See Faegri and van der Pijl (1971) for detailed discussion. Some species avoid the problems of pollen transfer entirely, through self-pollination (cleistogamy), consequently also losing the advantages of sexual reproduction discussed by Williams (1975).

LITERATURE CITED

Aisner, J., S. C. Schimpff, J. E. Bennett, V. M. Young, and P. H. Wiernik. 1976. Aspergillus infections in cancer patients: Association with fireproofing materials in a new hospital. J. Am. Med. Assoc. 235:411–412.

Ajello, L. 1971. Distribution of *Histoplasma capsulatum* in the United States. Chapter 15 (pages 103–122) *in* Histoplasmosis: Proceedings of the Second National Conference.

Alexakis, P. G., P. G. Feldon, M. Wellisch, R. E. Richter, and S. M. Finegold. 1976. West. J. Med 124:361–369.

AHA (American Hospital Assoc., Committee on Infections within Hospitals). 1974. Prevention and control of infection. Chapter 5 (pages 69–117) *in* Infection control in the hospital, 3rd ed. American Hospital Assoc., Chicago.

Anderson, K. 1959. *Pseudomonas pyocyanea* disseminated from an air-cooling apparatus. Med. J. Aust. 1:529.

Arlian, L. G., and G. W. Wharton. 1974. Kinetics of active and passive components of water exchange between the air and a mite, *Dermatophagoides pteronyssinus*. J. Insect Physiol. 20:1063.

Artenstein, M. S., W. S. Miller, T. H. Lomson, and B. L. Brandt. 1968. Large volume air sampling for meningococci and adenoviruses. Am. J. Epidemiol. 87: 471–483.

Ayer, H. E. 1971. Byssinosis. Chapter 2 (p. 207) *in* R. G. Bond and C. P. Straub, eds. Critical reviews in environmental control. CRC Press, Cleveland, Ohio.

Ayliffe, G. A. J., J. R. Babb, and B. J. Collins. 1974. Dispersal of *Staphylococcus aureus*. Lancet 2:1573.

Ayliffe, G. A. J., and B. J. Collins. 1967. Wound infections acquired from a dispenser of an unusual strain of *Staphylococcus aureus*. J. Clin. Pathol. 20:195–198.

Babudieri, B. 1953. Epidemiology, diagnosis and prophylaxis of Q fever. Pages 157–174 *in* Advances in the control of zoonoses. Monogr. Ser. 19, World Health Organ., Geneva.

Baer, H., C. J. Maloney, P. S. Norman, and D. G. Marsh. 1974. The potency and group 1 antigen content of six commercially prepared grass pollen extracts. J. Allergy Clin. Immunol. 54:157.

Banaszak, E. F., W. H. Thiede, and J. N. Fink. 1970. Hypersensitivity pneumonitis due to contamination of an air conditioner. N. Engl. J. Med. 283:271.

Barnes, E. H. 1964. Changing plant disease losses in a changing agriculture. Phytopathology 54:1314–1319.

Bate, J. G., and U. James. 1958. *Salmonella typhimurium* infection dustborne in a children's ward. Lancet 2:713–715.

Beitel, A., and W. J. Heideger. 1971. Surfactant effect on mass transfer. Chem. Eng. Sci. 26:711.

Belin, L. 1972. Separation and characterization of birch pollen antigens with special reference to the allergenic components. Int. Archiv. Allergy Appl. Immunol. 42:329.

Belin, L., J. Hoborn, F. Falsen, and J. André. 1970. Enzyme sensitization in consumers of enzyme-containing washing powder. Lancet 2:1153.

Bennett, J. E. 1974. Prevention of hospital-acquired systems mycoses. Prev. Med. 3:515–516.

Bernstein, I. L. 1972. Enzyme allergy in populations exposed to long-term, low-level concentrations of household laundry products. J. Allergy 49:219.

Bernstein, I. L., and R. S. Safferman. 1966. Sensitivity of skin and bronchial mucosa to green algae. J. Allergy 38:166.

Bernstein, I. L., and R. S. Safferman. 1970a. Viable algae in house dust allergies. Nature (Lond.) 227:851.

Bernstein, I. L., and R. S. Safferman. 1970b. Algae and house dust allergies. Excerpta Med. Int. Congr. Ser. 211:5.

Bernstein, I. L., and R. S. Safferman. 1971. Sampling living particulates in the atmosphere. Pages 106–108 *in* W. S. Benninghoff and R. L. Edmonds, eds. Aerobiology objectives in atmospheric monitoring. US/IBP Aerobiology Program Handb. 1. Univ. Michigan, Ann Arbor.

Bernstein, I. L., and R. S. Safferman. 1973. Clinical sensitivity to green algae demonstrated by nasal challenge and *in vitro* tests of immediate hypersensitivity. J. Allergy Clin. Immunol. 51:22.

Bernstein, I. L., G. V. Villacorte, and R. S. Safferman. 1969. Immunologic responses of experimental animals to green algae. J. Allergy 43:191.

Bernton, H. S., T. F. McMahon, and H. Brown. 1969. The German cockroach (*Blattela germanica-Linnaeus*)—An excitant of bronchospasm. J. Allergy 43:158.

Berrens, L. 1971. The chemistry of atopic allergens. Volume 7 *in* Monographs in allergy. S. Karger, Basel.

Berry, E. X. 1967. Cloud droplet growth by coalescence. J. Atmos. Sci. 24: 688–701.

Bice, D. E., J. Salvaggio, and E. Hoffman. 1974. Passive transfer of experimental hypersensitivity pneumonitis with lymphoid cells. J. Allergy Clin. Immunol. 55:71. (Abstr.)

Bigg, E. K., J. L. Brownscombe, and W. J. Thompson. 1969. Fog modification with long-chain alcohols. J. Appl. Meteorol. 8:75.

Blanchard, D. C. 1975. Bubble scavenging and the water-to-air transfer of organic material in the sea. Adv. Chem. Ser. 145:360–387.

Blanchard, D. C., and L. Syzdek. 1970. Mechanisms for the water-to-air transfer and concentration of bacteria. Science 170:626.

Bourdillion, R. B., and L. Colebrook. 1946a. Air hygiene in dressing-rooms for burns or major wounds. Lancet 1:561–565.

Bourdillion, R. B., and L. Colebrook. 1946b. Air hygiene in dressing-rooms for burns or major wounds. Lancet 1:601.

Brachman, P. S. 1971. Nosocomial infection—Airborne or not? Pages 189–192 *in* Proc. Int. Conf. Nosocomial Infect., August 1970, Atlanta, Ga. American Hospital Assoc., Chicago.

Brachman, P. S. 1974. Nosocomial respiratory infections. Prev. Med. 3:500–506.

Brown, R. M., Jr., D. A. Larson, and H. C. Bold. 1964. Airborne Algae: Their abundance and heterogeneity. Science 143:583–585.

Browne, A. J., E. A. Ryan, F. J. Glassow, C. J. Martin, and E. E. Shouldice. 1959. Staphylococcic wound infections: Study of wound infections in several thousand hernia cases. J. Am. Med. Assoc. 170:1274–1283.

Bruce, R. A. 1963. Bronchial and skin sensitivity in asthma. Int. Archiv. Allergy Appl. Immunol. 22:294.

Burgess, A. F. 1913. The dispersion of the gypsy moth. USDA Bur. Entomol. Bull. 119. 62 p.

Busse, W. W., C. E. Reed, and J. H. Hoehne. 1972. Where is the allergic reaction in ragweed asthma? II. Demonstration of ragweed antigen in airborne particles smaller than pollen. J. Allergy Clin. Immunol. 50:289.

Cain, S. A., and L. G. Cain. 1944. Size-frequency studies of *Pinus palustris* pollen. Ecology 25:229–232.

Cain, S. A., and L. G. Cain. 1948a. Size-frequency characteristics of *Pinus echinata* pollen. Bot. Gaz. 110:325–330.

Cain, S. A., and L. G. Cain. 1948b. Palynological studies at Sodon Lake. II. Size-frequency studies of pine pollen, fossil and modern. Am. J. Bot. 35:583–591.

Charnley, J. 1972. Postoperative infection after total hip replacement with special reference to air contamination in the operating room. Clin. Orthop. 87:167–187.

Charnley, J. 1973. Clean air in the operating room. Cleveland Clin. Q. 40:99–114.

Charnley, J., and N. Eftekhar. 1969. Postoperative infection in total prosthetic replacement arthroplasty of the hip joint, with special reference to the bacterial content of the air of the operating room. Br. J. Surg. 56:641–649.

Chiykowski, L. N., and R. K. Chapman. 1965. Migration of the six-spotted leafhopper, *Macrosteles fascifrons* (Stal.). 2. Migration of the six-spotted leafhopper in central North America. Wisc. Agric. Exp. Stn. Res. Bull. 261:21–45.

Clayton, Y. V., and W. C. Noble. 1966. Observations on the epidemiology of *Candida albicans*. J. Clin. Pathol. 19:76–78.

Cockroft, W. H., and F. R. C. Johnstone. 1964a. Asepsis in the operating theatre. Can. Med. Assoc. J. 90:659–662.

Cockroft, W. H., and F. R. C. Johnstone. 1964b. Asepsis in the operating theatre. Can. Med. Assoc. J. 90:1091.

Cohen, S. G., and C. B. Reif. 1953. Cutaneous sensitization to blue-green algae. J. Allergy 24:452.

Collins, C. W. 1917. Methods used in determining wind dispersion of the gypsy moth and some other insects. J. Econ. Entomol. 10:170–177.

Collins, C. W., and W. L. Baker. 1934. Exploring the upper air for wind-borne gypsy moth larvae. J. Econ. Entomol. 27:320–327.

Connell, J. T. 1969. Quantitative intranasal pollen challenges. III. The priming effect in allergic rhinitis. J. Allergy 43:33.

Cooke, W. B., and P. W. Kabler. 1956. The survival of *Histoplasma capsulatum* in water. Pages 261–264 *in* Public Health Tech. Monogr. 39, Proc. Conf. Histoplasmosis, 1952.

Cotton, W. R. 1972. Numerical simulation of precipitation development in supercooled cumuli—Part 1. Mon. Weather Rev. 100:757–763.

Couch, R. G., R. B. Douglas, Jr., K. M. Lindgren, P. J. Gerone, and V. Knight. 1970. Airborne transmission of respiratory infection with coxsackievirus A type 21. Am. J. Epidemiol. 91:78–86.

Couch, R. B., V. Knight, R. B. Douglas, S. H. Black, and B. H. Hamory. 1969. The minimal infectious dose of adenovirus type 4: The case for natural transmission by viral aerosol. Trans. Am. Clin. Climatol. Assoc. 80:205–211.

Craigie, J. H. 1945. Epidemiology of stem rust in western Canada. Sci. Agric. 25:285–401.

Cramer, H. H. 1967. Plant protection and world crop production. Pflanzenschutz Nachr. 20:524.

Cronquist, A. 1968. The evolution and classification of flowering plants. Houghton-Mifflin Co., Boston. 396 p.

D'Alessio, D. J., R. H. Heeren, S. L. Hendricks, P. Ogilvie, and M. L. Furcolow. 1965. A starling roost as the source of urban epidemic histoplasmosis in an area of low incidence. Am. Rev. Respir. Dis. 92:725–731.

Davidson, A. G., and R. M. Prentice, eds. 1967. Important forest insects and diseases of mutual concern to Canada, the United States, and Mexico. North Am. For. Comm., FAO. Minist. For. Rural Dev. Can. 248 p.

Davies, R. R., and W. C. Noble. 1962. Dispersal of bacteria on desquamated skin. Lancet 2:1295–1297.

Davies, R. R., and W. C. Noble. 1963. Dispersal of staphylococci on desquamated skin. Lancet 1:1111.

Davies, R. R., and L. P. Smith. 1973. Forecasting the start and severity of the hay fever season. Clin. Allergy 3:263.

Davis, D. D. 1973. Air pollution damages trees. USDA For. Serv., NE For. Exp. Stn., Upper Darby, Pa. 16 p.

Davis, D. J. 1972. NIAID initiatives in allergy research. J. Allergy Clin. Immunol. 49:323.

Delany, A. C., and S. Zenchelsky. 1976. The organic component of wind-erosion-generated soil-derived aerosol. Soil Sci. 121:146–155.

Delong, D. M. 1971. The bionomics of leafhoppers. Annu. Rev. Entomol. 16: 179–210.

Dickie, H. A., and J. Rankin. 1958. Farmers lung: An acute granulomatous interstitial pneumonitis occurring in agricultural workers. J. Am. Med. Assoc. 167: 1069.

Dineen, P. 1973. The role of impervious drapes and gowns in preventing surgical infection. Clin. Orthop. 96:210–212.

Dinger, J. E., H. B. Howell, and T. A. Wojciechowski. 1970. On the source and composition of cloud nuclei in subsident air mass over the North Atlantic. J. Atmos. Sci. 27:791–797.

Dingle, A. N. 1966. Pollens as condensation nuclei. J. Rech. Atmos. 2:231–237.

Dixon, R. E. 1973. The role of airborne bacteria in theatre-acquired surgical wound infection. Cleveland Clin. Q. 40:115–123.

Dobzhansky, T. 1973. Active dispersal and passive transport in *Drosophila* (Diptera:Drosophilidae). Evolution 27:565–575.

Dodge, H. J., L. Ajello, and O. Engelke. 1965. The association of a bird roosting site with infection of school children by *Histoplasma capsulatum*. Am. J. Public Health 55:1203–1211.

Dodge, H. J., O. K. Engelke, and K. M. Dohm. 1970. Observations on the decline of histoplasmal infections in school children: Milan, Michigan, 1958–1968. Am. J. Public Health 60:1442–1446.

Dorst, H. E., and E. W. Davis. 1937. Tracing long-distance movements of the beet leafhopper in the desert. J. Econ. Entomol. 30:948–954.

Ducksbury, C. F. J., and V. K. Dave. 1970. Contact dermatitis in home helps following the use of enzyme detergents. Br. Med. J. 1:537.

Durham, O. C. 1943. The volumetric incidence of atmospheric allergens. I. Specific gravity of pollen grains. J. Allergy 14:455–461.

Eames, A. J. 1959. The morphological evidence for a Paleozoic origin of the angiosperms. Recent Adv. Bot. 1:722–726.

Eckman, B. H., G. L. Schaefer, and M. Huppert. 1964. Bedside interhuman transmission of coccidioidomycosis via growth on fomites. Am. Rev. Respir. Dis. 89:175–185.

Edwards, J. H., J. T. Baker, and B. H. Davies. 1974. Precipitin test negative farmer's lung—Activation of the alternate pathway of complement by mouldy hay dusts. Clin. Allergy 4:379.

Ehrenkranz, J. J., and J. L. Kicklighter. 1972. Tuberculosis in a general hospital: Evidence for airborne spread of infection. Ann. Intern. Med. 77:377–382.

Eichenwald, H. F., O. Kotsevalov, and L. A. Fasso. 1961. Some effects of viral infection on aerial dissemination of staphylococci and on susceptibility to bacterial colonization. Bacteriol. Rev. 25:274–281.

Emmons, C. W., and W. R. Piggott. 1963. Eradication of *Histoplasma capsulatum* from soil. Mycologia 55:521–527.

Faegri, K., and L. van der Pijl. 1971. The principles of pollination ecology, 2nd ed. Pergamon Press, New York. 291 p.

Feder, W. A. 1973. Cumulative effects of chronic exposure of plants to low level of air pollutants. Adv. Chem. Ser. 122:21–30.

Feinberg, A. R., S. M. Feinberg, and C. Benaim-Pinto. 1956. Asthma and rhinitis from insect allergens. 1. Clinical importance. J. Allergy 27:437.

Feinberg, S. M. 1946. Allergy in practice, chap. 1, 2nd ed. Year Book Medical Publ., Inc., Chicago.

Feinberg, S. M., and P. L. Aries. 1932. Asthma from food odors. J. Am. Med. Assoc. 98:2280.

Ferry, B. W., M. S. Baddeley, and D. L. Hawksworth. 1973. Air pollution and lichens. Athlone Press, Univ. London, London. 389 p.

Figley, K. D. 1929. Asthma due to the may fly. Am. J. Med. Sci. 178:338.

Figley, K. D., and R. H. Elrod. 1928. Endemic asthma due to castor bean dust. J. Am. Med. Assoc. 90:79.

Finland, M. 1942. Recent advances in the epidemiology of pneumococcal infections. Medicine 21:307–344.

Fisher, A. A. 1973. Contact dermatitis, 2nd ed. Lea and Febiger, Philadelphia.

Flindt, M. L. H. 1969. Pulmonary disease due to inhalation of derivatives of *Bacillus subtilis* containing proteolytic enzyme. Lancet 1:1177.

Fontana, V. J., L. Indyk, and M. Zanjanian. 1974. Ragweed pollen challenges in a controlled environment. J. Allergy Clin. Immunol. 54:235.

Forbes, S. A. 1915. The insect, the farmer, the teacher, the citizen and the state. Bull. Ill. State Lab. Nat. Hist.

Foss, R. J., and S. Saslaw. 1971. Earth Day histoplasmosis: A new type of urban pollution. Arch. Intern. Med. 128:588–590.

Fowells, H. A. 1965. Silvics of forest trees of the United States. USDA Handb. 271. U.S. Dep. Agric., Washington, D.C. 762 p.

Frankland, A. W., and M. J. Hay. 1951. Dry rot as a cause of allergic complaints. Acta Allergol. 4:186.

Franz, T., K. D. McMurrain, S. Brooks, and I. L. Bernstein. 1971. Clinical, immunologic and physiologic observations in factory workers exposed to *B. subtilis* enzyme dust. J. Allergy 47:170.

Fresh, R. W. 1973. Microbial production of freezing nuclei from decomposing tree leaves. Rep. AR106. Dep. Atmos. Resour., Univ. Wyoming, Laramie. 15 p.

Fritts, H. C. 1966. Growth of trees: Their correlation with climate. Science 154:973–979.

Furcolow, M. L. 1965. Environmental aspects of histoplasmosis. Arch. Environ. Health 10:4–10.

Gadgil, M., and O. T. Solbrig. 1972. The concept of *r* and *K* selection: Evidence from wild flowers and some theoretical considerations. Am. Nat. 106:14–31.

Garrett, W. D. 1965. Collection of slick-forming materials from the sea surface. Limnol. Oceanogr. 10:602–605.

Gell, P. G. H., and R. R. A. Coombs. 1968. Clinical aspects of immunology, 2nd ed. Blackwell Scientific, Oxford, England.

Gibbs, J. A. 1969. Plant virus classification. Adv. Virus Res. 14:263–328.

Goldner, J. L., and B. L. Allen. 1975. Ultraviolet light in orthopedic operating rooms at Duke University: Thirty-five years experience, 1937–1973. Clin. Orthop. 96:195–205.

Goodwin, J. E., J. A. McLean, F. M. Hemphill, and J. M. Sheldon. 1957. Air pollution by ragweed: Medical aspects. Fed. Proc., Fed. Am. Soc. Exp. Biol. 18: 628.

Gordon, J. E. 1962. Chickenpox: An epidemiological review. Am. J. Med. Sci. 244:362–389.

Grant, V. 1950. The pollination of *Calycanthus occidentalis*. Am. J. Bot. 37: 294–297.

Graybill, J. R., L. W. Marshall, P. Charache, C. K. Wallace, and V. B. Melvin. 1973. Nosocomial pneumonia: A continuing major problem. Am. Rev. Respir. Dis. 108:1130–1140.

Greenbank, D. O. 1957. The role of climate and dispersal in the initiation of outbreaks of the spruce budworm in New Brunswick. II. The role of dispersal. Can. J. Zool. 35:385–403.

Gregory, P. H., E. D. Hamilton, and T. Sheeramulu. 1955. Occurrence of the alga *Gloeocapsa* in the air. Nature (Lond.) 176:1270.

Gregory, P. H., and J. M. Hirst. 1952. Possible role of basidiospores as airborne allergens. Nature (Lond.) 170:414.

Grzywacz, A. 1971. The influence of industrial air pollution on pathological fungi of forest trees. Sylwan 155:55–62.

Haddad, Z. H., D. G. Marsh, and D. H. Campbell. 1972. Studies on "allergoids" prepared from naturally occurring allergens. II. Evaluation of allergenicity and assay of antigenicity of formalinized mixed grass pollen extracts. J. Allergy Clin. Immunol. 49:197.

Hamburger, M., Jr., and M. J. Green. 1946. The problem of the "dangerous carrier" of hemolytic streptococci. IV. Observations upon the role of hands, of blowing the nose, of sneezing and coughing in the dispersal of these microorganisms. J. Infect. Dis. 79:33–44.

Hamburger, M., Jr., M. J. Green and V. G. Hamburger. 1945. The problem of the "dangerous carrier" of hemolytic streptococci. II. Spread of infection by individuals with strongly positive nose cultures who expelled large numbers of hemolytic streptococci. J. Infect. Dis. 77:96–108.

Hamburger, M., Jr., T. T. Puck, V. G. Hamburger, and M. A. Johnson. 1944. Studies on the transmission of hemolytic streptococcus infection. III. Hemolytic streptococci in the air, floor dust, and bedclothing of hospital wards and their relation to cross infection. J. Infect. Dis. 75:79–94.

Handel, S. N. 1976. Restricted pollen flow of two woodland herbs determined by neutron activation analysis. Nature (Lond.) 260:422–423.

Hanson, R. P., S. E. Sulkin, E. L. Buescher, W. McD. Hammon, R. W. McKinney, and T. H. Work. 1967. Arbovirus infections of laboratory workers. Science 158:1283–1286.

Hare, R. 1964. The transmission of respiratory infections. Proc. R. Soc. Med. 57:221–230.

Hargreave, F. E., J. Pepys, J. L. Longbottom, and D. G. Wraith. 1966. Bird breeder's (fancier's) lung. Lancet 1:445.

Harper, G. J. 1961. Airborne microorganisms: Survival tests with four viruses. J. Hyg. 59:479–486.

Harrington, J. B., Jr., and K. Metzger. 1963. Ragweed pollen density. Am. J. Bot. 50:532–539.

Hart, P. D., E. Russell, and J. S. Remington. 1969. The compromised host and infection. II. Deep fungal infection. J. Infect. Dis. 120:169–191.

Hasenclever, H. F., and W. R. Piggott. 1973. Air sampling for *Histoplasma capsulatum*. Pages 159–162 *in* R. L. Edmonds and W. S. Benninghoff, eds. Ecological systems approaches to aerobiology. II. Development, demonstration, and evaluation of models. US/IBP Aerobiol. Program Handb. 3. (NTIS no. AP-USIBP-H-73-3.) Univ. Michigan, Ann Arbor.

Hasenclever, H. F., and W. R. Piggott. 1974. Colonization of soil by *Histoplasma capsulatum:* Factors affecting its continuity in a given site. Health Lab. Sci. 11:197–200.

Hasenclever, H. F., M. H. Shacklette, R. V. Young, and G. A. Gelderman. 1967. The natural occurrence of *Histoplasma capsulatum* in a cave. 1. Epidemiological aspects. Am. J. Epidemiol. 86:238–245.

Heagle, A. S. 1973. Interaction between air pollutants and plant parasites. Annu. Rev. Phytopathol. 11:365–388.

Heck, W. W. 1973. Air pollution and the future of agricultural production. Adv. Chem. Ser. 122:118–130.

Heggestad, H. E. 1968. Diseases of crops and ornamental plants incited by air pollutants. Phytopathology 58:1089–1097.

Heise, H. A. 1949. Symptoms of hay fever caused by algae. J. Allergy 20:383.

Henson, W. R. 1951. Mass flights of the spruce budworm. Can. Entomol. 83:240.

Herbertson, S., J. Porath, and H. Colldahl. 1958. Studies of allergens from alder pollen (*Alnus glutinosa*). Acta Chem. Scand. 12:737.

Herxheimer, H., H. A. Hyde, and D. A. Williams. 1966. Allergic asthma caused by fungal spores. Lancet 1:572.

Herxheimer, H., H. A. Hyde, and D. A. Williams. 1969. Allergic asthma caused by basidiospores. Lancet 2:131.

Hill, J., A. Howell, and R. Blowers. 1974. Effect of clothing on dispersal of *Staphylococcus aureus* by males and females. Lancet 2:1131–1133.

Hindawi, I. J. 1970. Air pollution injury to vegetation. Natl. Air Pollut. Control Adm. (EPA) Publ. AP-71. U.S. Gov. Print. Off., Washington, D.C. 44 p.

Hodges, R. G., and C. M. MacLeod. 1946a. Epidemic pneumococcal pneumonia: I. Description of the epidemic. Am. J. Hyg. 44:183–192.

Hodges, R. G., and C. M. MacLeod. 1946b. Epidemic pneumococcal pneumonia: II. The influence of population characteristics and environment. Am. J. Hyg. 44:193–206.

Hodges, R. G., and C. M. MacLeod. 1946c. Epidemic pneumococcal pneumonia. IV. The relationship of nonbacterial respiratory disease to pneumococcal pneumonia. Am. J. Hyg. 44:231–236.

Hodges, R. G., and C. M. MacLeod. 1946d. Epidemic pneumococcal pneumonia: V. Final consideration of the factors underlying the epidemic. Am. J. Hyg. 44:237–243.

Hodges, R. G., C. M. MacLeod, and W. G. Bernhard. 1946. Epidemic pneumococcal pneumonia: III. Pneumococcal carrier studies. Am. J. Hyg. 44:207–230.

Hoehne, J. H., and C. E. Reed. 1971. Where is the allergic reaction in ragweed asthma? J. Allergy Clin. Immunol. 48:36.

Houk, V. N., D. C. Kent, J. H. Baker, and K. Sorensen. 1968. The epidemiology of tuberculosis infection in a closed environment. Arch. Environ. Health 16:26–35.

Howell, J. F., and A. E. Clift. 1974. The dispersal of sterilized codling moths (*Laspeyresia pomonella*) (Lepidoptera:Olethreutidae) released in the Wenas Valley, Washington. Environ. Entomol. 3:75–81.

Huff, F. H. 1963. Relation between leafhopper influxes and synoptic weather conditions. J. Appl. Meteorol. 2:39–43.

Hughes, R. B., and J. F. Stampfer, Jr. 1971. Enhanced evaporation of small, freely falling water drops due to surface contamination. Science 140:1244.

Hurst, V. 1960. Transmission of hospital staphylococci among newborn infants. II. Colonization of the skin and mucous membranes of the infants. Pediatrics 25:204–214.

Hyde, H. A., and D. A. Williams. 1945. Studies in atmospheric pollen. II. Diurnal variation in the incidence of grass pollen. New Phytol. 44:84–94.

Ibach, M. J., H. W. Larsh, and M. L. Furcolow. 1954. Isolation of *Histoplasma capsulatum* in the air. Science 119:71.

Ingold, C. T. 1971. Fungal spores—Their liberation and dispersal. Clarendon Press, Oxford, England. 302 p.

Ishizaki, T., T. Shida, T. Miyamoto, Y. Matsumara, K. Mizumo, and M. Tomaru. 1973. Occupational asthma from western red cedar dust in furniture factory workers. J. Occup. Med. 15:580.

Izraylet, L. I., and R. P. Feoktistova. 1970. Justification for health protection zones around some enzyme producing plants. Gig. Sanit. 6:80.

Jaenicke, R., and C. Junge. 1967. Studien zur obern Grenzgrosse des natürlichen aerosoles. Beitr. Phys. Atmos. 40:129–143.

Jensen, R. E., and J. R. Wallin. 1965. Weather and aphids: A review. USDC Weather Bur. Tech. Note 5. Agmet 1:1–19.

Jiusto, J. E., and T.-F. D. Lee. 1975. Aitken nuclei vs. cloud condensation nuclei. J. Rech. Atmos. 9:41–45.

Johnson, J. E., and P. J. Kadull. 1966. Laboratory-acquired Q fever: A report of fifty cases. Am. J. Med 41:391–403.

Johnson, P., and D. G. Marsh. 1965. "Isoallergens" from rye grass. Nature (Lond.) 206:935.

Johnson, W. L., W. H. Cross, J. E. Laggett, W. L. McGovern, H. C. Mitchell, and E. B. Mitchell. 1975. Dispersal of marked boll weevil: 1970–1973 studies (*Anthonomus grandis:* Coleoptera:Curculionidae). Ann. Entomol. Soc. Am. 68:1018–1022.

Jorde, W., and H. F. Linskens. 1974. Zur Persorption von Pollen und Sporen durch die intakte Darmschleimhaut. Acta Allergol. 29:165.

Junge, C. E. 1963. Air chemistry and radioactivity. Academic Press, New York. 382 p.

Junge, C. E. 1972a. The cycle of atmospheric gases—Natural and man made. Q. J. R. Meteorol. Soc. 98:711–729.

Junge, C. E. 1972b. Our knowledge of the physico-chemistry of aerosols in the undisturbed marine environment. J. Geophys. Res. 77:5183–5200.

Kaliner, M., S. I. Wasserman, and K. F. Austen. 1973. Immunologic release of chemical mediators from human nasal polyps. N. Engl. J. Med. 287:277.

Kammermeyer, J. K., and K. P. Mathews. 1973. Hypersensitivity to phenylglycine acid chloride. J. Allergy Clin. Immunol. 52:73–84.

Kawai, T., D. G. Marsh, L. M. Lichtenstein, and P. S. Norman. 1972. The allergens responsible for house dust. 1. Comparison of *Dermatophagoides pteronyssinus* and house dust extracts by assay of histamine release from allergic human leukocytes. J. Allergy Clin. Immunol. 50:117.

Kieckhefer, R. W., W. F. Lythe, and W. Spuhler. 1974. Spring movement of cereal aphids into South Dakota. Environ. Entomol. 3:347–350.

King, T. P., P. S. Norman, and J. T. Connell. 1964. Isolation and characterization of allergens from ragweed pollen: II. Biochemistry 3:458.

Knight, V. 1973. Airborne transmission and pulmonary deposition of respiratory viruses. Pages 1–9 *in* V. Knight, ed. Viral and mycoplasmal infections of the respiratory tract. Lea and Febiger, Philadelphia.

Knight, V., and J. A. Kasel. 1973. Influenza viruses. Pages 87–123 *in* V. Knight, ed. Viral and mycoplasmal infections of the respiratory tract. Lea and Febiger, Philadelphia.

Knox, R. B., J. Heslop-Harrison, and C. Reed. 1970. Localization of antigens associated with the pollen grain wall by immunofluorescence. Nature (Lond.) 225: 1066.

Kontou-Karakitsos, K., J. E. Salvaggio, and K. P. Mathews. 1975. Comparative nasal absorption of allergens in atopic and non-atopic subjects. J. Allergy Clin. Immunol. 55:241.

Kranz, J., ed. 1974. Epidemics of plant diseases: Mathematical analysis and modeling. Springer-Verlag, New York.

Kuribayashi, K., and A. Ichikawa. 1952. Studies on the forecasting of the rice blast disease. Spec. Rep. Nagano Agric. Exp. Stn. 13:1–229. (In Japanese.)

LaMarche, V. C., Jr. 1973. Holocene climatic variations inferred from treeline fluctuations in the White Mountains, California. Quat. Res. (N.Y.) 3:632–660.

Langmuir, A. D. 1961. Epidemiology of airborne infections. Bacteriol. Rev. 25:173–181.

Langmuir, A. D. 1964. Airborne infection: How important for public health? I. A historical review. Am. J. Public Health 54:1666–1668.

Large, E. C. 1940. The advance of the fungi. Henry Holt, New York. 488 p.

LeClerg, E. L. 1964. Crop losses due to plant diseases in the United States. Phytopathology 54:1309–1313.

Leedom, J. M. 1974. Q fever. Chapter 47 (p. 1–19) *in* T. C. Eickhoff, ed. Practice of medicine, vol. III. Harper and Row, New York.

Leighton, H. G., and R. R. Rogers. 1974. Droplet growth by condensation and coalescence in a strong updraft. J. Atmos. Sci. 31:271–279.

Leonof, L. F., P. S. Prokhorov, T. A. Efanova, and I. A. Zolotarev. 1971. Passivation of condensation nuclei by cetyl alcohol vapor. Adv. Aerosol Phys. 2:12.

Lerman, S. L., and E. F. Darley. 1975. Particulates. Pages 141–158 *in* J. B. Mudd and T. T. Kozlowski, eds. Responses of plants to air pollution. Academic Press, New York.

Leskowitz, S., J. E. Salvaggio, and H. J. Schwartz. 1972. An hypothesis for the development of atopic allergy in man. Clin. Allergy 2:237.

Levine, B. B. 1973. Genetics of atopic allergy and reagin production. Clin. Allergy 3:539.

Lichtenstein, L. M., K. Ishizaka, P. S. Norman, A. K. Sobotka, and B. M. Hill. 1973. IgE antibody measurements in ragweed hay fever: Relationship to clinical severity and the results of immunotherapy. J. Clin. Invest. 52:472.

Likens, G. E., F. H. Bormann, and N. M. Johnson. 1972. Acid rain. Environment 14:33–40.

List, R., and D. M. Whelpdale. 1969. A preliminary investigation of factors affecting the coalescence of colliding water drops. J. Atmos. Sci. 26:305–308.

Loosli, C. G., H. M. Lemon, O. H. Robertson, and M. Hamburger. 1952. Transmission and control of respiratory disease in army barracks. IV. The effect of oiling procedures on the incidence of respiratory diseases and hemolytic streptococcal infections. J. Infect. Dis. 90:153–164.

Loosli, C. G., M. H. D. Smith, J. Cline, and L. Nelson. 1950. The transmission of hemolyitic streptococcal infections in infant wards with special reference to "skin dispersers." J. Lab. Clin. Med. 36:342–359.

Lowbury, E. J. L. 1954. Air conditioning with filtered air for dressing burns. Lancet 1:292–294.

Lowbury, E. J. L. 1971. Evaluation of patient isolators. Pages 220–224 *in* Proc. Int. Conf. Nosocomial Inf., August 1970, Atlanta, Ga. American Hospital Assoc., Chicago.

Lowbury, E. J. L., and J. Fox. 1954. The epidemiology of infection with *Pseudomonas pyocyanea* in a burns unit. J. Hyg. 52:403–416.

McAndrews, J. H. 1968. Pollen evidence for the protohistoric development of the "Big Woods" in Minnesota (U.S.A.). Rev. Palaeobot. Palynol. 7:201–211.

MacArthur, R. H. 1972. Geographical ecology: Patterns in the distribution of species. Harper and Row, New York. 269 p.

McCune, D. C., L. H. Weinstein, D. C. MacLean, and J. S. Jacobsen. 1967. The concept of hidden injury in plants. AAAS Publ. 85:33–44.

McDonald, J. L., and W. R. Solomon. 1975. Effects of outdoor activity on aeroallergen levels in the human microenvironment. J. Allergy Clin. Immunol. 55:89. (Abstr.)

McLean, R. L. 1961. The effect of ultraviolet radiation upon the transmission of epidemic influenza in long-term hospital patients. Am. Rev. Respir. Dis. (Suppl.) 83:36–38.

McManus, M. L. 1973. A dispersal model for the larvae of the gypsy moth, *Porthetria dispar*. Pages 129–138 *in* R. L. Edmonds and W. S. Benninghoff, eds. Ecological systems approaches to aerobiology. II. Development, demonstration, and evaluation of models. US/IBP Aerobiol. Program Handb. 3. (NTIS no. AP-USIBP-H-73-3.) Univ. Michigan, Ann Arbor.

McMillan, C. 1967. Phenological variation within seven transplanted grassland community fractions from Texas and New Mexico. Ecology 48:807–813.

Maher, L. J., Jr. 1963. *Ephedra* pollen in sediments of the Great Lakes region. Ecology 45:391–395.

Maki, L. R., E. L. Galyan, M. C. Chien, and D. R. Caldwell. 1974. Ice nucleation induced by "*Pseudomonas syringae*." Appl. Microbiol. 28:456–459.

Manning, W. J. 1976. The influence of ozone on plant surfaces. Academic Press, New York. 669 p.

Marsh, D. G., Z. H. Haddad, and D. H. Campbell. 1970. A new method for determining the distribution of allergenic fractions in biological materials: Its application to grass pollen extracts. J. Allergy 46:107.

Martyn, E. B., and A. McIlwaine. 1951. Banana leaf spot disease control in Jamaica. Bull. Dep. Agric. Jamaica 46:30.

Mason, C. J., and R. L. Edmonds. 1974. An atmospheric dispersion model for the gypsy moth larvae. Pages 251–252 *in* Proc. AMS/WMC Symp. Atmos. Diffus. Air Pollut., Sept. 1974, Santa Barbara, Calif.

May, K. R., and N. P. Pomeroy. 1973. Bacterial dispersion from the body surface. Pages 426–432 *in* J. F. P. Hers and K. C. Winkler, eds. Airborne transmission and airborne infection. Oosthoek Publ. Co., Utrecht, The Netherlands.

Mayron, L. W., and R. J. Loiselle. 1972. Ragweed pollen: Morphological changes on incubation with saliva from ragweed allergic and non-allergic humans. Clin. Sci. 42:25.

MRC (Medical Research Council). 1954. Air disinfection with ultraviolet irradiation: Its effect on illness among school-children. Spec. Rep. Ser. no. 283. Her Majesty's Stationery Office, London.

Metcalf, C. L., W. P. Flint, and R. A. Metcalf. 1951. Destructive and useful insects. McGraw-Hill, New York. 1071 p.

Middleton, J. T., and A. O. Paulus. 1956. The identification and distribution of air pollutants through plant response. AMA Arch. Ind. Health 14:526–532.

Miller, P. R. 1973. Oxidant-induced community change in a mixed conifer forest. Adv. Chem. Ser. 122:101–117.

Miller, P. R., and J. R. McBride. 1975. Effects of air pollutants on forests. Pages 195–235 *in* J. B. Mudd and T. T. Kozlowski, eds. Responses of plants to air pollution. Academic Press, New York.

Mitchell, N. J., and D. R. Gamble. 1974. Clothing design for operating-room personnel. Lancet 2:1133–1136.

Mitchell, W. F., G. W. Wharton, D. G. Larson, and R. Modic. 1969. House dust, mites and insects. Ann. Allergy 27:93.

Miyamoto, T., S. Oshima, T. Ishizaki, and S. Sato. 1968. Allergenic identity between the common floor mite (*Dermatophagoides farinae* Hughes, 1961) and house dust as a causative antigen in bronchial asthma. J. Allergy 42:14.

Morris, R. F., C. A. Miller, D. O. Greenbank, and D. G. Mott. 1958. The population dynamics of the spruce budworm in eastern Canada. Proc. 10th Int. Congr. Entomol. 4:137–149.

Mortimer, E. A., Jr., P. J. Lipsitz, E. Wolinsky, A. J. Gonzaga, and C. H. Rammerlkamp, Jr. 1962. Transmission of staphylococci between newborns: Importance of the hands of personnel. Am. J. Dis. Child. 104:289–295.

Mortimer, E. A., Jr., E. Wolinsky, A. J. Gonzaga, and C. H. Rammerlkamp, Jr. 1966. Role of airborne transmission in staphylococcal infections. Br. Med. J. 1:319–322.

Mudd, J. B. 1973. Biochemical effects of some air pollutants in plants. Adv. Chem. Ser. 122:31–47.

Mudd, J. B., and T. T. Kozlowski, eds. 1975. Responses of plants to air pollution. Academic Press, New York. 381 p.

Mufson, M. A., H. E. Mocega, and H. E. Krause. 1973. Acquisition of parainfluenza 3 virus by hospitalized children. I. Frequencies, rates, and temporal data. J. Infect. Dis. 128:141–147.

Mullan, N. A. 1973. Immunological properties of carbohydrate and protein fractions of timothy grass proteins. Int. Archiv. Allergy Appl. Immunol. 45:43.

Nadel, J. A. 1974. Parasympathetic nervous control of airway smooth muscle. Ann. N.Y. Acad. Sci. 221:99.

Naegale, J. A., ed. 1973. Air pollution damage to vegetation. Advan. Chem. Ser. 122. Am. Chem. Soc., Washiongton, D.C. 137p.

NAS (National Academy of Sciences). 1975. Forest pest control. Volume IV *in* Pest control: An asseessment of present and alternative technologies. Print. Publ. Off., NAS, Washington, D.C.

NAS/NRC (National Academy of Sciences/National Research Council). 1964. Postoperative wound infections: The influence of ultraviolet irradiation of the operating room and of various other factors. Ann. Surg. 160 (Suppl.2):1–192.

Nichols, J. O. 1961. The gypsy moth in Pennsylvania. Pa. Dep. Agric. Misc. Bull. 4404. 82 p.

Nickiparick, W. 1965. The aerial migration of the six-spotted leafhopper and the spread of the virus disease, aster yellows. Int. J. Bioclimatol. Biometeorol. 9: 219–227.

Noble, W. C. 1962. The dispersal of staphylococci in hospital wards. J. Clin. Pathol. 15:552–558.

Nobel, W. C. 1971. Dispersal and acquisition of microorganisms. Pages 193–197 *in* Proc. Int. Conf. Nosocomial Inf., August 1970, Atlanta, Ga. American Hospital Assoc., Chicago.

Nobel, W. C. 1975. Dispersal of skin microorganisms. Br. J. Dermatol. 93: 447–485.

Norman, P. S. 1969. A rational approach to desensitization. J. Allergy. 44: 129.

Ohman, J. L., F. C. Lowell, and K. J. Bloch. 1973. Allergens of mammalian origin: Characterization of allergen extracted from cat pelts. J. Allergy Clin. Immunol. 52:229.

Ou, S. H. 1972. Rice diseases. Commonw. Mycol. Inst., Kew, England. 368 p.

Parish, W. E. 1967. Release of histamine and slow reacting substance with mast cell changes after challenge of human lung with reagin *in vitro*. Nature (Lond.) 215:738.

Parlato, S. J. 1929. A case of coryza and asthma due to sand flies (caddis fly). J. Allergy 1:35.

Patterson, R., and J. F. Kelly. 1974. Animal models of the asthmatic state. Annu. Rev. Med. 25:53.

Patterson, R., and I. M. Suszko. 1974. Polymerized ragweed antigen E. III. Differences in immune response to three molecular weight ranges of monomer and polymer. J. Immunol. 112:1855.

Paugam, J. Y. 1975. Formation de noyaux Aitken dans l'air au-dessus de la zone littorale. J. Rech. Atmos. 9:67–75.

Pearl, R. 1930. The biology of population growth. A. A. Knopf, New York.

Pepys, J. 1969. Hypersensitivity diseases of the lungs due to fungi and organic dusts. Volume 4 *in* Monographs in allergy. Karger, Basel.

Pepys, J., C. Pickering, and H. London. 1972. Asthma due to inhaled chemical agents—Piperazine dihydrochloride. Clin. Allergy 2:189.

Perkins, J. E., A. M. Bahlke, and H. F. Silverman. 1947. Effect of ultraviolet irradiation of classrooms on spread of measles in large rural central schools. Am. J. Public Health 37:529–537.

Perlman, F. 1961. Insect allergens: Their interrelationship and differences. J. Allergy 32:93.

Pianka, E. R. 1972. *r* and *K* selection or *b* and *d* selection? Am. Nat. 106:581–588.

Pielou, E. C. 1975. Ecological diversity. John Wiley & Sons, New York.

Pienkowski, R. L., and J. T. Medler. 1964. Synoptic weather conditions associated with long-range movement of the potato leafhopper, *Empoasca fabae,* into Wisconsin. Ann. Entomol. Soc. Am. 57:588–591.

Pimentel, D. 1976. World food crisis: Energy and pests. Bull. Entomol. Soc. Am. 22:20–26.

Podzimek, J., and A. N. Saad. 1975. Retardation of condensation nuclei growth by surfactant. J. Geophys. Res. 80:3386–3392.

Poole, R. W. 1974. An introduction to quantitative ecology. McGraw-Hill, New York. 532 p.

Portnoy, B., H. L. Eckert, B. Hanes, and M. A. Salvatore. 1966. Multiple respiratory virus infections in hospitalized children. Am. J. Epedimiol. 82:262–272.

Poston, F. L., and L. P. Pedigo. 1975. Migration of plant bugs and the potato leafhopper in a soybean-alfalfa complex. Environ. Entomol. 4:8–10.

Proctor, M., and P. Yeo. 1972. The pollination of flowers. Taplinger Publ., New York. 418 p.

Purseglove, J. W. 1968. Tropical crops: Dicotyledons. John Wiley & Sons, New York. 719 p.

Quinn, J. A. 1969. Variability among high plains populations of *Panicum virgatum*. Bull. Torrey Bot. Club 96:20–41.

Quinn, J. A., and R. T. Ward. 1969. Ecological differentiation in sand dropseed (*Sporobolus cryptandrus*). Ecol. Monogr. 39:61–78.

Ramazzini, B. 1713. *In* De morbis artificum diatriba. Geneva. (Transl. by W. C. Wright, 1940. Univ. Chicago Press, Chicago.)

Rasool, S. I. 1973. Chemistry of the lower atmosphere. Plenum Press, New York. 335 p.

Raynor, G. S., J. V. Hayes, and E. C. Ogden. 1972b. Dispersion and deposition of corn pollen from experimental sources. Agron. J. 64:420–427.

Raynor, G. S., J. V. Hayes, and E. C. Ogden. 1973. Dispersion of pollens from low-level, crosswind line sources. Agric. Meteorol. 11:177–195.

Raynor, G. S., E. C. Ogden, and J. V. Hayes. 1970. Dispersion and deposition of ragweed pollen from experimental sources. J. Appl. Meteorol. 9:885–895.

Raynor, G. S., E. C. Ogden, and J. V. Hayes. 1972a. Dispersion and deposition of timothy pollen from experimental sources. Agric. Meteorol. 9:347–366.

Reinert, R. A., A. S. Heagle, and W. W. Heck. 1975. Plant responses to pollutant combinations. Pages 159–177 *in* J. B. Mudd and T. T. Kozlowski. eds. Responses of plants to air pollution. Academic Press, New York.

Riemensnider, D. K. 1967. Reduction of microbial shedding from humans. Pages 242–255 *in* Proc. Sixth Annu. Tech. Meet. Exh. AACC.

Riley, H. D., Jr. 1969. Hospital-associated infections. Pediatr. Clin. North Am. 16:701–734.

Riley, R. L. 1961. Airborne pulmonary tuberculosis. Bacteriol. Rev. 25:243–248.

Riley, R. L. 1972. The ecology of indoor atmospheres: Airborne infection in hospital. J. Chronic Dis. 25:421–423.

Riley, R. L. 1974. Airborne infection. Am. J. Med. 57:466–475.

Riley, R. L., C. C. Mills, F. O'Grady, L. U. Sultan, F. Wittstadt, and D. N. Shivpuri. 1962. Infectiousness of air from a tuberculosis ward. Am. Rev. Respir. Dis. 85:511–525.

Riley, R. L., and F. O'Grady. 1961. Airborne infection, transmission and control. Macmillan Co., New York.

Ritchie, J. C., and S. Lichti-Federovich. 1967. Pollen dispersal phenomena in arctic-subarctic Canada. Rev. Palaeobot. Palynol. 3:255–266.

Ritter, C., and R. L. Culp. 1956. Studies of *Histoplasma capsulatum* experimentally introduced into tap water. Pages 261–264 *in* Public Health Monogr. 39, Proc. Conf. Histoplasmosis, 1952.

Roberts, R., F. Wenzel, and D. Emanual. 1975. The incidence of precipitins to thermophilic actinomycetes and *Aspergillus* species in a Wisconsin farm population. J. Allergy Clin. Immunol. 55:70. (Abstr.)

Robertson, C. 1904. The structure of flowers and the mode of pollination of the primitive angiosperms. Bot. Gaz. 37:294–298.

Robertson, P. A., and R. T. Ward. 1970. Ecotypic differentiation in *Koeleria cristata* (L.) Pers. from Colorado and related areas. Ecology 51:1082–1087.

Rogers, K. B. 1951. The spread of infantile gastroenteritis in a cubicled ward. J. Hyg. 49:140–151.

Rose, H. D., and B. Varkey. 1975. Deep mycotic infection in the hospitalized adult: A study of 123 patients. Medicine (Baltimore) 54:499–507.

Rosenstern, I. 1948. Control of airborne infections in a nursery for young infants. Am. J. Dis. Child. 75:193–202.

Rosinski, J., and F. Parungo. 1966. Terpene-iodine compounds as ice nuclei. J. Appl. Meteorol. 5:119–123.

Rotsettis, J., J. A. Quinn, and D. E. Fairbrothers. 1972. Growth and flowering of *Danthonia sericea* populations. Ecology 53:227–234.

Rummel, D. R., J. R. White, and L. J. Wade. 1975. Late season immigration of boll weevils into an isolated cotton plot (*Anthonomus grandis:* Col. Curculionidae). J. Econ. Entomol. 68:616–618.

Salvaggio, J. E., J. J. A. Cavanaugh, F. C. Lowell, and S. Leskowitz. 1964. A comparison of the immunologic responses of normal and atopic individuals to intranasally administered antigen. J. Allergy 35:62.

Salvaggio, J. E., J. H. Seabury, H. A. Buechner, and V. G. Jundur. 1967. Bagassosis: Demonstration of precipitins against extracts of thermophilic actinomycetes in the sera of affected individuals. J. Allergy 39:106.

Salvaggio, J. E., J. H. Seabury, and E. A. Schoenhardt. 1971. New Orleans asthma. V. Relationship between charity hospital asthma admissions rates, semiquantitative pollen and fungal spore counts, and total particulate aerometric sampling data. J. Allergy Clin. Immunol. 48:96.

Sanford, J. P., and A. K. Pierce. 1971. Current infection problems—Respiratory. Pages 77–81 *in* Proc. Int. Conf. Nosocomial Inf., August 1970, Atlanta, Ga. American Hospital Assoc., Chicago.

Sarosi, G. A., J. D. Parker, and F. E. Tosh. 1971. Histoplasmosis outbreaks: Their patterns. Chapter 16 (p. 123–128) *in* Proc. 2nd Natl. Conf. Histoplasmosis.

Saunders, P. J. W. 1966. The toxicity of sulfur dioxide to *Diplocryon rosae* Wold causing blackspot of roses. Ann. Appl. Biol. 58:103–114.

Schachter, J., M. Sung, and K. F. Meyer. 1971. Potential danger of Q fever in a university hospital environment. J. Infect. Dis. 123:301–304.

Schaffner, W., L. B. Lefkowitz, J. S. Goodman, and M. G. Koenig. 1969. Hospital outbreak of infections with group A streptococci traced to an asymptomatic anal carrier. N. Engl. J. Med. 280:1224–1225.

Schimpff, S. C., W. H. Greene, V. M. Young, C. L. Fortner, L. Jepsen, N. Cusack, J. B. Block, and P. H. Wiernik. 1975. Infection prevention in acute nonlymphocytic leukemia: Laminar air flow room reverse isolation with oral, nonabsorbable antibiotic prophylaxis. Ann. Intern. Med. 82:351–358.

Schlichting, H. E., Jr. 1964. Meteorological conditions affecting the dispersal of airborne algae and Protozoa. Lloydia 27:64.

Schlichting, H. E., Jr. 1974. Ejection of microalgae into the air via bursting bubbles. J. Allergy Clin. Immunol. 53:185.

Schnell, R. C., and G. Vali. 1972. Atmospheric ice nuclei from decomposing vegetation. Nature (Lond.) 236:163–165.

Schnell, R. C., and G. Vali. 1976. Biogenic ice nuclei: I. Terrestrial and marine sources. J. Atmos. Sci. 33:1554–1564.

Schulman, J. L. 1967. Experimental transmission of influenza virus infection in mice. IV. Relationship of transmissibility of different strains of virus and recovery of airborne virus in the environment of infector mice. J. Exp. Med. 125:479–488.

Schulman, J. L. 1968. The use of an animal model to study transmission of influenza virus infection. Am. J. Public Health 58:2092–2096.

Schulman, J. L., and E. D. Kilbourne. 1968. Experimental transmission of influenza in mice. 1. The period of transmissibility. J. Exp. Med. 128:256–266.

Schwartz, H. J., and A. I. Terr. 1971. The immune response of allergic and normal subjects to pneumococcal polysaccharide. Int. Arch. Allergy Appl. Immunol. 40:250.

Sciple, G. S., D. K. Riemensnider, and C. A. J. Scheyler. 1967. Recovery of microorganisms shed by humans into a sterilized environment. Appl. Microbiol. 15:1388–1392.

Sevitt, S. 1949. Source of two hospital-infected cases of tetanus. Lancet 2:1075–1078.

Sevitt, S. 1953. Gas-gangrene infection in an operating theatre. Lancet 2:1121–1123.

Shacklette, M. H., and H. F. Hasenclever. 1968. The natural occurrence of *Histoplasma capsulatum* in a cave. 3. Effect of flooding. Am. J. Epidemiol. 88:210–214.

Shacklette, M. H., H. F. Hasenclever, and E. A. Miranda. 1967. The natural occurrence of *Histoplasma capsulatum* in a cave. 2. Ecologic aspects. Am. J. Epidemiol. 86:246–254.

Shafiee, A., E. J. Staba, and Y. T. Abul-Hajj. 1973. Partial purification of antigen E from mixed stems and leaves of short ragweed plant. J. Pharmaceut. Sci. 62:1654.

Shapiro, R. S., and B. C. Eisenberg. 1971. Sensitivity to proteolytic enzymes in laundry detergents. J. Allergy 47:76.

Shooter, R. A., G. W. Taylor, G. Ellis, and J. P. Ross. 1956. Postoperative wound infection. Surg. Gynecol. Obstet. 103:257–262.

Simpson, K., R. C. Tozzer, and W. A. Gillespie. 1960. Prevention of staphylococcal sepsis in a maternity hospital by means of hexachlorophane. Br. Med. J. 1:315–317.

Small, V. A. 1952. Increasing castor bean allergy in southern California due to fertilizer. J. Allergy 23:406.

Smith, C. D., M. L. Furcolow, and F. E. Tosh. 1964. Attempts to eliminate *Histoplasma capsulatum* from soil. Am. J. Hyg. 79:170–180.

Smith, F. B. 1944. The occurrence and distribution of algae in soils. Proc. Fla. Acad. Sci. 7:44.

Snead, C. C., and J. T. Zung. 1968. The effects of insoluble films upon the evaporation kinetics of liquid droplets. J. Colloid Interface Sci. 27:25.

Solberg, C. O. 1965. A study of carriers of *Staphylococcus aureus*. Acta Med. Scand. 178:Suppl. I.

Solomon, A. M. 1975. A rational model for interpretation of pollen samples from arid regions. Paper presented at 8th Annu. Meet., Am. Assoc. Strat. Palynol., October 1975, Houston, Texas. (Abstr.)

Solomon, W. R. 1969. Experimental ragweed pollinosis induced during normal respiration in a test chamber: Initial observations. J. Allergy 43:181. (Abstr.)

Solomon, W. R. 1974. Fungus aerosols arising from cold-mist vaporizers. J. Allergy Clin. Immunol. 54:222.

Solomon, W. R. 1975. Assessing fungus prevalence in domestic interiors. J. Allergy Clin. Immunol. 56:235.

Sommer, H. E., and J. Stokes. 1942. Studies on airborne infection in a hospital ward. I. The effect of ultraviolet light on cross infection in an infants' ward. J. Pediatr. 21:569–576.

Sosman, A. A., D. P. Schleuter, J. N. Fink, and J. J. Barboriak. 1969. Hypersensitivity to wood dust. N. Engl. J. Med. 281:977.

Soulage, F. B. 1957. Les noyaux de congelation de l'atmosphere. Ann. Geophys. 13:103–134.

Stebbins, G. L. 1966. Processes of organic evolution. Prentice-Hall, Englewood Cliffs, N.J. 191 p.

Steinberg, P., R. Ehtessabian, L. M. Ford, and N. K. Bayne. 1975. Role of IgE-mediated reactions in immunity: Effect of intranasal ragweed extract on vascular and mucosal permeability. J. Allergy Clin. Immunol. 55:119. (Abstr.)

Stotzky, G., and A. H. Post. 1967. Soil mineralogy as a possible factor in geographic distribution of *Histoplasma capsulatum*. Can. J. Microbiol. 13:1–7.

Stover, R. H. 1962. Intercontinental spread of banana leaf spot (*Mycosphaerella musicola* Leach). Trop. Agric. (Trinidad) 39:327–338.

Stover, R. H. 1972. Banana, plantain and abaco diseases. Commonw. Mycol. Inst. Kew, Surv., England. 316 p.

Taylor, L. R. 1965. Flight behavior and aphid migration. Proc. North Cent. Branch Entomol. Soc. Am. 20:9–19.

Taylor, O. C. 1973. Acute responses of plants to aerial pollutants. Adv. Chem. Ser. 122:9–20.

Tosh, F. E., I. L. Doto, S. B. Beecher, and T. D. Y. Chin. 1970. Relationship of starling-blackbird roosts and endemic histoplasmosis. Am. Rev. Respir. Dis. 101: 283–288.

Tosh, F. E., I. L. Doto, D. J. D'Alessio, A. A. Medeiros, S. L. Hendricks, and T. D. Y. Chin. 1966a. The second of two epidemics of histoplasmosis resulting from work on the same starling roost. Am. Rev. Respir. Dis. 94:406–413.

Tosh, F. E., R. J. Weeks, F. R. Pfeiffer, S. L. Hendricks, and T. D. Y. Chin. 1966b. Chemical decontamination of soil containing *Histoplasma capsulatum* Am. J. Epidemiol. 83:262–270.

Tosh, F. E., R. J. Weeks, F. R. Pfeiffer, S. L. Hendricks, D. L. Greer, and T. D. Y. Chin. 1967. The use of formalin to kill *Histoplasma capsulatum* at an epidemic site. Am. J. Epidemiol. 85:259–265.

Tourville, D. R., W. I. Weiss, P. T. Wertlake, and G. M. Laudemann. 1972. Hypersensitivity pneumonitis due to contamination of home humidifier. J. Allergy Clin. Immunol. 49:245.

Treshow, M. 1975. Interactions of air pollutants and plant disease. Pages 307–334 *in* J. B. Mudd and T. T. Kozlowski, eds. Responses of plants to air pollution. Academic Press, New York.

Tse, K. S., P. Warren, D. S. McCarthy, and R. N. Cherniack. 1973. Respiratory abnormalities in workers exposed to grain dust. Arch. Environ. Health 27:74.

Turk, A., L. M. Lichtenstein, and P. A. Norman. 1970. Nasal secretory antibody to inhalant allergens in allergic and nonallergic patients. Immunology 19:85.

Twomey, S. 1971. The composition of cloud nuclei. J. Atmos. Sci. 28:377–381.

USPHS (U.S. Dep. Health, Education, and Welfare, Public Health Serv.). 1974. Morbidity and mortality 22(53). U.S. Public Health Serv., Atlanta, Ga.

Vali, G. 1968. Ice nucleation relevant to the formation of hail. McGill Univ. Stormy Weather Group, Sci. Rep. MW-58. 51 p.

Vali, G., M. Christensen, R. W. Fresh, E. L. Galyan, L. R. Maki, and R. C. Schnell. 1976. Biogenic ice nuclei. II. Bacterial sources. J. Atmos. Sci. 33:1565–1570.

van Bronswijk, J. E. M. H., and R. N. Sinha. 1971. Pyroglyphid mites (Acari) and house dust allergy. J. Allergy 47:31.

Van der Plank, J. E. 1963. Plant diseases: Epidemics and control. Academic Press, New York. 349 p.

Van der Plank, J. E. 1968. Disease resistance in plants. Academic Press, New York. 206 p.

Van der Plank, J. E. 1975. Principles of plant infection. Academic Press, New York. 210 p.

Van der Werff, P. J. 1958. Mould fungi and bronchial asthma. Stenfert Kroese, Leiden.

Van Toorn, D. E. 1970. Coffee worker's lung: A new example of extrinsic allergic alveolitis. Thorax 25:399.

Varga, D. F., and A. White. 1961. Suppression of nasal, skin and aerial staphylococci by nasal application of methicillin. J. Clin. Invest. 40:2209–2214.

Verhulst, P. F. 1838. Notice sur la loi que la population suit dans son accroissement. Corresp. Math. Phys. 10:113–121.

Von Maur, R. K., T. A. E. Platt-Mills, K. Ishizaka, L. M. Lichtenstein, and P. S. Norman. 1975. IgA and IgG antibodies in nasal secretions. J. Allergy Clin. Immunol. 55:120. (Abstr.)

Voorhorst, R., T. T. M. Spieksma, H. Varekamp, M. J. Leupen, and A. W. Lyklema. 1967. The house-dust mite (*Dermatophagoides pteronyssinus*) and the allergens it produces: Identity with house-dust allergy. J. Allergy 39:325.

Wallin, J. R., and D. V. Loonan. 1971. Low-level jet winds, aphid vectors, local weather, and barley yellow dwarf virus outbreak. Phytopathology 61:1068–1070.

Wallin, J. R., D. Peters, and L. C. Johnson. 1967. Low-level jet winds, early cereal aphid and barley yellow dwarf detection in Iowa. Plant Dis. Rep. 51:517–530.

Walter, C. W., R. B. Knudsen, and M. M. Brubaker. 1963. The incidence of airborne wound infection during operation. J. Am. Med. Assoc. 186:908–913.

Wardlaw, C. W. 1934. Banana diseases. 9. The occurrence of sigatoka disease (*Cercospora musae* Zimm.) on bananas in Trinidad. Trop. Agric. (Trinidad) 11:173–175.

Wardlaw, C. W. 1972. Banana diseases including plantains and abaca, 2nd ed. Longmans, London. 878 p.

Watson, M. A., and R. T. Plumb. 1972. Transmission of plant-pathogenic viruses by aphids. Annu. Rev. Entomol. 17:425–452.

Way, M. J. 1976. Entomology and the world food situation. Bull. Entomol. Soc. Am. 22:125–129.

Weeks, R. J., and F. E. Tosh. 1971. Control of epidemic foci of *Histoplasma capsulatum*. Chapter 24 (p. 184–189) *in* Proc. 2nd Nat. Conf. Histoplasmosis.

Wehrle, P. F., J. Posch, K. H. Richter, and D. A. Henderson. 1970. An airborne outbreak of smallpox in a German hospital and its significance with respect to other recent outbreaks in Europe. Bull. W.H.O. 43:669–679.

Weil, H., L. C. Waddell, and M. Ziskind. 1971. A study of workers exposed to detergent enzymes. J. Am. Med. Assoc. 217:425.

Wellington, W. G. 1945. Conditions governing the distribution of insects in the free atmosphere. IV. Distributive processes of economic significance. Can. Entomol. 77:69–74.

Wellman, F. L. 1972. Tropical American plant disease. Scarecrow Press, Metuchen, N.J. 989 p.

Wells, P. V. 1970. Postglacial vegetation history of the Great Plains. Science 167:1574–1582.

Wells, W. F., M. W. Wells, and T. S. Wilder. 1942. The environmental control of epidemic contagion. I. An epidemiologic study of radiant disinfection of air in day schools. Am. J. Hyg. 35:97–121.

Went, F. W. 1964. The nature of Aitken condensation in the atmosphere. Proc. Natl. Acad. Sci. 51:1259–1267.

Went, F. W., D. V. Slemons, and H. N. Mozingo. 1967. The organic nature of atmospheric condensation nuclei. Proc. Natl. Acad. Sci. 58:69–74.

Wharton, G. W. 1970. Mites and commercial extracts of house-dust. Science 167:1382.

White, A. 1961. Relation between quantitative nasal cultures and dissemination of staphylococci. J. Lab. Clin. Med. 58:273–277.

White, A., J. Smith, and D. T. Varga. 1964. Dissemination of staphylococci. Arch. Intern. Med 114:651–656.

Whitehead, D. R. 1969. Wind pollination in the angiosperms: Evolutionary and environmental considerations. Evolution 23:28–35.

Wickman, B. E., R. R. Mason, and C. G. Thompson. 1973. Major outbreaks of the Douglas-fir tussock moth in Oregon and California. USDA For. Serv. Tech. Rep. PNW 3. USDA For. Serv., Corvallis, Oreg. 18 p.

Wilkie, B., B. Pauli, and M. Gygaz. 1973. Hypersensitivity pneumonitis: Experimental production in guinea pigs with antigens of *Micropolyspora* fungi. Pathol. Microbiol. 39:393.

Willard, J. R. 1973. Wandering time of the crawlers of California red scale *Aonidiella aurantii* (Mask.) (Homoptera:Diaspididae) on citrus. Aust. J. Zool. 21:217–229.

Williams, G. C. 1975. Sex and evolution. Monogr. Pop. Biol. 8. Princeton Univ. Press, Princeton, N.J. 200 p.

Williams, R., E. D. Williams, and D. E. Hyams. 1960. Cross-infection with *Pseudomonas pyocyanea*. Lancet 1:376–379.

Williams, R. E. O. 1966. Epidemiology of airborne staphylococcal infection. Bacteriol. Rev. 30:660–672.

Wilson, A. F., H. S. Novey, R. A. Berke, and E. L. Suprenant. 1973. Deposition of inhaled pollen and pollen extract in human airways. N. Engl. J. Med. 288:1056.

Wiseman, R. D., W. G. Woodin, H. C. Miller, and M. A. Myers. 1959. Insect allergy as a possible cause of inhalant sensitivity. J. Allergy 30:191.

Wodehouse, R. P. 1939. Weeds, waste and hayfever. Nat. Hist. 63:150–163; 178.

Wolinsky, E., P. J. Lipsitz, E. A. Mortimer, Jr., and C. H. Rommelkamp, Jr. 1960. Acquisition of staphylococci by newborns: Direct versus indirect transmission. Lancet 2:620–622.

Wood, F. A. 1968. Sources of plant-pathogenic pollutants. Phytopathology 58:1075–1084.

Woodcock, A. H. 1955. Bursting bubbles and air pollution. Sewage Ind. Wastes 27:1189.

Woodwell, G. M. 1970. Effects of pollution on the structure and physiology of ecosystems. Science 168:429–433.

Yunginger, J. W., and G. J. Gleich. 1972. Measurement of ragweed antigen E by double antibody radioimmunoassay. J. Allergy Clin. Immunol. 50:326.

Zeterberg, J. M. 1973. A review of respiratory virology and the spread of virulent and possibly antigenic viruses via air conditioning systems. II. Ann. Allergy 31:291–299.

Zuskin, E., R. L. Wolfson, G. Harpel, J. W. Wellborn, and A. Bouhuys. 1969. Byssinosis in carding and spinning workers. Arch. Environ. Health 19:666.

7. MODELING OF AEROBIOLOGICAL SYSTEMS

J. R. Burleigh, R. L. Edmonds, J. B. Harrington, Jr., B. Lighthart, R. E. McCoy, C. J. Mason, G. H. Quentin, A. M. Solomon, and J. R. Wallin

GENERAL APPROACHES TO MODELING AEROBIOLOGY SYSTEMS*

OBJECTIVES OF MODELING

The major objectives for modeling aerobiological systems are (1) to enable the prediction of ecological impacts and (2) to evaluate the major factors influencing these impacts. By understanding the ultimate impact upon a given ecosystem, it is envisioned that man might avoid deleterious results by (1) altering conditions at some point in the aerobiological pathway or (2) using control measures at the point of impact. An example of such objectives would be the prediction and control of crop diseases, which has been a long-range goal of plant pathologists. Currently, prediction of disease epidemics is based primarily upon mortality studies, which represents a post facto analysis and does not give a great deal of forewarning to permit adequate preventive control measures. The corn blight epidemic that struck the United States in 1970 destroyed a sizable portion of the annual crop and portrayed both the serious results that aerobiological phenomena can cause and the inability of current methods to predict such occurrence (Wade 1972).

The task of modeling aerobiological systems to achieve a predictive capability requires the integration of diverse information regarding the many stages and factors in a given system cycle into a comprehensive descriptive package. Early attempts to formulate predictive models have been confronted by deficiency of information describing some of the important stages in the aerobiological cycle. This in itself has helped to shape the directions taken in the development of the models. One approach has been to construct statistical models directly relating overall results with several major causative factors, in

*G. H. Quentin

effect ignoring the main functional relationships describing the stages between cause and effect, e.g., regression models. The alternate approach has been the formulation of simulation models that are deterministic in nature, attempting to represent a complete aerobiological system cycle as a sequence of interacting events that include all discernible functional relationships. These efforts have been confronted by the necessity of supplying data for many relationships that heretofore have been studied very little and consequently are poorly understood. As a result, the modeling efforts have provided new directions for experimental research. There has also been the necessity to incorporate interacting effects into simulation models, which frequently require different types of mathematical representation from other segments of a model.

As an example of the latter case, it is possible to consider the spore cycle in a fungal plant disease in terms of the sequence of events concerning the germination of spores on the host organism. The mechanisms of spore release and deposition between the host and the atmosphere are critical to the evolution of the disease, however, and they depend greatly on external factors such as meteorological conditions as much as they do on the spore-host relationships.

Previous chapters have covered many of the interrelated facets of system modeling such as sources of materials, atmospheric release, transport, deposition, and ecological impact. The latter portion of this chapter is devoted to examples of aerobiological system models developed to date. Three major approaches are discussed: (1) simulation models, (2) statistical regression models, and (3) synoptic models. These examples show the state of development and the future directions that must be taken to accomplish the overall objective, achievement of a predictive model.

CONCEPTUALIZATION OF SYSTEMS

In the development of a model of any type, the first step recommended is to construct a conceptual model of the system to be studied. The purpose of the conceptual model is to assist in the planning and organization of the project leading to a successful working model. The objectives of such a modeling project are frequently presented in terms of difficult questions that ecological system managers and planners (including everyone from farmers to forest rangers to health specialists) would like to have answered. An example of the type of questions a modeling project might answer for white pine blister rust caused by *Cronartium ribicola* is presented in Table 7.1 (Edmonds and Benninghoff 1973b). An example of a basic conceptual model is presented in Figure 7.1 for white pine blister rust.

The construction of a conceptual model, of course, would not answer the questions posed at the outset. Rather, a conceptual model attempts to depict (1) the sequence of events in a complete system cycle, (2) the significant factors (parameters and variables) that affect the system, and (3) the interrela-

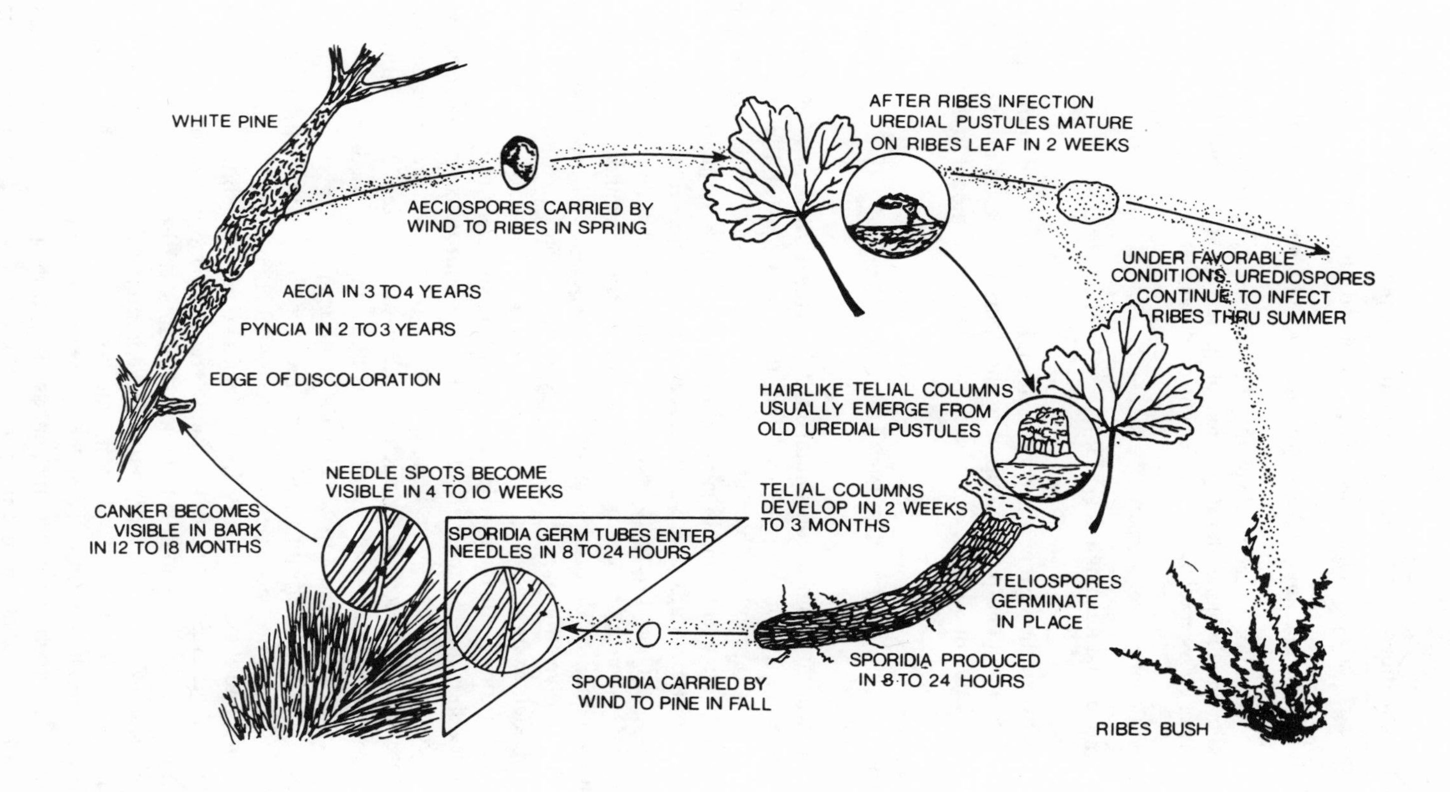

Figure 7.1. Life cycle of *Cronartium ribicola* (after Boyce 1961).

Table 7.1. Questions to be asked of an overall white pine blister rust model.[a]

1. How far can white pine blister rust spores travel and still cause infection?
2. What is the apparent limit to the geographic range of *Cronartium ribicola?*
3. Will an eradication program involving the alternate host (*Ribes*) work if we can predict the dispersal and deposition of spores? That is, if we increase the efficiency of *Ribes* eradication can we control blister rust?
4. What additional control measures are feasible?

[a]From Edmonds and Benninghoff (1973b).

tionships between the system and other external systems. Based upon the time and resources available to the group or agency pursuing the answers, the conceptual model aids in the selection of the ultimate type of model to be formed.

MODELING APPROACHES

In proceeding from a conceptual model to the formation of a working model, there are several alternatives. First, there is the possibility of a deterministic simulation model, which attempts to simulate each event occurring in a system with an exact functional relationship. In such simulation models, each separate step usually involves a separate mathematical description; or, in certain cases, a single differential equation might properly describe a number of events in sequence. An example of the latter instance would be the partial differential equation describing atmospheric dispersion phenomena. As a second alternative, the overall aerobiology system might be described by a synoptic model, using statistical methods based upon the relative frequency of occurrence of important factors. Here it would be sufficient to gather experimental data showing the final effects relative to major causative factors. By statistical correlation techniques it is possible to evaluate whether the relationship with each factor is truly significant, and to what degree. Then with this knowledge in hand, statistical regression methods might be used to determine a suitable empirical formula to describe the relationship.

The major considerations governing the selection of the type of model include: (1) scope and accuracy of the desired final results, (2) information requirements and the currently available data, and (3) availability of time and resources. In general, a simulation model requires more information (i.e., data), time, and resources to develop than a synoptic statistical model. Simulation models, however, hold more hope for comprehensive final results with better degrees of definition. To demonstrate how the factors mentioned above can affect the modeling effort, consider the example of the white pine blister rust cycle, for which the conceptual model is shown in Figure 7.1. In this system cycle, note that the pycniospore and aeciospore stages on the white pine host are shown to incubate for several years each, prior to spore appearance. Such stages are certain to present a formidable obstacle for a deterministic simulation model, particularly in terms of required knowledge concerning the sporulation process. Because of the extensive cycle time frame for these stages, both the host and germinating spores may be subjected to numerous beneficial or adverse effects. To accurately predict the outcome of these stages of the cycle, a model would require detailed information on all the potential factors affecting the stages. This might predicate a great deal of experimental research, and undoubtedly would constitute a large number of model calculations and hence computer time. If these stages are not fully defined, a crude approximation might be used to fill the gap until adequate research results are available. Another likelihood is that a statistical analysis of available data might yield an empirical formula for these stages in an otherwise deterministic model.

It should be pointed out that statistical methods can be quite valuable in a deterministic modeling effort, and one should not consider these two approaches as mutually exclusive. Of course, in the white pine blister rust example a statistical approach might be chosen as the most direct avenue to a working model of the overall system. It must be recognized that such models, which attempt to span the gap from cause to overall effect ignoring the interim stages, are limited to rather general predictions within the range of the given data. Climatic weather models for example, based on previous history, can provide a general estimate about weather expectations. Such predictions would be within the realm of prior history, i.e., they could not accurately foresee anomalous weather occurrences. A deterministic simulation model, on the other hand, would be more reliable for predicting anomalous patterns because of the more exact representation of the intermediate occurrences in the system cycle.

VALIDATION OF THE MODEL

The discussion of anomalous behavior points out the need for validation of a model. That is, given an actual experimental set of observations, how well can a working model anticipate the final outcome? It is essential that those responsible for model development take into account the necessity of validation and the means to accomplish it. The image that comes to mind at the mention of the term *validation* is that after a complete working model is established, someone must "take it out and test it." While the final working model should be verified for its capability, it is important to do the utmost to ensure validity along the way.

The following series of recommendations is offered not only to assist in model validation, but generally to improve the chance for success in planning, development, and ultimate acceptance of the working model (particularly one involving simulation):

1. Start with a conceptual model. Though it is implicit that one must start with a basic definition of the phenomena involved, it is usually quite informative to portray the system graphically.
2. Set up a series of objectives leading to a final working model. Such goals should be measureable to permit an assessment of progress along the way, that is, allow for validation of the working model at a number of intermediate points.
3. Stage the tasks according to degree of difficulty and size. That is, simple tasks should precede the more comprehensive ones. Build smaller submodels, make them workable, and then incorporate them into a larger model.
4. Plan for generation of vital experimental data at the outset, particularly where needed to determine functional relationships.
5. Consider the task of translation of data into a working model, and the generation of a usable form of output from the computer.
6. Allow for evaluation and selection of possible alternative methods of simulation and computation.
7. Generalize and streamline the model to make it easy for other less sophisticated users to operate.

EXAMPLES OF MODELS

AIR POLLUTION DISPERSION MODELS*

An air pollution dispersion model relates meteorological elements to air pollution levels. Input data usually include weather conditions and total emissions from pollution sources; output data constitute a prediction of pollutant concentration as it varies in space and time. The physical and chemical processes that "transform" the source emissions into concentrations observed at sampling sites are represented by mathematical algorithms.

Air pollution dispersion models can be generalized to all materials transported in the atmosphere. Thus many of the aerobiological models discussed in this chapter include air pollution dispersion models.

Components of an Air Pollution Model

An air pollution model contains three essential components. The first is a source inventory; what materials, how much, from what location, and at what

---

*Conrad J. Mason

rate are they being injected into the atmosphere are questions addressed by it. The second is a meteorological data network to gather the necessary weather data. The third is the mathematical algorithm referred to above. Figure 7.2 illustrates the relation of the model to the physical system being modeled.

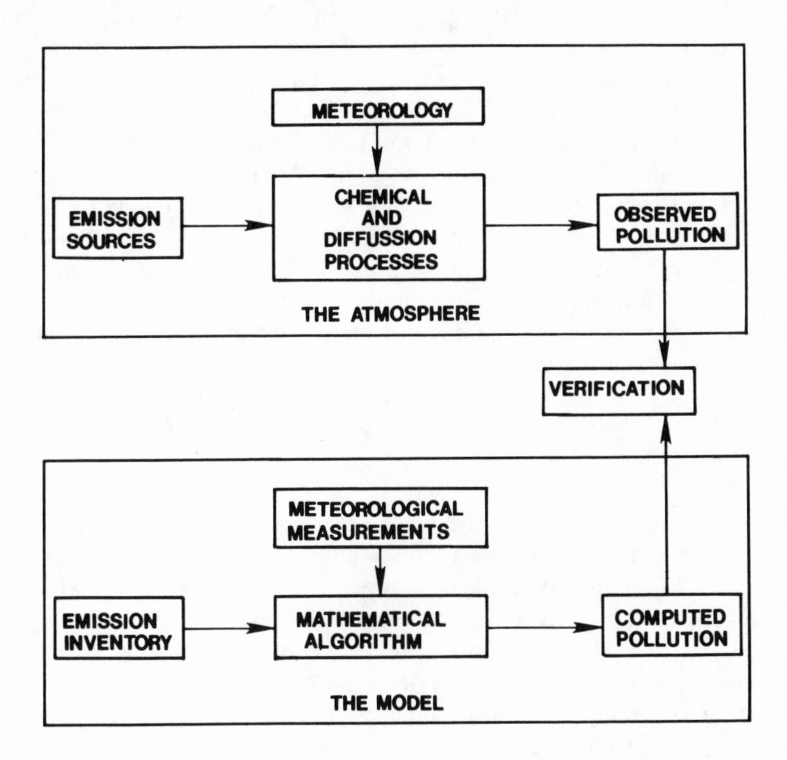

Figure 7.2. Schematic diagram of an air pollution model.

The source inventory

An air pollution model usually provides for two types of sources: (1) point sources, such as the stack of a large foundry or public utility; and (2) distributed sources, both area and line, such as the homes in a residential area or the highways that may pass through it. A pollution-source grid is frequently constructed, on which point sources are specifically located and distributed sources are mapped as to location, extent, and intensity. Source location and emission rate as a function of time must be known. Hourly emission rate data are desirable, but usually only data with much coarser time resolution are available. Knowledge of seasonal variations in hourly emission rates is, of course, also essential.

Usually detailed source inventory data are lacking. In some instances the sparseness of the data forces one to make certain assumptions to fill in what is missing. In many cases these assumptions do go hand in hand with the built-in limitations of the model. For example, suppose a prediction of the downwind concentration of pollen from a large field is required. An exact solution would require knowledge of the pollen release rate for each plant in the field. Obviously, such numerous measurements are not possible; instead, the release rates of several individual plants are measured and, assuming they are representative of the whole, an average release rate to be applied to the entire plant population is determined. In this instance the procedure is in accord with the dispersion modelers' technique of treating such a field as a uniform area source rather than as a conglomeration of point sources. Conversely, a knowledge of model limitations can allow one to avoid getting data that are too detailed and, consequently, data that cannot be handled adequately by the dispersion model. As in the preceding example, one could conceivably measure the pollen release rate of every plant in the field. The modeler might just as well assume a uniform area source; however, his predictions would not be improved by such a precise and detailed knowledge of the emission rates because of other approximations in the model.

Meteorological measurements

Wind speed and direction are among the most important meteorological measurements required. Finding representative sites at which to make these measurements is frequently difficult, however, since the horizontal wind field is heterogeneous especially during light winds. Both terrain features and structures that act as obstacles to flow introduce local perturbations in the wind field. Usually, National Weather Service wind speed and direction data (from local airports) are used by most modelers, although in special cases private towers and stations are erected.

An important characteristic of the atmosphere with regard to dispersion, in addition to the dilution effect of wind speed, is its *turbulent intensity,* a measure of the energy in the turbulent eddies that exist at any given time. In general, turbulent intensity depends upon three factors. The first is the roughness of the terrain over which the wind is blowing. The second factor is vertical wind shear; that is, the change of wind speed with height. The third factor, and perhaps the most important, is the vertical temperature structure of the atmosphere, i.e., the variation in temperature of the atmosphere as altitude increases. The temperature variation is referred to as the *temperature lapse rate* or, simply, the *lapse rate*. The lapse rate may be measured by temperature sensors mounted at various heights on appropriate towers or carried aloft by tethered or free balloons. A determination of the turbulent intensity is essential either by its direct measurement with special instrumentation or by an estimation of it from lapse rate measurements or other techniques described below. In a complete study, precipitation, solar radiation, and net radiation flux also may be measured.

Often the meteorological input to a model may be very sparse or lacking in detail. In these cases the failure of a model to produce reasonable results is not because it necessarily is a poor model; instead, its performance is masked by the uncertainties in the measurement of the wind and temperature fields. All too often the modeler is forced to generate additional wind and temperature field models, to serve as inputs to his dispersion model, on the basis of too few observations from too few stations.

The mathematical algorithm

The mathematical algorithm, the backbone of any air pollution model, can be conveniently divided into three major sections: (1) source emissions, (2) chemical kinetics, and (3) diffusion, the last of which includes the meteorological measurements. Each of these components can be treated as an independent entity for the analysis of an existing model, but in the development of a model, one must "size" the various components and generate subroutines of compatible complexity and precision.

Source emissions

The source emissions subroutine describes in mathematical terms the emissions of the various sources as a function of time. The subroutine is usually based on the pollution-source grid, which assigns coordinates to each identified source. Source strengths as functions of time and location are thus made available as input data to the remainder of the model. Emission data are obtained from the source emission inventory discussed previously.

Chemical kinetics

The removal or modification of pollutants as they are dispersing is affected by the chemical reactions they undergo with the ambient atmosphere and with each other. The formalism developed by the modeler to describe these processes is the chemical kinetics subroutine.

In the most general case the mathematical system of equations to be solved consists of a set of continuity or mass balance equations, one for each specific chemical species included (Lamb and Neiburger 1971).

Considerable simplification of the general problem can be effected if chemical reactions are not included and all variables and parameters are assumed to be time independent (steady-state solution). Although this assumption may seem contradictory to our intent and an oversimplification, it applies to any pollutant that has a long residence time (i.e., long compared with the time span for which the model is valid, at most a few hours) in the atmosphere as, for example, carbon monoxide. If chemical reactions are not included a solution is obtained that forms the basis for most diffusion models—the normal bivariate or Gaussian distribution for the downwind diffusion of effluents from a continuous source. Because of its widespread application, this equation and its numerous variations are discussed in the next section.

The second level of chemical-kinetic complexity assumes that first-order or pseudo-first-order reactions are responsible for the removal of a particular pollutant; as a result, its concentration decays exponentially with time. In this case a characteristic residence time or half-life describes the temporal behavior of the pollutant. Consequently, the removal of pollutants by chemical action can be included by simply multiplying the solution to the diffusion equation by an exponential decay term of the form $\exp(-t/T)$, where T represents the half-life of the pollutant under consideration (Roberts et al. 1971, Martin 1971). Since chemical reaction rates are markedly influenced by temperature, e.g., a doubling for every 10°C rise in temperature, the residence time must also be a function of temperature if chemical reactions are involved. It should be noted here that chemical reactions are not the only removal mechanism for pollutants; some other processes contributing to their disappearance may be absorption by plants, soil bacteria action, impaction on surfaces, and washout. To the extent that these processes are simulated by or can be fitted to an exponential decay, the above approximation proves useful and valid.

If specific chemical interactions between pollutants need be considered, then a set of reaction equations to describe the chemistry for the entire system must be written. For complex chemical systems, the number of equations is reduced by resorting to a "lumped-parameter" stratagem. In the lumped-parameter representation, typical reaction rate equations (usually selected from the rate-determining reactions) are employed with adjusted rate constants derived from appropriate verification data. Lumped-parameter subroutines simulate atmospheric conditions with a simplified chemical kinetic scheme to reduce computing time (Hecht and Seinfeld 1972, Friedlander and Seinfeld 1969).

Diffusion

The diffusion model describes the diffusion and dispersion of pollutants in the atmosphere. A wide array of plausible assumptions regarding the actual diffusion process forms the basis for the many kinds of diffusion models in existence. Thus, for example, we have fixed-cell models, moving-cell models, multiple-box models, particle-in-cell models, and so on (Johnson 1972). Some of these model types are illustrated below, where we discuss specific models developed by various investigators.

The earliest models are characterized by a plethora of assumptions and approximations whose purpose was to simplify their application. Recently, models have become more "rigorous" as modelers have tried different approaches to eliminate some of the poorer approximations. The wide-scale availability of large computing systems has made some of these attempts feasible. Of late, however, several authors (Hanna 1971, Milford et al. 1971) have called for a return to simple models. Their major contention is that in many instances "over-modeling" occurs; that is, sophisticated models are employed that are not warranted by the relative crudeness of the input data.

Dispersion Models

As discussed previously in this chapter, there are two major approaches to modeling aerobiological systems: simulation models and statistical models. We give several examples of each of these types of models as applied to atmospheric dispersion modeling in what follows. Considerable attention is paid to the Gaussian plume model because of its widespread and frequent use; an example of a simple model developed by Hanna (1971) is included.

Simulation models

The Gaussian plume model

The Gaussian plume model is perhaps the most widely used dispersion model. With reference to a coordinate system whose x-axis is aligned in the direction of the mean wind, with the z-axis along the vertical and the y-axis as the crosswind axis, the equation that gives the concentration as a function of distance from a point source is:

$$\chi = \left(\frac{Q}{U}\right)\frac{1}{\sqrt{2\pi}\sigma_y} \exp\left[-\left(\frac{1}{2}\right)\left(\frac{y}{\sigma_y}\right)^2\right]\frac{1}{\sqrt{2\pi}\sigma_z} \exp\left[-\left(\frac{1}{2}\right)\left(\frac{z}{\sigma_z}\right)^2\right] \qquad (7.1)$$

where the symbols are defined subsequently. It is written in this form to emphasize that it is the product of two normal distributions, i.e., the bivariate normal distribution.

A Gaussian point-source plume from an elevated source is depicted in Figure 7.3. If the origin of the previously defined coordinate system is located at the source (at the top of the stack in the figure), then Equation (7.1) holds. Note that either a vertical or a horizontal slice through the plume shows a normally distributed concentration pattern in accord with Equation (7.1). The cross section of the plume is an ellipse as illustrated.

Consider now the physical interpretations of the factors in Equation (7.1). First, the quantity Q refers to the strength of the source; that is, the amount of material emitted per unit time such as the number of pollen grains per second. Here Q is assumed to be constant in time; there is no temporal variation in emission. Second, the factor $1/U$, where U is the mean wind speed, is the *dilution factor,* so-called because the concentration is universely proportional to wind speed, as expected.

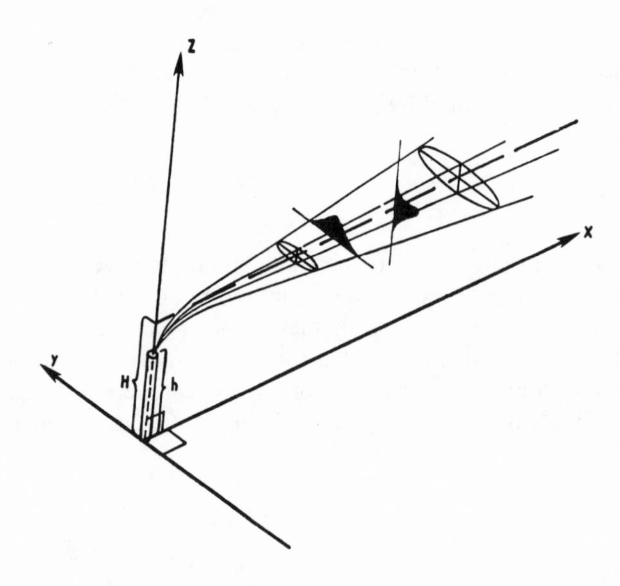

Figure 7.3. A Gaussian plume (from Turner 1969).

The dispersion coefficients σ_y and σ_z are analagous to the standard deviation of a normal distribution and so are measures of the lateral (y) and vertical (z) dimensions of the plume. They vary with the turbulent intensity of the atmosphere, the distance from the source, the sampling time, and characteristics of the terrain. To characterize atmospheric turbulence and turbulent intensity, one can use a completely theoretical treatment that relates the dispersion coefficients to the wind-vector fluctuations; details of this development are given in Slade (Slade 1968). It is sufficient to note that the Gaussian plume idea works well because of the statistical nature of turbulent diffusion. In order to use Equation (7.1), empirical values for the dispersion coefficients can be determined for certain special cases. In fact, this has been done and the results appear as Figures 7.4 and 7.5. These figures are plots of the horizontal and vertical dispersion coefficients versus downwind distance from a point source with "atmospheric stability" as a parameter. Practically, one can establish atmospheric stability categories of turbulence if the surface wind speed and the amount of incoming solar radiation are known. Table 7.2 lists these; note there are six categories (A through F) ranging from extremely unstable conditions to moderately stable conditions. The first column refers to surface wind speed, the second to the amount of solar radiation or daytime insolation. In the presence of clear skies, insolation is determined only by solar altitude; that is, the insolation is "strong" if the sun is at an angle greater than 60 deg with respect to the horizon, "moderate" if it lies between 35 and 60 deg, "slight" if it is greater than 15 deg but less than 35 deg. Nighttime conditions are assumed to prevail from 1 h before sunset to 1 h after sunrise. Clouds modulate the amount of incoming solar radiation; in the presence of broken

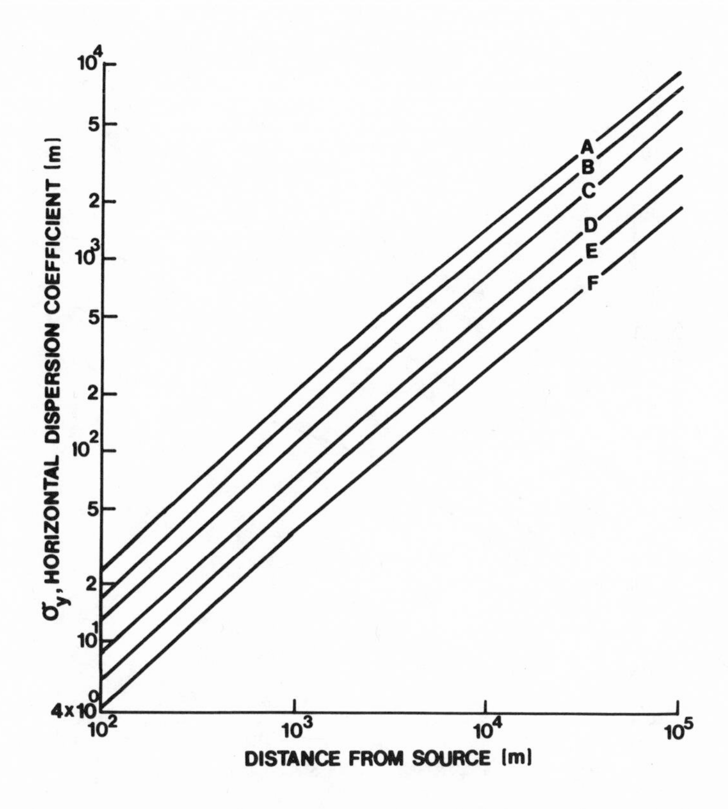

Figure 7.4. Lateral dispersion versus downwind distance from a source (after Slade 1968). Stability classes as follows: (A) extremely unstable, (B) moderately unstable, (C) slightly unstable, (D) neutral, (E) slightly stable, (F) moderately stable; see Table 7.2.

Table 7.2. Key to stability categories.[a,b]

Surface wind speed at 10 m (m/sec)	Day: Incoming solar radiation			Night	
	Strong	Moderate	Slight	Thinly overcast or ⩾4/8 low cloud	⩽3/8 low cloud
<2	A	A—B	B		
2—3	A—B	B	C	E	F
3—5	B	B—C	C	D	E
5—6	C	C—D	D	D	D
>6	C	D	D	D	D

[a]The neutral class, D, should be assumed for overcast conditions day or night.
[b]After Pasquill (1962) as quoted by Turner (1969).

middle clouds, "strong" becomes "moderate," whereas broken low clouds modulate "strong" radiation to "slight." To determine the values of the dispersion coefficients that are appropriate to a given stability class, refer to Figures

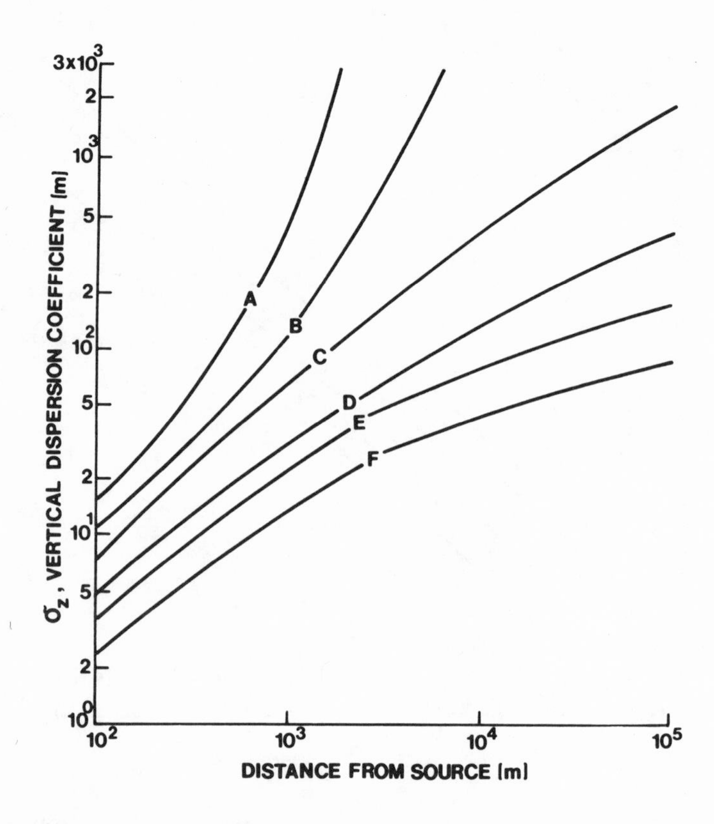

Figure 7.5. Vertical dispersion versus downwind distance from a source (after Slade 1968). Stability classes as follows: (A) extremely unstable, (B) moderately unstable, (C) slightly unstable, (D) neutral, (E) slightly stable, (F) moderately stable; see Table 7.2.

7.4 and 7.5. If these values are substituted in Equation (7.1), then the concentration (χ = number/volume) at the chosen point from a source emitting material at a given rate Q can be determined. Figure 7.6 shows a typical surface concentration pattern from an elevated source. Note that the maximum concentration is not at the base of the source but at some downwind point determined by the precise values of the dispersion coefficients and the mean wind speed.

The actual values assigned to the dispersion coefficients depend on the sampling time, that is, the time period over which the concentration measurements are effectively averaged. Figure 7.7 shows the outlines of a plume as a function of sampling time. On the left, the approximate outlines of a plume observed first instantaneously, then averaged over 10 min, and finally over 2 h are shown. On the right, the cross-plume concentration patterns are given; note how the width of the plume increases as the sampling time increases. Since the dispersion coefficients are a measure of this width, then they too

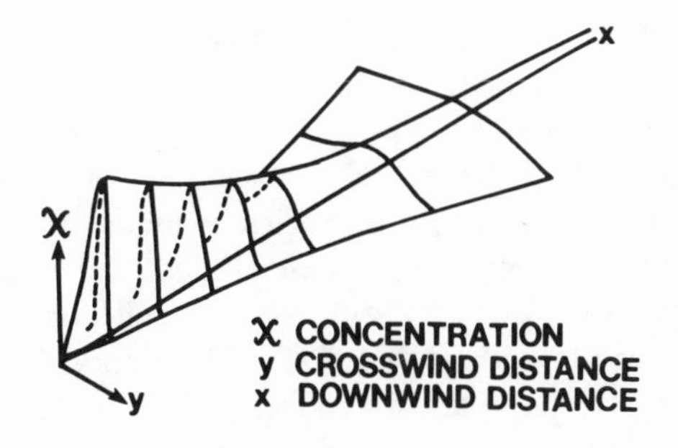

Figure 7.6. Concentration at surface, downwind from an elevated point source (from Slade 1968).

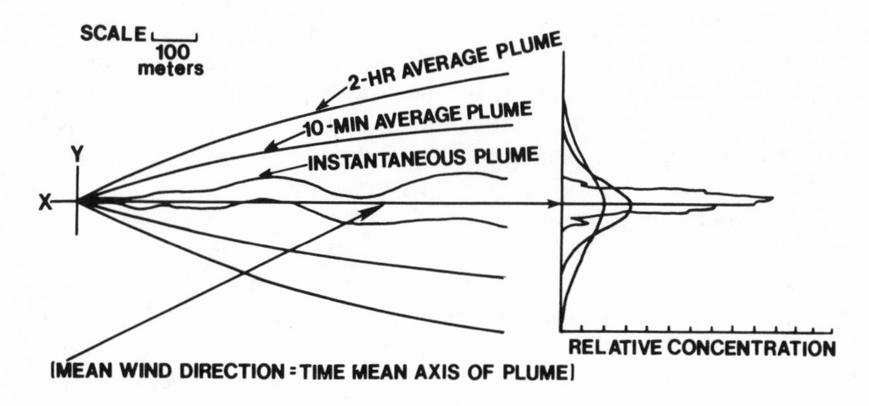

Figure 7.7. Approximate plume outlines averaged over various time intervals (left); corresponding cross-plume concentration (right); (from Slade 1968).

must increase as the sampling time increases. Figures 7.4 and 7.5 used to determine σ_y and σ_z are based on 10-min sampling times. To determine the average concentration for some other time period, empirical relationships exist that allow one to calculate these averages while still using the same figures (Turner 1969).

Finally, the last factor affecting the dispersion is the character of the terrain. Where the dispersion is occurring—on grasslands, in a heavily forested area, or perhaps in an urban environment (with many tall buildings and obstructions to flow)—is obviously going to affect the rate at which the plume spreads. Figures 7.4 and 7.5 pertain to grasslands. Some studies have been conducted to measure the dispersion coefficients in other locales as, for example, mountainous terrain (Start et al. 1975).

Many variations are possible on the basic model and sometimes, in fact, they are mandatory. For example, the presence of inversions (a more stable layer of air overlying the surface air layer) acts as a lid that effectively prevents any further upward diffusion. In the same way, the plume eventually intercepts the earth's surface, which acts as a barrier to any further downwind dispersion. Somehow our model must treat both of these boundaries. The easiest case to consider is that of a gaseous plume; here, one assumes that perfect reflections occur at both bounding surfaces, i.e., at the inversion height and at the ground. In this case, far enough downwind from the source, the model predicts a uniform concentration of effluent independent of altitude. For particles and aerosols, one can still assume "reflection" from the inversion, but for the surface bound one must consider deposition—a loss mechanism discussed at length in chapter 4, and treated briefly below. Other complications such as the presence of area sources or line sources require further modification of the Gaussian plume model (Turner 1969).

Because of their size, dry deposition (in contrast to deposition brought about by rain) is the most important loss mechanism for particulates of biological origin. These particulates are on the order of 10–100 μm in diameter; they have large gravitational fall speeds, i.e., terminal velocities of about 0.3–30 cm/s. Under these conditions and with the assumption of horizontal transport, one can assume the particles disperse as predicted by the Gaussian plume model while, at the same time, they settle with their respective terminal velocities. Consequently, for an elevated source, this settling tilts the centerline of the wind plume downward. We can effect this mathematically by replacing the variable z in Equation (7.1) with the expression $(z - V_g \cdot t) = (z - V_g \cdot [x/U])$, where V_g is the gravitational settling velocity of a particle of a given size, t is the transit time to the point x, x is the downwind distance, and U is the mean wind speed. As a result, we can calculate the surface concentration for particles of a particular size. If the source emits particles of different sizes, then a separate calculation must be carried out for each size. Note that the surface concentration multiplied by the settling velocity $(\chi \cdot V_g)$ is equal to the downward flux of material into the surface, that is, the amount removed per unit time per unit area—a quantity of some interest to biologists.

For very small particles, the observed dry-deposition rates are greater than those predicted by gravitational-settling theory alone. We can treat these very small particles in the manner outlined above by defining an *effective deposition velocity* (Slade 1968).

The Gaussian plume model is valid only for a region close to the source and for a time period during which no significant changes in any important parameters occur. These are very severe limitations. For example, suppose we wished to predict the diurnal variation of the pollen concentration at an urban sampling station from sources located in rural areas. The Gaussian plume model would not yield accurate estimates because the sources are far removed from the sampling point. While the pollen was in transit, significant changes in the dispersion parameters could occur because of a change in the wind direction, the amount of solar radiation, or the terrain characteristics. These changes would alter the character of the plume in a fashion not amenable to analysis using the framework provided by the Gaussian plume model.

Advecting Gaussian puff model

An example of what can be done to overcome some of these limitations is given by an advecting Gaussian puff model (Roberts et al. 1971). Here, the emission from a source is treated as a series of puffs, each of which undergoes Gaussian diffusion while its center moves with the mean wind. At any point in space, the concentration is determined by the cumulative effect of successive puffs. Thus the purely diffusive model, as in the Gaussian plume, is replaced by a trajectory-diffusion model. Moreover, changes in any model parameters can occur "between" puffs. First, the source strength can change from puff to puff, i.e., time-varying emission sources are allowed. Second, the stability class can change while the puff is in motion; that is, the puff disperses at one rate characterized by certain values of the dispersion coefficients prior to the change and at another rate (with different values of the dispersion coefficients) after the change. Recall that such a change in stability class (or turbulence level) can occur either from a change in the weather or a change in the character of the terrain. Consequently, diffusion from an urban area to a rural area or vice versa can be remodeled. Third, since the time history of a puff can be followed for a considerable period, the model is no longer restricted to a localized region. For example, the puff can be considered as moving with the winds associated with mesoscale weather systems. Thus, by this one strategem, many of the limitations of the Gaussian plume model have been removed. The one important remaining problem is the experimental determinations of the dispersion coefficients under all of the conditions for which the model may be applied.

Deposition can be handled in a manner similar to that used in the Gaussian plume approach: one can associate an effective deposition velocity with a particle of a given size and density; a puff of these particles then disperses, moves with the mean wind, and settles with that velocity simultaneously. If there is a spectrum of particle sizes, then a family of puffs is needed, each characterized by the appropriate settling velocity.

An adaptation of the advecting Gaussian puff model that incorporates a settling velocity term has been successfully applied to the dispersion of gypsy moth larvae (Mason 1975, Mason and Edmonds 1974, Mason and McManus 1975).

The "multi-box" model

MacCracken et al. (1971) have developed a "multi-box" model based on an Eulerian (fixed) grid approach in which both transport and photochemical processes are treated simultaneously. Because of the multiple-box approach, complex geographic regions can be modeled since both the meteorological and the pollutant concentrations can vary in space and time. A large set of coupled rate equations, one equation per pollutant per box, is produced. In order to simplify this set, MacCracken et al. (1971) confine their attention to the transport of a single nonreacting pollutant. Further simplification is affected by treating only a single-layer box model, in which the bottoms are formed by the local topographic surface while the tops are the time- and space-varying inversion height. Since the model yields only average vertical concentration under these circumstances, the authors have assumed a logarithmic pollutant profile with height that allows them to calculate the average surface concentration. It was necessary to generate inversion and wind fields on a grid for input to the model; this process was accomplished in the usual way, namely, the smoothing of available meteorological network data.

The model was applied to the San Francisco Bay area, a very complex geographic region. The diurnal variation of carbon monoxide levels was predicted and compared with observations at three urban sampling stations separated by over 64 km (San Francisco, Oakland, and San Jose). For two of the stations, the agreement is very good; the discrepancy for the third station (Oakland) is attributed to the complex terrain in its vicinity as well as the 6-h averaging of meteorological data. In addition, the model correctly reproduces the log-normal frequency distribution with the same slope that Larsen found (Larsen 1971).

The Hanna-Gifford model

Hanna has proposed the use of a very simple yet physically realistic model to calculate surface concentrations from area sources in urban environments (Hanna 1971). He assumes that the various area sources are fairly uniform in source strength (which is, in fact, the case) and shows that the surface concentration in a grid block (χ) is proportional to the source strength of that grid block Q and inversely proportional to wind speed U, i.e., $\chi = CQ/U$, where C is a proportionality factor that is only a weak function of stability. The predictions of this model are compared with those of a more complex model for natural gas concentrations in the Los Angeles basin; the predicted $\chi U/Q$ ratios for the two models agree with statistical uncertainties. Similarly, in a Chicago-based study, SO_2 isopleths are constructed and compared with those of a complex model; in this instance, the forms of the two sets of isopleths are similar although the simpler model overestimates the concentrations by a factor of two. Yet, in this case, the entire calculation took only 30 min by hand. Hanna concludes that his model could be used in those instances where the slight improvement in predicted concentrations resulting from the use of a more complex model does not justify the effort. He advocates its use as a baseline estimator during the evaluation of a more complex model: if the proposed model does not produce a significant improvement over the simple model, there is no *practical* justification for its use.

Statistical models

The linear regression model

An example of a statistical model is that developed by Roberts and Croke (Croke et al. 1968). Here, for any given sampling site, a linear relation between the concentration and various sources is assumed, i.e.,

$$\chi = C_0 + C_1Q_1 + C_2Q_2 + \sum_{i=1}^{n} k_iQ_i \quad (7.2)$$

where C_0 represents the background level of the pollutant, Q_1 = a particular type of source such as commercial and industrial emissions, Q_2 = emissions from home heating and the k_iQ_i terms are large individual point sources. The coefficients C_1, C_2, and k_i represent the $1/\sigma_y\sigma_z$ term as well as the contribution of the exponential factor in the Gaussian diffusion equation (7.1).

Multiple discriminant analysis techniques for individual sampling sites can be used to determine the probability that pollutant concentrations fall within a given range or that they exceed a given critical value. Meteorological variables such as temperature, wind speed, and stability are used as the independent variables in the discriminant function.

The tabulation prediction technique

This method consists of developing an ordered set of combinations of relevant meteorological variables and presenting the percentile distribution of SO_2 concentrations for each element in the set (Moses 1970). Obviously, the technique can be applied to any pollutant for which an adequate data set is available. Since the independent meteorological variables are ordered, it is possible to look up any combination of these variables to find the predicted distribution. The advantages of the tabulation prediction method are: (1) it is easy to use, (2) it provides predictions of pollution concentrations rapidly, and (3) it provides a percentile distribution of predicted pollutant concentration. Its disadvantages are (1) at least 2 yr of meteorological data for a given area are necessary; and (2) adding, removing, or modifying important pollution sources invalidates the model.

PLANT PATHOLOGY MODELS*

In the epidemic development of plant pathogens several events occur in sequence: (1) spores are produced and released, (2) they are dispersed, (3) they are deposited on susceptible hosts, and (4) they germinate and infect plants. That sequence is repeated several times during an epidemic and the outcome of the cycle can be measured as crop loss. That sequence of biological events in disease development has been called the *aerobiology pathway* (Edmonds 1972) and aims to describe plant disease epidemics in a systematic manner. Although that sequence is perfectly logical, many researchers are interested in only one event or another and fail to view the study of epidemics as a system. There are notable exceptions, however. The studies of Edmonds (1973), Waggoner et al. (1972), Kranz et al. (1973), McCoy (1971), and Shrum (1975) clearly emphasize a systems approach and simulation modeling. Simulation models attempt to quantify biometeorological interactions between host, pathogen, and environment affecting the rate of disease development. In that regard, successful simulators mimic disease progress and permit rapid measurement of the cumulative effects of changes of environmental variables on individual events in the disease cycle such as spore production, release, germination, penetration, infection, and incubation period. Successful simulators also might be good predictors of disease but that is not their primary intent.

Another approach to disease modeling that has enjoyed considerable use is linear regression. Whereas simulation of plant disease epidemics describes in a logical fashion the effects of biological and meteorological variables on all events in the aerobiology pathway, linear regression describes mathematical relationships among one or more variables affecting spore production, release, dispersal, deposition, and infection. Functional relationships expressed by regression lines are empirically fitted curves; so, unlike simulation, the functions simply represent the best mathematical fit to an observed set of data and do not always have a clearly interpretable biological meaning. Such curves may have real merit as predictors of events in the aerobiology pathway, but they fall short of a precise description of the effects of biological-meteorological events on each step.

Early attempts to predict plant disease development used neither simulation nor regression techniques but relied heavily on observation of the interactions between pathogens and environmental factors favoring rapid disease development. The efforts of Wallin and Waggoner (1950), Hyre (1957), and Bourke (1957) were among the first to relate epidemic development of plant pathogens to the frequency of disease-favorable days. Each day was evaluated on the basis of temperature and moisture requirements for fungal activity and, after a prescribed number of consecutive disease-favorable days passed, a forecast was issued. It is noteworthy that among the many attempts to develop

*J. R. Burleigh, R. E. McCoy, and J. R. Wallin

plant disease forecasts this rather simple approach has given plant pathologists one of the few workable disease prediction systems. Based on Hyre's and Wallin's studies, Krause et al. (1975) wrote a computer program for potato late blight that gives regular, timely, local forecasts. That system is operative and has proved to be successful by issuing forecasts at a time when disease incidence and severity permit control.

In the next few pages we present specific examples of models and mathematical representations used to describe the events in the aerobiology pathway and discuss advantages and disadvantages of each approach in achieving stated goals.

Simulation Models

The simulation of plant disease development by a computer is a new concept of unparalleled importance in plant pathology. The importance of the plant-pathogen-environment triangle has been stressed for many years, but not until the development of the plant disease simulator did it become possible to numerically analyze this system in toto, adding the dimension of time as well. The application of this concept to plant pathology originated with Waggoner (1968), who used data from the literature to construct a computer simulator of the *Phytophthora* late blight of potato. This simulator successfully imitated late blight epidemics in North America. A year later Waggoner and Horsfall (1969) followed up with EPIDEM, a simulator of *Alternaria* early blight of tomato, which successfully followed five years of early blight epidemics as well as allowing some basic new insights into the interactions within the tomato-*Alternaria*-weather-time system.

Since these first attempts, plant disease simulators have been developed for southern corn leaf blight (Waggoner et al. 1972, Massie 1973), apple scab (Kranz et al. 1973), *Fomes* butt rot of conifers (Edmonds 1973), hop downy mildew (Waggoner 1974), and *Mycosphaerella* blight of chrysanthemum (McCoy 1971). In addition, a general-purpose simulator adaptable to a number of plant diseases has been developed by Shrum (1975).

McCoy (1971), in investigating the life cycle of *Mycosphaerella ligulicola,* a pathogen of chrysanthemum, incorporated data found in controlled environment studies and in the literature into a simulation model of this disease system. This model, entitled MYCOS, will serve as an example in this section. Program MYCOS is a computer program that estimates disease development each day of the growing season in light of the plant-pathogen-environment triangle. Each day is divided into four 6-h periods for which means of temperature, relative humidity, wind, and sky cover are evaluated for their effects on pathogen and disease development. Other factors directly evaluated are amounts and duration of rainfall or irrigation, and dew. A period-by-period, day-by-day account of the host and pathogen factors intrinsic to disease development is then calculated until, at the end of the season, a graph of lesion area is plotted against time.

Life cycle

The systematic analysis of *Mycosphaerella* blight is based on the life cycle (Figure 7.8) of *M. ligulicola*. The aerobiology pathway is plainly evident in this cycle, which is divided into two phases for convenience. Each phase, infection and dispersal, is of equal importance in the development and spread of the pathogen. Infection concerns the series of events occurring on the host plant, i.e., spore germination and ingress, development of lesions, and production of reproductive structures. Dispersal includes the liberation, transportation, and deposition of spores from their source to their ultimate infection sites.

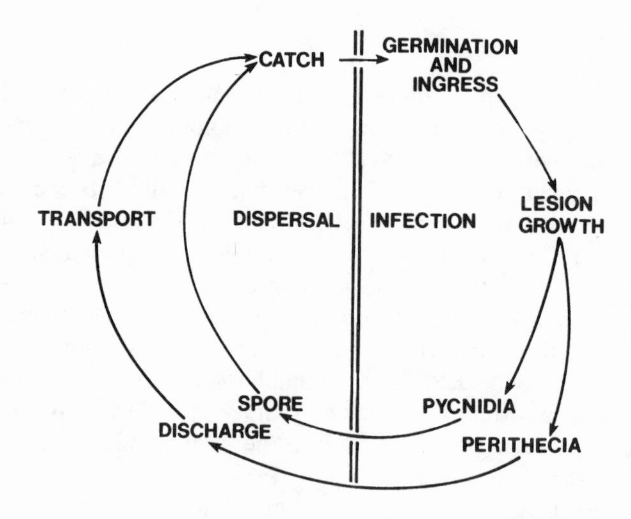

Figure 7.8. Life cycle of *Mycosphaerella ligulicola*.

Each step of the disease cycle was examined separately and numerical parameters were placed on the biological processes involved. The numerical parameters were found either in the literature or in controlled-environment studies of pathogen and disease development (McCoy et al. 1972, McCoy and Dimock 1972). The dispersal phase, particularly transport and catch, is based on the physical processes of turbulent atmospheric diffusion. The simplified flowchart given in Figure 7.9 depicts the logical sequence of operations performed by MYCOS in calculating disease progress.

Lesion growth

Once established, an infection results in lesion development provided a favorable environment is present. Temperature and relative humidity (RH; Figure 7.10-A) were found to be the controlling factors regulating the rate of lesion growth in chrysanthemum (McCoy and Dimock 1972). Temperature data are incorporated in MYCOS in the function

Equation $$LESN = [\exp(TEMP/7) - 1](30 - TEMP)Z \qquad (7.3)$$

where Z is a constant related to lesion growth. This same function is also used

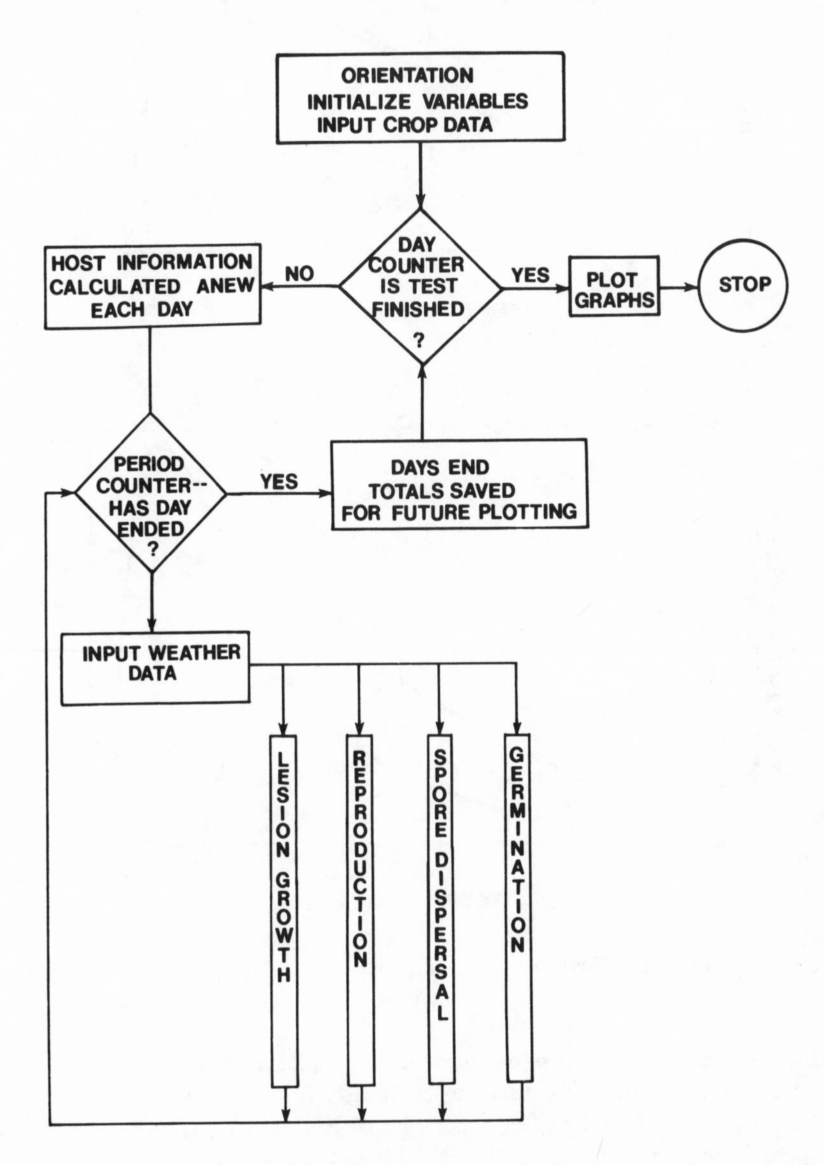

Figure 7.9. Simplified flow chart of operations performed by program MYCOS in calculating disease progress.

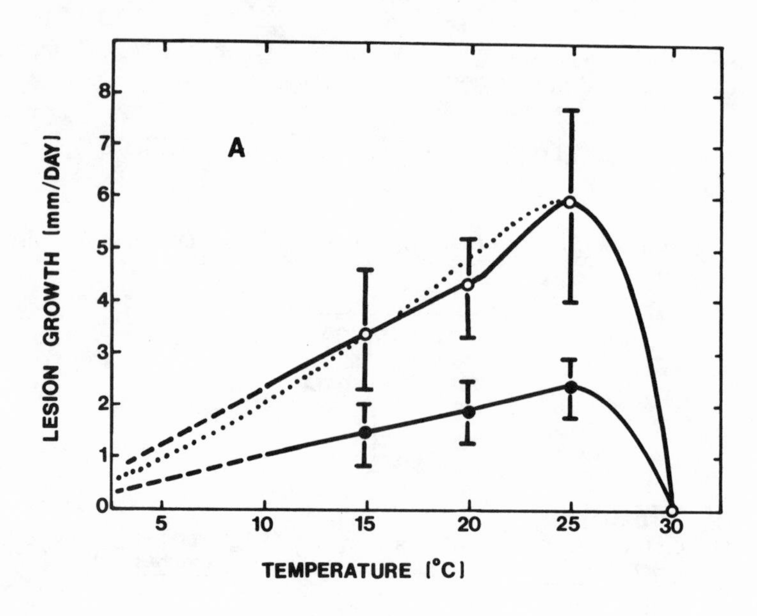

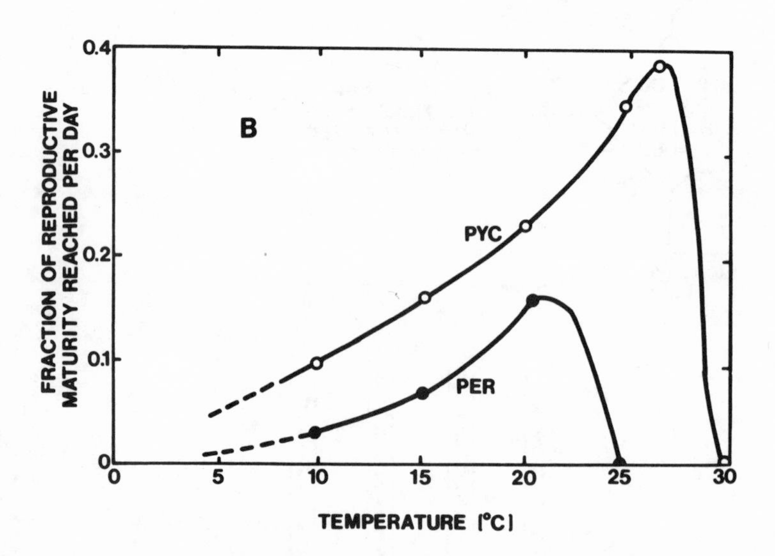

Figure 7.10 Lesion growth (A) and reproductive development (B) of *Mycosphaerella ligulicola*.

to calculate the rates of spore germination and the rate of pycnidium maturation; a different constant Z is used for each case.

Since RH profoundly affects the rate of lesion growth, the LESN value obtained is then multiplied by the RH for that period, i.e., an RH of 50% halves the calculated value. Multiplying this lesion value by the number of infections present gives the increment in lesion growth in centimeters for the crop as a whole for the particular day and period being considered. This information is stored in a table (array) for future use in calculating reproductive development.

Reproductive development

Depending on environment, reproductive structures (pycnidia and/or perithecia) may or may not be produced on *M. ligulicola* lesions. McCoy et al. (1972) reported that pycnidia and perithecia have different temperature optima as well as different rates of development. Pycnidia are produced over a wider temperature range, have a higher temperature optimum, and require less time for maturation than perithecia (Figure 7.10-B). Increments of pycnidial (PYC) and perithecial (PER) maturation are calculated for each period for each lesion increment. When mature, the numbers of pycnidia and perithecia are calculated and used to estimate the number of spores available for initiating subsequent infections.

To complicate matters, it was found that conidia produced at low temperatures (10°–15°C) were large and septate, whereas conidia produced at high temperatures (25°C) were small and unicellular (Blakeman and Hadley 1968), although many more spores were produced at high than at low temperatures. In evaluating virulence of these two conidium types, Blakeman and Fraser (1969) found the larger conidia to have a greater pathogenic potential than the smaller, unicellular conidia.

Information regarding the production of these two conidium types has been programmed into MYCOS, thereby allowing calculation of the percentages of each type (UNIC = % of unicellular conidia, SEPC = % septate conidia) from past weather data. Direct evaluation of the potential of the two conidium types has been made possible by including these data in MYCOS.

Spore liberation

Mycosphaerella ligulicola produces two types of spores: conidia, which are water dispersed, and ascospores, which are carried by air currents. Conidia are oozed from pycnidia whenever there is sufficient moisture to form a water film on the tissue surface. The conidia, once exuded into the water film, are admirably suited for dispersal by splashing water.

The number of conidia liberated at any one time (LIBC) is calculated by multiplying the number of mature pycnidia (NPYC) by the number of spores that an average 150-μm-diam pycnidium holds. This amounts to 1.5×10^4 unicellular conidia, or 2.6×10^3 septate conidia. The formula

$$LIBC = NPYC[(1.5 \times 10^4)(UNIC) + (2.6 \times 10^3)(SEPC)] \quad (7.4)$$

is used for mixtures of the two conidium types.

Ascospores are explosively discharged from their perithecia when moisture from dew, rain, or irrigation accumulates to the extent of 0.1 mm or more (McCoy and Dimock 1973). The ascospores are discharged to distances up to 6 mm (McCoy 1973), sufficient to ensure their passage through the boundary layer of air above the tissue surface. Once airborne, the ascospores are subject to the laws of atmospheric diffusion.

Microscopic examination of perithecia indicates that ascospores in all stages of maturity are present at any one time, although 64 is the usual number of mature spores present (McCoy 1971). Therefore the number of ascospores discharged in any one period (LIBA) is calculated by multiplying the number of perithecia (NPER) by 64.

Spore transport

The transport and subsequent deposition of spores is treated in MYCOS as a model within a model. Virtually no quantitative data on spore dispersal are available. Yet, rather than use empirical estimates of dispersion, it was desired to place dispersal on a physical basis. The incorporation of an untested dispersal model in MYCOS is bound to create some quantitative errors, but these are not expected to be qualitative in nature. Also, in creating the model to fit as closely as possible the physical principles responsible for dispersal, it is felt that the errors are not likely to be large. As dispersal equations are improved, their incorporation into MYCOS will simply be a minor modification and will not affect the basic structure of the simulator.

Splash dispersal

Conidia of *M. ligulicola* are dispersed by splashing water, although spread may also occur by handling with hands or equipment; MYCOS considers splash dispersal only. Gregory et al. (1959) determined that the distance to which a falling raindrop hitting a surface would splash was dependent on the kinetic energy contained in the drop. Larger drops splashed farther on impact than did smaller drops falling at a lower velocity. Program MYCOS attempts to relate the distance of splash from a disease locus to the rate of rainfall (RRAIN). A heavy rain splashes spores farther than a light drizzle. In addition, wind blows splash droplets downwind thereby increasing the area covered by water splashing from any disease locus. This area (DSPC) in square meters is determined from wind velocity (WIND) and RRAIN.

Once conidia have been splashed from their source to an infection site, they are subject to washing if rainfall continues. Information on the washing of spores from infection sites by rain is scant. The WASH factor used in MYCOS, therefore, is the same factor used by Waggoner and Horsfall (1969) modified to denote the percentage of spores remaining per site after rain.

Airborne dispersal in the greenhouse

In a field situation discharged spores are carried downwind in a spore cloud, which becomes diluted as it moves farther from its source. Some of the spores eventually escape from the field. In the greenhouse we essentially have a closed system in which airborne spores are trapped and cannot escape. Venting the greenhouse, however, will allow the escape of some spores.

Program MYCOS assumes that ascospores liberated within a greenhouse are distributed evenly throughout the house. The number of infection sites receiving ascospores settling evenly over all horizontal surfaces within the house is calculated from Equation (7.5).

$$\text{CATA} = \frac{\text{LIBA}}{\text{AREA}(1.1 + \text{LAR})}(0.14 \times 10^{-4}) \times (\text{WASH})(\text{SITE})(\text{AREA}) \tag{7.5}$$

The number of ascospores liberated (LIBA) divided by floor area plus leaf area gives the number of spores present/m^2 of horizontal surface. Multiplying by the exposed surface area of one infection site (0.28 cm^2/2) then gives the probability of any one site's receiving a spore. In the event that overhead watering is used, this probability is multiplied by WASH. Multiplying the resultant probability by the number of available infection sites in the greenhouse (SITE $\times$ AREA) gives a probable number of ascospore inoculations. If conditions are suitable for germination and ingress, this number is transferred to an account for ascospore infections.

Airborne dispersal in the field

The primary factor affecting the transport of airborne spores in the field is the degree of atmospheric stability. Stable atmospheric conditions tend to dampen or suppress turbulent transport, whereas an unstable atmosphere accelerates it. Spores released into a stable atmosphere remain close together and tend not to be diluted as they move downwind, while spores released into an unstable atmosphere are diluted rapidly in both the lateral and vertical directions with increasing downwind distance.

The field dispersal model used in MYCOS is based on Pasquill's (1962) theory for turbulent transport. Depending on atmospheric stability, Pasquill's equations predict the downwind dispersion of particles released from a point source. The distances of horizontal and vertical dispersion for stable, neutral, and unstable atmospheric conditions are graphed in Figure 7.11-A,B. If these two graphs are matched so the x-, y-, and z-axes correspond to the downwind, crosswind, and vertical directions, respectively, we can graph the dispersal cones seen in Figure 7.11-C.

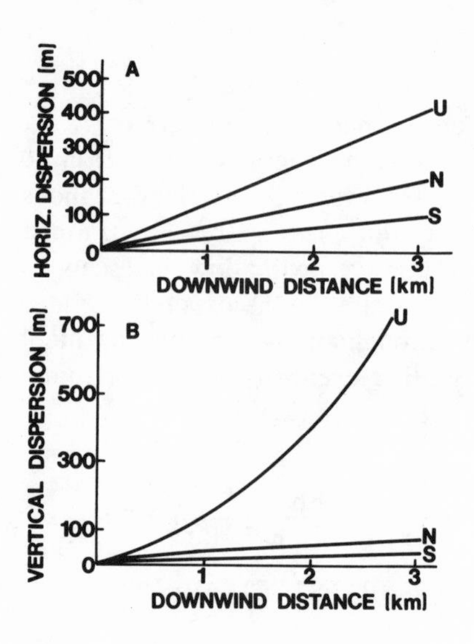

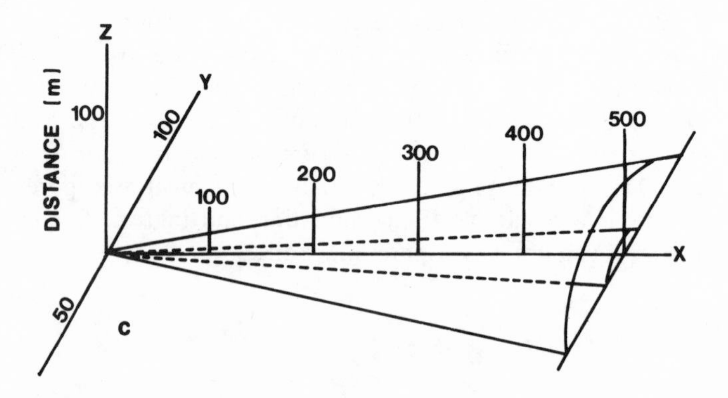

Figure 7.11 Distances of horizontal (A) and vertical (B) dispersion of particles for stable, neutral, and unstable atmospheric conditions; and (C) hypothetical dispersion cone achieved from matching (A) and (B).

These dispersal cones represent the time-averaged dispersal pattern emerging from a point source such as a disease focus. The boundaries of the cones are actually the standard deviations of the limits of dispersal, but their use allows a more accurate evaluation of the ascospore dose to which the crop within the dispersal sector is exposed. This is because MYCOS assumes an even distribution of spores in the dispersal cone, while in nature the spores would be most concentrated along the downwind axis. Using the standard deviation helps to correct the discrepancy. Later dispersal models may attempt to simulate dispersal cones with crosswind variations in spore concentration, as well as the downwind variations simulated by MYCOS.

The catch of ascospores on infection sites involves two processes: impaction and sedimentation. Spores traveling in an airstream may be impacted onto objects jutting into the airflow. The efficiency of impaction (EFF) depends on spore size, target area, and wind velocity. The probability of catch by impaction is calculated by multiplying the flux rate of spores through the crop (FLUX) at any one distance from the spore source by the frontal area of one infection site, and by EFF and WASH. The number of resulting inoculations is then calculated by multiplying by the number of infection sites exposed to impaction.

Catch by sedimentation is calculated by determining the total plant and ground surface area exposed to any one value of FLUX and incorporates the fall rate of the airborne ascospores. The resultant probability is multiplied by WASH, and by the number of infection sites exposed.

Since FLUX decreases continuously with downwind distance, MYCOS must integrate ascospore catch (CATA) from the spore source to the edge of the field. As MYCOS integrates CATA down the sector of dispersal FLUX continuously decreases, but the area of the crop subjected to FLUX steadily increases. Then CATA is summed over 10 integration intervals using mean values of FLUX for each interval. In order to eliminate positional effects, MYCOS assumes that the inoculum source is in the center of the field. In doing so, MYCOS provides an average CATA value for calculating ascospore infections.

Germination and ingress

Once spores are dispersed from their source to an infection site, germination will occur if moisture is present. It is necessary for MYCOS to know the rates of germination of spores in order to determine if the duration of the wet period is great enough for the spores to germinate and enter the host before drying commences. Equation (7.3) calculates hourly increments of the percentage of the germination process completed for the two conidium types and ascospores; a different value of Z is used for each. These calculations are made for each hour of leaf wetness. When the cumulative total reaches 1.5, ingress is presumed to have occurred and subsequent drying will not stop the infection. Blakeman and Fraser (1969) found that conidia produced at 15°C germinate more rapidly than conidia produced at 26°C. Their data were integrated with Equation (7.3) by McCoy (1971) and the relationship was determined between the rates of germination of septate and nonseptate conidia and the temperature at which germination is occurring. Blakeman and Fraser also found that septate, but not unicellular, conidia germinate at RH greater than 96%. This information is also incorporated into MYCOS.

Host

Disease development depends not only on the fungus and its physical environment, but is vastly influenced by the condition of the host crop. As the host plants grow, they present greater numbers of targets for inoculation. Program MYCOS has been programmed to calculate new host parameters each day of the growing season. These calculations are based on the measured development of chrysanthemum cultivar Fred Shoesmith grown under normal greenhouse conditions. Information for simulating other varieties or cultural conditions can easily be programmed into MYCOS.

The total host plant surface area in the crop is calculated from

$$\mathrm{HOST} = (\mathrm{LAP})(\mathrm{DENS})(\mathrm{AREA}) \tag{7.6}$$

where LAP is leaf area per plant, DENS is the number of plants/m^2, and AREA is field size in square meters. When disease is progressing through the crop, the accumulated lesion area (TLES) is subtracted from HOST each day to produce AVHOST, the remaining amount of susceptible host tissue. The AVHOST value is then used in the next day's round of calculations determining disease progress.

Output

So far as possible, the plant disease simulator attempts to mimic all aspects of the plant disease cycle concurrently, which is of immediate value in assessing the disease potentials induced by differing sets of environmental conditions. Output from MYCOS consists of a set of graphs depicting percentage of disease incidence in the crop, TLES/HOST (Figure 7.12), and the logarithmic transformation, ln (TLES/AVHOST), which are used in calculations of the apparent rate of spread of disease as defined by Van der Plank (1963). In addition, a weekly summary of major host and pathogen parameters such as host area, lesion area, number of infections, and numbers and types of fungal reproductive structures that have been calculated throughout the season is presented. If additional detail is requested, a period-by-period, day-by-day summary of these facts can be printed.

Uses, potentials, and problems

The most immediate use of a plant disease simulator is to determine the accuracy or completeness of our knowledge of a plant disease system. If the simulator cannot accurately follow disease development in the field, a gap in our knowledge exists. When the missing information is found and the simu-

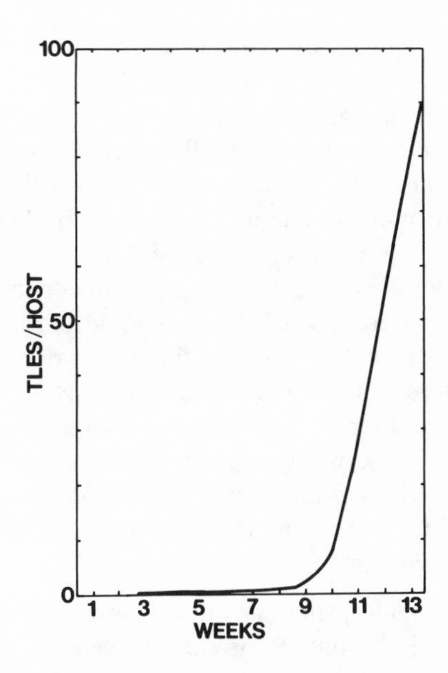

Figure 7.12. Percentage of disease incidence in the crop, TLES/HOST, as charted by program MYCOS.

lator is modified accordingly, both the simulator itself and our knowledge of the disease process are benefited. In addition, the construction of a simulator often reveals large gaps in our knowledge that must be incorporated either through direct experimentation or by inserting into the model an assumption that can be verified later.

The greatest potential use for plant disease simulators lies in their propensity to determine disease potential. The possibilities for disease prediction and the subsequent recommendation of control measures are limitless. Various control measures either singly or integrally may be compared under different sets of environmental conditions and the degree of control can be determined. It is even possible to compare the costs of applying these control measures to the economic benefits derived entirely through simulation. Through simulation it will ultimately be possible to develop optimized control programs based on the calculated disease pressure brought about by actual weather conditions.

One of the most important properties of simulation models is their flexibility. Changing parameters for particular situations can be achieved by simply changing one or two statements in the program. Such changes might concern shifts in the virulence of pathogen populations monitored from year to year as discussed in the case of race T of *Helminthosporium maydis* by Waggoner (1974). Differences in host resistance can be accounted for similarly, as in the case of *M. ligulicola,* which produces rapidly extending lesions on chrysanthemum C. V. Fred Shoesmith, but only slowly expanding lesions on C. V. Yellow Delaware. These different rates of lesion extension were incorporated in MYCOS by McCoy (1971) and the effect of host resistance was simulated accordingly.

The flexibility of simulation models is a property not shared by either multiple regression models or by simple differential equation models. Simulation attacks from within the problem of understanding the interaction between plant disease and environment. The disease cycle is broken down into discrete steps, each of which is thoroughly investigated before its reassembly into the logical whole, or model. Regression attempts to understand these interactions from without; the interaction itself is observed and the important driving factors are determined statistically. The resulting regression equation or model is inflexible in that different host or pathogen parameters not specifically analyzed in establishing the model cannot be incorporated later. The differential equations describing disease progress (Van der Plank 1963) cannot incorporate the wealth of variables present in a simulation model, nor can they account for the day-by-day variations in weather, which are the driving forces behind plant disease epidemics.

A major problem facing simulator developers, as discussed by Waggoner (1974), lies in relating the output produced by the simulator to the actual amounts of disease present in the field. Program EPIDEM (Waggoner and Horsfall 1969) skips this problem entirely by calculating continuously increasing numbers of lesions without regard to the quantity of susceptible tissue remaining in the crop. Program EPIMAY (Waggoner et al. 1972) partially corrects the problem by multiplying the possible number of infections produced by any given number of spores by the probability of a spore's landing on susceptible tissue. Schrum (1975) attacks this problem in the simulator EPIDEMIC by attempting to determine the spatial distribution of disease in any particular field, then calculating the actual number of infections in the field. This same factor is accounted for in MYCOS (McCoy 1971) by using only the remaining amount of susceptible host tissue (AVHOST) to calculate subsequent disease progress. Program MYCOS attempts to calculate absolute numbers in the form of total lesion area (TLES) and total host area (HOST) in a crop of a given size. As a result it is possible to calculate percentage of infection (TLES/HOST) and use these values in calculating rate of spread. The accuracy of these absolute numbers, however, is governed by the accuracy of the dispersal section of the simulator. Both MYCOS and EPIDEMIC contain involved models of dispersal based on aerobiological principles; however slight changes in the parameters involving airborne dispersal or spore catch may cause large changes in the outcome of the simulated epidemic (McCoy 1971). This points up the importance of the aerobiological pathway and the necessity of improving our knowledge of airborne spore dispersal in the future if plant disease simulators are to do more than serve as simple calculators of disease potential.

Regression Models

The concept of regression in aerobiology

An excellent treatise on the nature and utility of regression analysis in disease epidemiology is given by Butt and Royle (1974). Their comments are directly applicable to aerobiological concepts and need not be repeated here. We believe, however, that a brief description of the mechanics of regression will be useful to our readers to aid in the interpretation of studies reviewed herein.

In aerobiology we are concerned with relations between pairs of variables hypothesized to be in a cause-and-effect relationship, e.g., number of spores produced per unit of time, number of spores caught per unit of distance from a source, number of infections per unit number of spores caught, and crop loss as a function of number of lesions. The philosophical requirements for establishing whether the relationship between two variables is really cause and effect are not our concern here. We assume established procedures of scientific inquiry are followed so we can direct our attention to the form and significance of functional relationships between two variables. A function is a mathematical relationship enabling us to predict what values of a variable Y correspond to given values of a variable X. That relationship is usually written as $Y = f(X)$.

The simplest type of regression follows the equation $Y = X$, which is illustrated in Figure 7.13 and shows the relation between amount of disease and amount of inoculum. The figure shows that whatever the value of X (units of inoculum), the value of Y (units of disease) is of corresponding magnitude. Therefore 10 units of inoculum produce 10 units of disease. When $X = 0$, Y also equals zero and the line describing the function goes through the origin of the coordinate system. This means that with no inoculum there is no disease; a perfectly logical assumption. It is clearly seen that with the relationship $Y =$

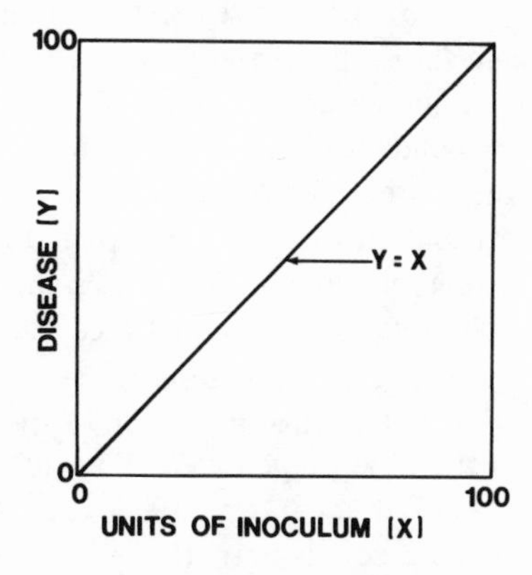

Figure 7.13. Regression of inoculum on percentage of disease severity.

X we can predict accurately the amount of disease, given the amount of inoculum. Of course we must assume that conditions for infection remain stable continuously, otherwise our comparisons would be invalid. We call the variable Y the *dependent variable,* while X is called the *independent variable.* The magnitude of Y depends on the magnitude of X and can be predicted from the independent variable, which is free to vary and is assumed to be measured without error.

Another functional relationship is illustrated in Figure 7.14 and is described by the equation $Y = bX$. Here the independent variable X is multi-

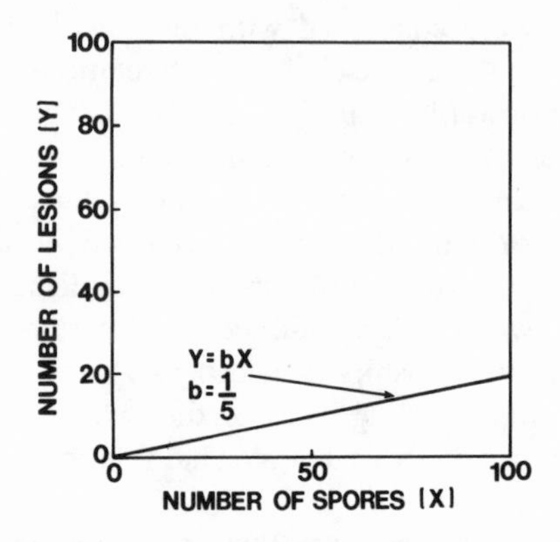

Figure 7.14. Regression of number of spores on number of lesions (after Van der Plank 1975).

plied by a coefficient, b. In our example, taken from Van der Plank (1975:75), the number of lesions produced on a host plant is related to the number of spores used in inoculation. Here we find that it takes five spores to produce one lesion. This means that for an increase of five units of X there must be an increase of one unit of Y; thus the slope factor (b) equals 1/5. Therefore 10 spores can be expected to cause 2 lesions; 15 spores, 3 lesions; 100 spores, 20 lesions. For any fixed scale the slope of the function line depends on the size of b, the slope factor. As b increases, the slope of the line becomes steeper. Again when $X = 0$, the dependent variable also equals zero. When no spores are applied to the plant, there are no lesions.

Finally, we turn to a situation such as that depicted in Figure 7.15, which illustrates the effect of time on sporulation of *Helminthosporium maydis* (Waggoner et al. 1972). The relationships illustrated can be expressed by the formula $Y = a + bX$. The regression line depicts the relationship $Y = 15 + 3X$. From the equation given, it is easy to calculate the percentage sporulation to be expected within the time range given. Thus after 6 h the percentage sporulation would be $Y = 15 + 3(6)$, or 33%. Note that, by this formula, when the independent variable equals zero the dependent variable does not equal zero but equals a. When $X = 0$ this function predicts 15% sporulation.

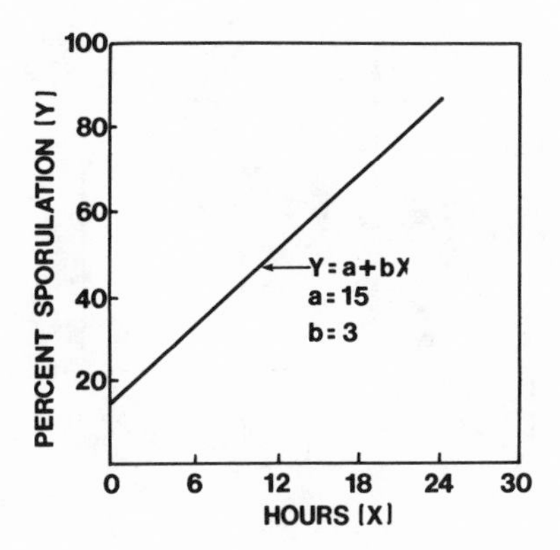

Figure 7.15. Regression of time on percentage of sporulation of *Helminthosporium maydis* (after Waggoner et al. 1972).

In this example some sporulation occurred during an hour's drying time before the fungus was placed in a moist atmosphere for testing. The magnitude of a can be obtained by solving the equation after substituting zero for X or by examining the figure for the intersection of the function line with the Y-axis. That point gives the magnitude of a, which is called the Y-intercept.

So far we have considered only simple linear regression. In aerobiology we frequently are interested in the effects of two or more independent variables on a dependent variable, e.g., the effects of temperature, light, and relative humidity on spore release. Generally, if we suspect several variables (light, temperature, and RH) as being functionally related to Y (spore release), we try to regress Y on all simultaneously. This leads to a multiple regression equation, which has the form

$$Y = a + b_1X_1 + b_2X_2 + b_3X_3 + \cdot \cdot \cdot + b_nX_n \qquad (7.7)$$

in which the X_1, X_2, X_3, ..., X_n variables refer to separate, independent variables that need not be uncorrelated with each other. A regression coefficient such as b denotes the regression coefficient of Y on variable X_1 when the other variables in the regression equation, X_2, X_3, are held constant. It is called a *partial regression coefficient,* and by its use we generally are able to make a more accurate prediction of Y. Figure 7.16 illustrates multiple regression with two independent variables, X_1 and X_2. Note that the line of simple linear regression has now become a plane. The multiple linear relationship cannot be illustrated easily for more than two independent variables, but the principle remains the same for any number of independent variables.

A relation between two variables may be approximately linear when studied over a limited range but markedly curvilinear when a broader range is considered. For example, such can be the case in studying the effects of tem-

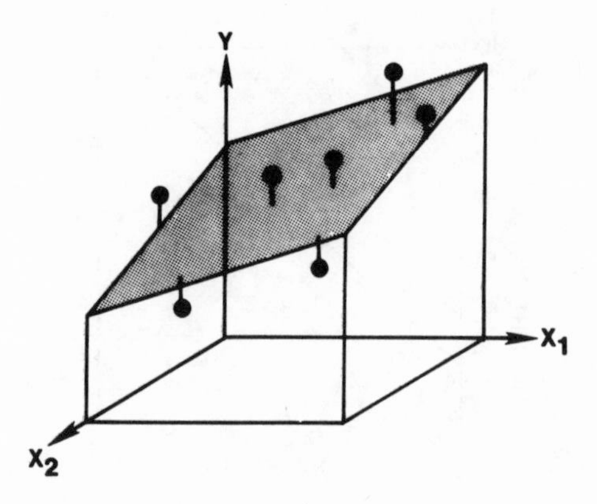

Figure 7.16. Multiple regression of Y on X_1 and X_2 (after Sokal and Rohlf 1969).

perature on spore germination. Temperature effects are linear when only temperatures between the minimum and optimum are considered, but distinctly curvilinear when considered between the minimum and maximum. Such considerations lead to two types of solutions. We can decide upon the curve to be fitted, then fit the observed data to that curve, or we can transform the data so that an easier fitting procedure can be used. Curvilinear regression seldom is used in aerobiology. In their work on rust epidemiology in the midwestern United States, Burleigh et al. (1969) showed that increase of *Puccinia recondita* (Rob. ex Desm.) f. sp. *tritici* (on winter wheats), when considered for the entire growing season, follows a cubic polynomial function rather than a linear function. When only that increase that occurred during the spring and summer was considered, however, the relation was decidedly linear. The cubic polynomial ($Y = a + bX + cX^2 + dX^3$) implies that wheat leaf rust increased in severity, decreased, then increased again (Figure 7.17). Polynomials have

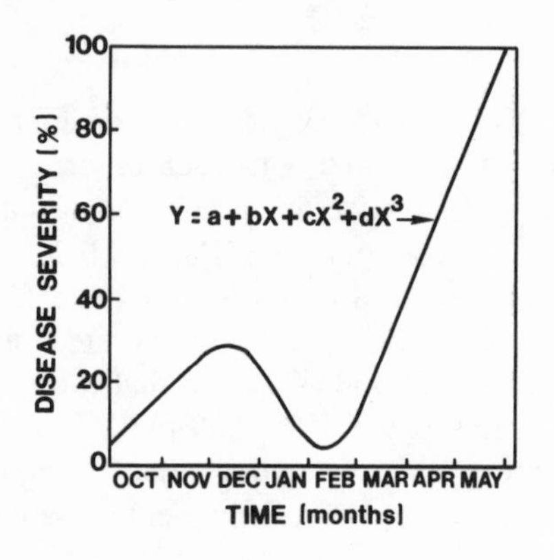

Figure 7.17. Nonlinear regression of time on percentage of disease.

peaks and depressions numbering one less than the highest exponent. In this case the polynomial gives a more accurate description of the disease cycle than the linear function of the relation between amount of diseased tissue and time. Infected tissue is destroyed by freezing and thawing in the early spring months; new tissue emerges and becomes infected, and then with the coming of warm temperatures disease increase becomes exponential. Although this example is not strictly an aerobiological one, it does illustrate the utility of the polynomial to describe some biological events. Since spore production is a function of amount of disease, we would expect the polynomial to predict spore numbers with equal accuracy during the same time period.

The second solution (data transformation) is widely used in aerobiology and specifically in the study of spore and infection dispersal gradients (Gregory 1973). Distance usually is plotted on the x-axis and spore numbers on the y-axis. Plotted on a linear scale, dispersal gradients commonly show a hollow curve, decreasing rapidly near the source, then flattening out and becoming parallel with the x-axis (Figure 7.18). When both distance and spore numbers

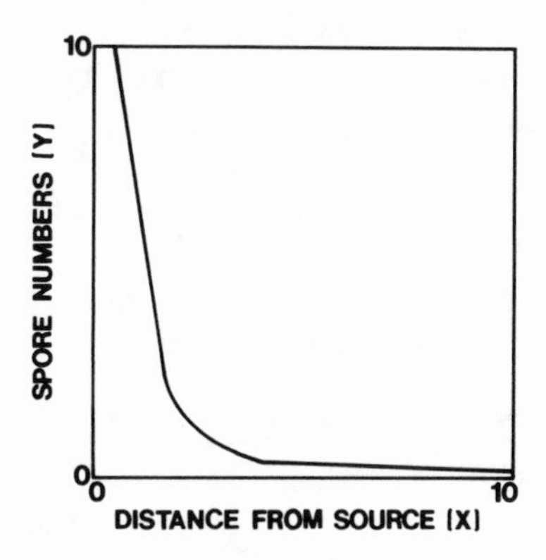

Figure 7.18. Typical hollow curve of spore dispersal gradient from a point source plotted with linear scales of X and Y (after Gregory 1968).

are transformed to logarithms, linear regression can be applied in the form $\log y = \log a + b \log x$, and the gradient becomes a straight line (Figure 7.19). The regression coefficient (b) is a measure of the gradient and can be used to compare dispersal gradients with one another.

Schrödter (1964) solved the problem of nonlinear responses of fungi to temperature by using a $\sin^2$ transformation of a temperature-growth function that has a linear relationship to temperature. By definition, the temperature function (x) can lie only between 0 and 1 and is calculated by

$$x = \frac{t - t_n}{t_x - t_n}(100) \qquad (7.8)$$

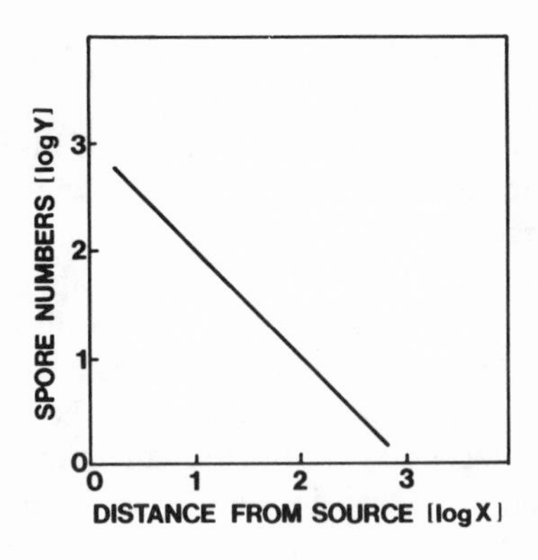

Figure 7.19. Spore dispersal gradient plotted on log-log scale (after Gregory 1968).

where t = observed temperature, t_n = minimum temperature at which growth occurs, and t_x = maximum temperature for growth. The x values are plugged into the polynomial $y = \sin^2 (a_1x + a_2x^2 + a_3x^3)$ and temperature growth values are generated. Since x can lie only between 0 and 1 or 0 and 180 deg, y also lies between 0 and 1 and therefore is never negative. In that way linearization of a nonlinear response is achieved.

Analysis of variance as applied to regression fits a straight line to data points so that distance is at a minimum between the regression line and the data. In the language of regression, the sums of squares of the deviation of the observed points from the straight line is at a minimum. Total variation in observed values of the dependent variable Y (total sums of squares from their mean $\bar{Y} = \Sigma \hat{y}^2$) is made up of: (1) the sum of squares (SS) due to regression, i.e., the SS of the deviations of predicted values $\hat{Y}$ from the mean $\bar{Y} = \Sigma \hat{y}^2$; plus (2) the residual variation due to differences in SS between observed Y values and predicted $\hat{Y}$ values. *Residual variation* is an error term and is a measure of the inadequacy of the regression model to explain all the variability in Y. It is used to calculate variance ratios (F values), which in turn are used to test the significance of the reduction in sums of squares attributable to regression. The coefficient of determination (R^2) is the ratio $\Sigma \hat{y}^2 / \Sigma y^2$, R^2 is the proportion of the total sums of squares attributable to regression, and a test of significance of R^2 is equivalent to the F test.

Uses of regression models in aerobiology

The aerobiology pathway described by Edmonds (1972) gives a convenient and logical sequence of the events involved in disease development. It is apparent that inoculum (spores) must be produced, released, dispersed, and deposited before infection can occur and plant health can be affected. Because that pathway is important to the full understanding of aerobiology, we use it as an outline for our description of regression models in aerobiology.

It is reasonable to expect that the numbers of spores of a pathogen per unit volume of air is a function of the amount of disease per unit area of crop and meteorological factors such as temperature, rainfall, relative humidity, surface wetness duration, and light, among others. If functional relationships can be established between disease severity, climatic interactions, and spore numbers, powerful predictive equations would result so that disease incidence and disease severity could be estimated from spore numbers. Since development of disease caused by many plant pathogens can proceed exponentially during favorable periods, the rapid detection of disease and determination of rates of disease development becomes particularly critical if control is to be attempted. Because spore trapping equipment is available and can operate continuously without constant attention, airborne spore numbers usually are more rapidly assessed than disease severity estimates, especially if the area being monitored is large. Spore numbers then would appear to be an effective measure of disease. Several studies have been made based on that premise, using regression analysis. Kerr and Shanmuganathan (1966) predicted the sporulation of *Exobasidium vexans* (cause of tea blister blight) with two equations:

$$y = 2.5824 - 0.6169x_1 + 0.06x_2 \tag{7.9}$$

$$y = 3.1411 - 0.9867x_1 \tag{7.10}$$

where y = log spores per blister, x_1 = log blisters per 100 shoots, and x_2 = mean daily sunshine. (Notice that the regression coefficients associated with x_1 are negative, indicating that there are more spores released per blister when infection is low than when it is high.) Equation (7.9) was effective for the period April to December, excluding August, and Equation (7.10) was for August. By calculating y, converting to spores per blister, and multiplying by the number of blisters per unit area of tea, the number of spores/m^3 was estimated. Although the authors did not calculate correlation coefficients for observed and predicted spore numbers, it is obvious from their data that estimated spores/m^3 closely approximated observed values. Those data were used in a later study (Kerr and Rodrigo 1967a,b) to predict disease incidence as a function of spores/m^3.

Royle and Thomas (1972) used multiple linear regression (MLR) to predict spore release of *Pseudoperonospora humuli* (M.Y. and Tak.), the cause of hop downy mildew. They used as independent variables only meteorological factors thought to cause spore liberation, i.e., temperature, relative humidity, rainfall quantity and duration, leaf surface wetness duration, and sunshine.

Sporangia/m^3 was the dependent variable. Values for each independent variable were calculated for four 6-h periods per day, for each of 3 days preceding the day of spore release. Although the number of possible combinations of independent variables was large, the authors stipulated that no more than four independent variables were to be used in any one regression analysis, as they thought that the inclusion of additional variables would give little additional information. Their analysis showed that several combinations of independent variables gave significant R^2 values (0.482–0.554). For prediction, however, they used the equation with the highest R^2 (=0.554). That equation was constructed from three independent variables (duration of rain 2.2–1.1 days before spore release, mean temperature 1.4–0.1 days before spore release, and the mean vapor pressure deficit 1.3 days before spore release) and closely approximated numbers of sporangia/m^3 of air. When predictions were made for periods during those years (1967, 1968) and measures of the independent variables were made, the equation satisfactorily estimated spore numbers, but predictions usually were exaggerated when applied to data from 1969 and 1970. Royle and Thomas (1972) considered the inaccuracy of their predictions to be caused by extremes in temperature and rainfall, which resulted in overestimates and underestimates, respectively. If additional weather measures (e.g., wind) and important biological variables were incorporated into existing equations, the accuracy of predictions would be improved.

Inaccurate prediction is a problem common to regression when extrapolations are made beyond the limits of the independent variables. The predictability of a regression equation is only as good as the data used in its generation. If an equation is expected to predict spore numbers from measurements of weather outside the limits of data used to build the equation, overestimates and underestimates will occur. Therefore it is necessary to construct regression equations from data representative of the range of conditions expected to be encountered if accurate predictions are expected. Even then, unacceptable estimates will occur if the most important independent variables are not represented in the equation. Royle and Thomas (1972) thought wind data would increase the accuracy of their predictions.

Edmonds (1973) expressed spore release as a function of time of day as well as a function of weather. Since most fungi exhibit circadian rhythms of spore release (Ingold 1971) it seems reasonable to expect that a periodicity function would be a powerful variable to include in the regression model for spore release. Edmonds did that for *Fomes annosus* and derived two equations to predict spore release:

$$Y = 0.278 + 0.22t + 0.030T \quad (7.11)$$

$$Y = 3.38 - 0.143t + 0.030T \quad (7.12)$$

where Y = number of spores released/cm^2 of sporophore area per second per 2-h period, t = time (from 0 to 8 or 8 to 16) and is the middle hour of each 2-h period, and T = temperature (0°C) and is the average of each 2-h period.

Equation (7.11) predicted spore release from 1600 to 2400 (0–8 on time scale) and Equation (7.12) from 2400 to 1600 (8–16 on time scale). The value for Y was multiplied by the surface area of the sporophore to derive the source strength term applied in the dispersion model of the disease. Edmonds and Driver (1974) did not test the accuracy of Edmonds' spore release predictions by conventional methods; however, his dispersion model successfully predicted spore numbers up to 60 m from the source, which suggests that his release function was adequate.

Workers who use regression models to study sporulation usually are interested in prediction. Consequently, data to construct equations are obtained from the field; and the amount of variation explained by regression, although statistically significant, usually ranges from 50% to 70%. Values of R^2 greater than 0.70 seldom are obtained, indicating that other independent variables not used in the analysis also are functional in affecting sporulation. Massie et al. (1973) used regression techniques to quantify the effects of dew duration and dew temperature under controlled conditions on sporulation of *Helminthosporium maydis* (cause of southern corn leaf blight), with the plan to use that functional relation in a simulation of the disease cycle. When the authors transformed the dependent variable (number spores/mm^2 of lesion area) to the $\sqrt[4]{\ }$, R^2 values of 0.94–0.99 were obtained, which indicates a negligible difference between observed and expected sporulation. That use of regression is new and heretofore unconventional but opens new vistas to the use of regression as a tool to understand the complete cycle of a disease.

Spore dispersal

The dispersal of spores is a complex subject that does not lend itself to simple solutions. Some understanding has been achieved by applying variations of the Gaussian plume model, discussed in this chapter by C. J. Mason. Regression models, on the other hand, have received little attention here because they are descriptive, not interpretive. A regression equation might adequately predict the number of spores trapped downwind from a source of known strength given constant wind direction, topography, and source geometry, but might be entirely inadequate with substantive changes in those variables. Where regression techniques have been used, the purpose of the work was to predict spore numbers: (1) some distance downwind from a source (Roelfs 1972, Rowe and Powelson 1973, Gregory 1968), (2) perpendicular to wind direction (Roelfs 1972), and (3) on annuli from a point source (Roelfs 1972).

Gregory (1968) has made considerable use of regression to describe dispersion gradients as a function of: (1) source strength, (2) geometry of source,

(3) proximity to source, (4) dilution rate, and (5) wind direction. If log spores (Y) are plotted against log distance (X), the regression coefficient (b) is a measure of the dispersal gradient, is usually negative, and indicates that spore numbers are inversely proportional to some power of the distance. Referring back to our discussion of the mechanics of regression, the constant a was defined as the Y-intercept, i.e., the point where the regression line or plane crossed the Y-axis. Therefore it is understandable that a is affected by source strength. Very simply, a varies directly with the number of spores emitted from a source.

Dispersion gradients (regression coefficients) commonly range from $b = -1$ to $b = -4$ (Gregory 1968). Near-zero values of the coefficients suggest the presence of other sources, background contamination, or proximity to a large-area source. Steep gradients ($b = -3, -4$) suggest that spores are dispersed only a short distance from the source. Usually an area source gives a flatter dispersion gradient than do line or point sources because the total number of spores being emitted from an area source is greater than that from line or point sources even though the number of spores/m^3 emitted might be the same regardless of source geometry.

Dispersion gradients also are a function of wind direction (Gregory 1968). Values for b downwind from a source usually are smaller than for mean wind direction. That relationship is consistent particularly for short sampling periods. Wind direction fluctuates and there is less difference between b values for downwind and mean wind as length of the sampling period increases.

Roelfs' work (1972) with urediospore dispersal of wheat leaf and stem rust supports the concepts of Gregory (1968). He successfully used regression analysis to predict the number of urediospores downwind ($\log Y = a + bx$) from an area source and on annuli ($\log y = \log a + b \log x$) around an area source. Spore numbers predicted downwind closely approximated numbers trapped. His equation predicted that 10% of the spores released at a source should be trapped 10 m downwind. Roelfs trapped 10.36% and 9.62% for stem rust and leaf rust, respectively. Those values agree closely with Gregory's theoretical model. Numbers predicted on annuli 36 m and 72 m from the source, however, were only 44% and 41% of those expected, respectively, for leaf rust and 32% and 25% for stem rust. It was assumed that differences between observed and predicted numbers were due to losses by deposition and vertical diffusion. Regression coefficients for spore numbers impacted on an annulus were $b = -0.4953$ and $b = -0.6002$ for leaf rust and stem rust, respectively. Those values indicate that spore numbers vary inversely as approximately one-half the first power of the distance. In comparison, Rowe and Powelson's (1973) data for conidia dispersal of *Cercosporella herpotrichoides* Fron. (cause of *Cercosporella* foot rot of wheat) show a regression coefficient for spore-carrying rain droplets of $b = -1.06$, indicating that spore dispersal varies inversely as the first power of the distance. Both Roelfs (1972) and Rowe and Powelson (1973) used the same regression model ($\log y = \log a +$

$b \log x$) to calculate spore dispersal gradients so their data are comparable. Rust spores principally are windborne and therefore easily dispersed for long distances, while conidia of *C. herpotrichoides* are dispersed by raindrops only 0.9–1.2 m from their source (Rowe and Powelson 1973). If we assume log $a = 1$ and log $x = 1$, then for *C. herpotrichoides* log $y = 0$ and $y = 1$, since $b = 1$. For *Puccinia recondita tritici* and *P. graminis tritici,* log $y = 0.05$ and $y = 3.162$ as $b = 0.5$. Therefore one would expect to find approximately three times more rust spores than foot rot conidia per unit distance from a source given equal source strength for both fungi.

The difference between regression coefficients also suggests a basic difference in the relative number of spores expected per unit distance from a source and is reflective of the different mechanism of spore dispersal exhibited by the two fungi.

Disease severity and rate of epidemic development

The greatest use of regression in aerobiology has been to predict disease severity and crop loss, and it is in those studies that the predictive value of regression is seen best. Early studies (Schmitt et al. 1959, Underwood et al. 1959) with wheat stem rust (*P. graminis tritici*) were designed to compare rates of disease spread from foci with different levels of initial infection. Schmitt et al. found that rates of spread (Table 7.3) and, indeed, final severi-

Table 7.3. Regression coefficients for rust spread within circles of 6.1 and 12.2 m radius around foci with different levels of initial inoculum (from Schmitt et al. 1959).

	Initial inoculum level		
Circle radius (m)	30 uredia	900 uredia (concd)	900 uredia (dispersed)
6.1	0.417	0.398	0.252
12.2	0.371	0.357	0.388

ties varied little between epidemics initiated with 30 and 900 uredia. Even when the focus was dispersed, not concentrated, rate of disease development within a distance of 12.2 m from the focus was similar to rates from a concentrated focus of similar intensity and a concentrated focus of 30-fold lower intensity. The similarity in rates of spread may be a function of the focus intensities chosen rather than being descriptive of a principle of rust intensification. Rate differences might have been obtained with foci of 1 and 30 or 10 and 300 uredia. Certainly that would be true if time were limiting; however, if time were not limiting, the data of Schmitt et al. suggest that epidemics of similar magnitude could occur even though severalfold differences exist in initial inoculum levels.

It is the extremely rapid rate of disease development of the rusts that has alarmed and excited workers over the years. Severe epidemics seem to arise *ex nihilio* and progress at uncontrollable rates. Tenfold increases occur in 4–5 days given favorable moisture and temperatures. It is that characteristic of the rusts perhaps, more than others, that has prompted attempts to predict disease so that control strategies could be employed. Linear regression has proved to be an extremely useful tool to associate epidemic development with biological and climatological variables. In a preliminary study of historical data, Romig and Dirks (1966) used MLR to show that there was a linear relationship between the cumulative numbers of airborne urediospores of *P. recondita tritici* and *P. graminis tritici* with time in days and suggested that spore numbers might serve as a measure of epidemic development. In a later study the same authors (Dirks and Romig 1970) again used MLR to predict cumulative spore numbers 14, 28, and 42 days after the date of prediction. Of the 11 independent variables used, 1–6 gave significant regression coefficients depending on the predictive period. In all cases, significant R^2 values were obtained only when biological or climatological and biological variables were entered into the equations. Climatological variables alone were nonsignificant. Numbers of spores of *P. graminis tritici* were predicted 14 days before crop ripeness in winter wheats ($R^2 = 0.728$) but 42 days before ripeness with spring wheats ($R^2 = 0.758$). Spores of *P. recondita* were predicted 42 days before ripeness in winter wheats ($R^2 = 0.643$) and spring wheats ($R^2 = 0.614$). Those results indicate that accurate predictions of spore numbers could be made 42 days before ripe stage or 2 wk before the crop headed.

Spore numbers were used successfully by Kerr and Rodrigo (1967a) to predict disease incidence of *Exobasidium vexans* (tea blister blight) in Sri Lanka (Ceylon). They used MLR to construct an equation to predict the number of blisters per 100 shoots of *E. vexans* on tea bushes. Independent variables were number of spores/m^3 of air and mean daily sunshine. Equation (7.13)

$$\hat{Y} = 33 + 0.3145x_1 - 0.03725x_1x_2 \tag{7.13}$$

required the estimation of disease incidence about 3 wk after the date of prediction. Although R^2 values are not given, their data show little difference between predicted and observed severities. In spite of their success with spore numbers as a predictor of disease, Kerr and Rodrigo (1967b) attempted to simplify their forecasting procedures so that predictions could be made on individual tea estates, as the mountainous terrain of Sri Lanka made impossible the issuance of a single accurate forecast for the entire country. They found that disease incidence could be estimated on leaves picked for processing and those estimates used in a MLR equation to predict disease 2–3 wk hence. Their equation (Equation 7.14)

$$\hat{Y} = 1.8324 + 0.8439x_1 + 0.9665x_2 - 0.1031x_3 \tag{7.14}$$

required the estimation of disease at the beginning (t_1) and end (t_2) of a 3-wk period, and transformation of the severity estimates to the log $\sqrt{\%\ \text{severity}}$, x_1 = log $\sqrt{\%\ \text{severity}}$ at the end of the 3-wk period (t_2), and x_2 = [log ($\sqrt{\%\ \text{severity}}\ t_2$) − log ($\sqrt{\%\ \text{severity}}\ t_1$], and x_3 = mean daily sunshine for a 7-day period preceding t_2. A calculating device was designed so that the equation could be solved easily by growers to provide an accurate and practical forecasting procedure.

In contrast to the above, Burleigh et al. (1969) showed a high, positive correlation between spore numbers and disease severity for *P. graminis tritici* and *P. recondita tritici* and suggested that spore numbers could serve as predictors of disease as well as disease severity estimates. When spore numbers were used with other biological and climatological variables to predict disease, however, coefficients of determination were never as high as when disease severity estimates replaced spore numbers (Eversmeyer and Burleigh 1970, Burleigh et al. 1972a, Eversmeyer et al. 1973). With disease severity estimates the regression model explained 57%–70% of the variation in epidemic development, but 10%–40% less was explained when spore numbers replaced disease severity in the model.

Royle (1973) used MLR to predict severity of *Pseudoperonospora humuli* (hop downy mildew) as a function of spore numbers/m^3 of air (x_1), leaf surface wetness (x_2), and quantity of rainfall (x_3). When data from 3 yr were combined Equation (7.15) resulted:

$$\hat{Y} = 0.037 + 0.023x_1 + 0.066x_2 + 0.002x_3 \tag{7.15}$$

where $\hat{Y}$ = percentage of leaves infected and x_1, x_2, and x_3 are as before. Over 70% of the variation in percentage of infection was explained by that equation. When spore numbers were purposely omitted from the analysis, the significant independent variables were RH (number of hours ≥ 90%) and quantity of rainfall. The R^2 value associated with that equation was 0.636, only slightly lower than 0.70. Nevertheless, only Equation (7.15), with spore numbers as a variable, was used to predict disease. It would be interesting to see if omission of spore numbers seriously reduced the accuracy of disease predictions.

Simple linear regression has been used extensively to predict the severity of rice blast disease (*Piricularia oryzae,* Ono 1965). Percentage of severity has been expressed as a function of sunshine, temperature, precipitation, crop growth stage, starch distribution in sheath tissue, number of silicated cells in the top leaf of the rice plant, and number of spores. Regression equations built on those variables are used throughout Japan to warn growers of impending epidemics.

Related to our present discussion but nevertheless unique is the application of regression to the study of dosage-response relations in host vs. plant-pathogenic bacteria. The quantification of plant-pathogenic bacteria interactions is a field of study ripe for research as little information is available to evaluate pathogen virulence and host susceptibility because of the lack of analytical tools to quantify dose-response curves. Civerolo (1975), however, in an interesting study with *Xanthomonas pruni,* offers a solution to that question that might prove useful with other plant pathogenic bacteria as well. Using simple linear regression, Civerolo described lesion numbers as a function of colony-forming units (CFU; Figure 7.20). The quotient of CFU divided by the mean number of lesions was defined as lesion-forming units (LFU) or the number of *X. pruni* cells required to initiate a single lesion. From Figure 7.20, LFU values of 9–23 can be calculated. The mean value is 16. Therefore an estimated 16 CFU are required to induce a single lesion at each infection site. When numbers of lesions were transformed to probits and regression was applied to the dose-response interaction, regression coefficients for inoculations on peach, apricot, plum, and sweet cherry were nonsignificantly different and less than 2, which suggests that *X. pruni* cells act independently to cause infection. This procedure might be used to evaluate virulence and host susceptibility in other host vs. plant-pathogenic bacterial combinations.

Most studies discussed so far have relied heavily on both biological and meteorological variables to predict spore numbers or disease severity (Kerr and Rodrigo 1967a,b, Dirks and Romig 1970, Eversmeyer and Burleigh

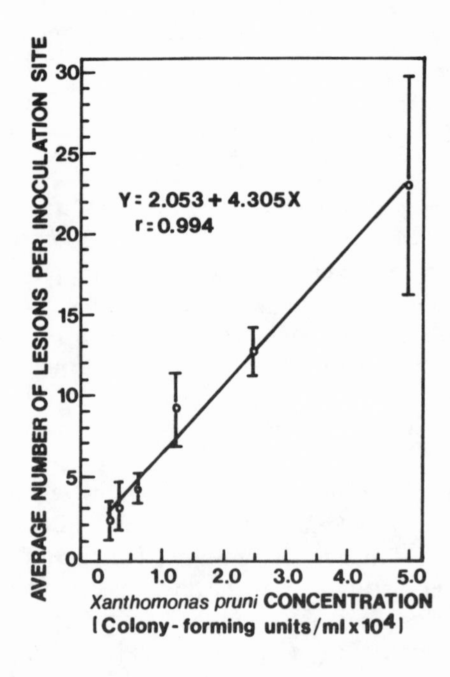

Figure 7.20 Proportional relationship between average number of lesions per inoculation site and concentration of *Xanthomonas pruni* in the inoculum. Each point is the average number of lesions at six inoculation sites in each of the three experiments. One standard error from each mean is indicated by the brackets (after Civerolo 1975).

1970, Burleigh et al. 1972a). In all cases regression equations based solely on meteorological events did not predict disease with accuracy equivalent to equations constructed from biometeorological data. Accurate forecasts required a measure of disease either as spore numbers or percentage severity. Measures of disease for forecasting can be obtained, but the reader should be aware of the difficulties associated with gathering standardized data over a large geographic region. Standard weather stations from which measures of rain, temperature, and humidity can be obtained usually are readily available, however. Therefore a practical forecasting scheme based only on meteorological events would permit rapid evaluation with standardized data and the speedy issuance of forecasts—a task somewhat more cumbersome when disease measures are required as well.

One of the more thoughtful studies using MLR to build a practical disease forecasting system is that of Schrödter and Ullrich (1965) and Schrödter (1967) for potato late blight. Their work typifies the power of MLR to quantify functional relations between phases of the disease cycle and weather parameters and then use only weather measures for prediction. They used standard MLR techniques to formulate the equation:

$$Y = C_1 y(K_D) + C_2 y(S_D) + C_3 y(M) + C_4 y(U) \tag{7.16}$$

where Y = rate of infection, $C_1 + C_2 + C_3 + C_4 = 1$, $y(K_D)$ = effect of temperature and moisture on spore germination and infection, $y(S_D)$ = effect of temperature and moisture on sporangiophore and sporangia formation, $y(M)$ = effect of temperature on mycelial growth, and $y(U)$ = effect of intervening dry periods on spread of the epidemic. Each independent variable is a functional expression of the relevant weather measures on a component of the disease cycle. So with weather data alone the magnitude of Y can be calculated, thereby expediting a timely forecast. When Schrödter and Ullrich compared, observed, and calculated rates of infection (Figure 7.21), they obtained a correlation coefficient of 0.756 ($R^2 = 0.57$). Although that R value is lower than others from similar studies discussed in this chapter, it was considered acceptable given the large error expected from assumptions made regarding the nature of disease spread in plots, growth of the host, and host susceptibility.

The interest of epidemiologists rightly has been directed to the prediction of disease and evaluation of variety resistance in homogenous stands of crop varieties since plantings of mixed stands of susceptible and resistant lines is relatively new and not widespread. As multilines increase in popularity, however, there will be needed a technique to evaluate their effect on disease development. Leonard (1969) proposed a simple but workable procedure to make such an evaluation for oat stem rust (*Puccinia graminis avenae*). Using simple, linear regression he found that the rate of stem rust increase in a mixed stand could be predicted from estimates of rust increase in a pure stand of the sus-

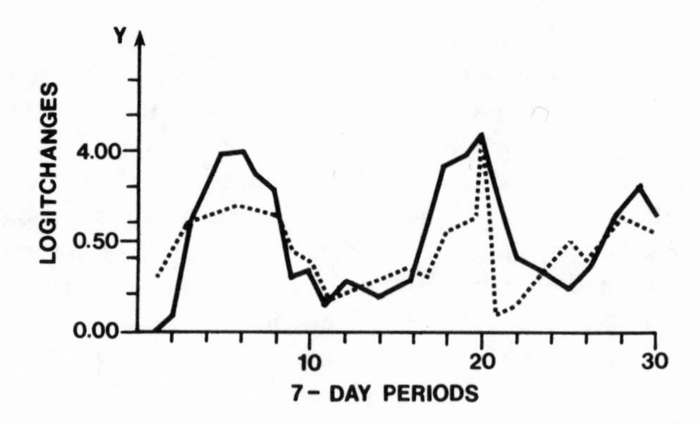

Figure 7.21. Comparison between observed (———) and calculated (▪▪▪▪▪▪) rates of infection on the basis of multiple regression equations (after Schrödter and Ullrich 1965).

ceptible line and the proportion of susceptible plants in the mixture. By combining those variables into a regression model, the following equation was developed:

$$r_m = r_s + \frac{n}{t} \ln M \tag{7.17}$$

where r_m = rate of stem rust increase in a mixed stand, r_s = rate of stem rust increase in a pure stand of susceptible plants, n = number of rust generations, t = number of days of rust increase, and M = proportion of susceptible plants in the host mixture.

Disease incidence

Often the purpose of MLR is not to predict disease but simply to identify those variables that might explain most of the variation observed in disease incidence. If the relationships are strong, prediction might be the outcome but not the stated goal. For example, Kincaid et al. (1970) used MLR to relate the incidence of tobacco black shank (*Phytophthora parasitica* var. *nicotianae*) and root-knot nematode (*Meloidogyne incognita acrita*) to pH, and pounds per acre of oxides of potassium, calcium, and magnesium. They found a positive, linear relation between pH 5 and 6 and disease incidence of black shank and a quadratic relationship between K_2O and root-knot incidence.

In an interesting study made to understand the vagaries of fusiform rust (*Cronartium fusiforme,* Hedge and Hunt ex Cumm.) distribution and incidence in Florida and Georgia, Schmidt et al. (1974) used MLR with longitude, latitude, tree age, and year planted as independent variables. Longitude and latitude explained over 50% of the variaton in rust incidence (Table 7.4)

Table 7.4. Independent variables, regression equations, and percent of fusiform rust variability associated with independent variables for slash pine plantations.[a]

Eq. no.	Independent variables	Regression equation	% rust variability assoc. with independent variables[b]	Number of plantations
1	Long. & lat. Florida & Georgia	% rust = 54.4396 − 2.8477 (long) − 45.0556 (lat + 0.6259 (long) (lat)	50.1**	1286
2	Long. & lat. Florida	% rust = −2065.42 + 18.4173 (long) + 18.7739 (lat)	54.8**	854
3	Long. & lat. Georgia	% rust = 2.9733 − 4.2444 (long) + 12.0648 (lat)	8.6**	432
4	Age of tree & year planted;[c] Florida & Georgia	% rust = −29.2902 + 7.6823 (age) − 0.5898 (yr-1950) − 0.3053 $(age)^2$ + 0.0992 $(yr\text{-}1950)^2$ + 0.1868 (yr-1950) • (age)	9.8**	1286

[a]From Schmidt et al. (1974). [b]Statistical significance: ** = 1%. [c]Year of planting is indexed as (yr [planted]-1950).

while tree age and year planted explained 9.8%. Both values appear small, but they are statistically significant. They showed that rust increases along a vector running southeast to northwest and suggested that this increase might be the result of higher precipitation and humidity along the coast. Also, their equation (4), (Table 7.4), showed that rust increases in plantations up to 10–11 yr of age and then declines (Figure 7.22). They offer several possible explanations for that relationship but decline to make any conclusions without further observations.

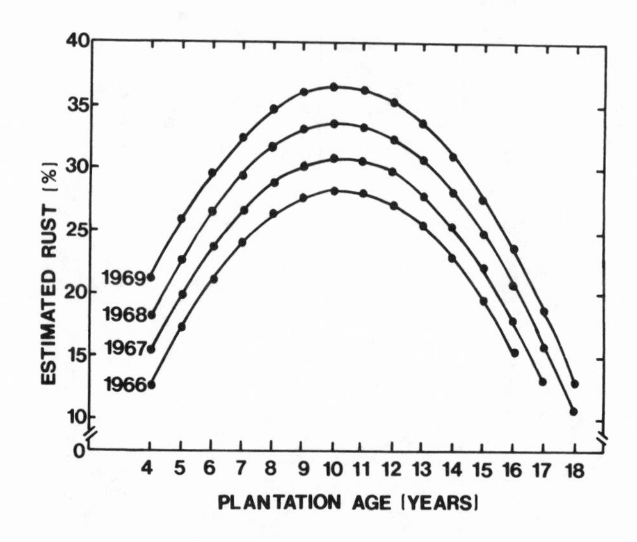

Figure 7.22. Percentage of rust estimated as a function of plantation age and year of establishment. Numbers at left of curves are years the data were collected (after Schmidt et al. 1974).

Relation of disease incidence to disease severity

One of the most sticky problems in the use of regression to predict disease is the questionable accuracy of disease estimates used as independent variables. Model I regression requires independent variables to be measured without error. That is seldom the case in the instance of disease severity estimates, particularly at the onset of epidemic development when little disease is present. James and Shih (1973) attempted to solve that problem by quantifying the relation between disease incidence (number of infected leaves) and disease severity (percentage of leaf infected) for wheat leaf rust (*Puccinia recondita tritici*) and powdery mildew (*Erysiphe graminis tritici*). Incidence measurements were regarded as error-free since they were based on presence or absence of disease. Using the linear regression equation $Y = bX$, where Y = severity, X = incidence, and b = regression coefficient (0.013), James and Shih successfully predicted severity from incidence when $X \leqslant 65\%$ but found that large errors were associated with predictions when $X > 65\%$. The regression equation explained 92% and 96% of the variation in percentage of severity in mildew and leaf rust, respectively. Therefore severity estimates could be calculated from incidence measurements with a high degree of accuracy.

Kranz (1970) also offers a solution to the problem of obtaining accurate disease severity estimates. He made 25 models of apple leaves and pasted on them round spots or parts of brown paper to represent intended percentage of infection from 0% to 100%. In a random sequence the models were presented to 200 persons who were asked to determine percentage of infection of each leaf. Kranz found that percentage infection between 0.3% and 96% could be divided into 10 classes that could be estimated with the least amount of error (Figure 7.23). When regression was applied to expected and estimated values for these 10 classes, a regression coefficient of 0.87 was obtained. A value of 1 would indicate perfect fit.

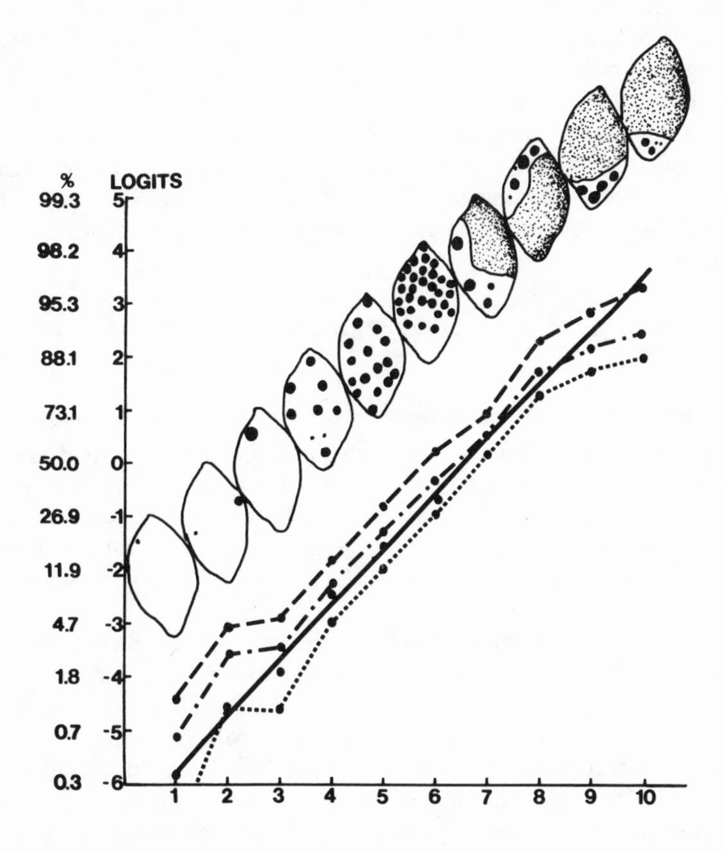

Figure 7.23. Severity estimates on apple leaf models that have known "infected" surface areas: mean of 200 estimates (▬▬▬), confidence limits (5%) for the mean (▪▪▪▪▪▪▪) and (▪ ▬ ▪), and standard error for each estimate (▬ ▬); (after Krantz 1970).

As stated at the beginning of this section, the systems analysis approach to aerobiology encourages study of the complex but natural forces that link spore production, release, dispersal, and deposition to impact. In plant pathology, *impact* is a euphemism for *disease loss,* and to that area we now turn our attention.

Few but significant contributions have been made to the study of disease-yield loss models. Kranz (1974) and James (1974) give excellent summaries of those studies. It is noteworthy that all studies reviewed by Kranz and James used regression to express crop loss as a function of disease severity or incidence. Studies using simple, linear regression (Romig and Calpouzos 1970, Mogk 1973, Mundy 1973, Jenkyn and Bainbridge 1974, Eyal and Ziv 1974) described loss or crop yield as a function of a single estimate of percentage of disease severity. Those studies that used MLR (James et al. 1972, Burleigh et al. 1972b) also described crop loss as a function of disease severity, but rather than a single estimate of disease, loss was related to multiple severity estimates at different crop growth stages.

Perhaps the best examples of studies that show the accuracy of crop loss estimates that can be achieved by simple linear regression and by MLR are those of Romig and Calpouzos (1970) and James et al. (1972), respectively. Romig and Calpouzos estimated yield loss of wheat due to wheat stem rust by the equation:

$$Y = -25.53 + 27.17 \ln X \qquad (7.18)$$

where Y = % yield loss and X = % disease severity. The coefficient of determination associated with that equation was 0.99, indicating that almost all of the variation in yield was attributable to amount of disease. Similar accuracy was achieved by James and his colleagues. They estimated tuber weight loss caused by *Phytophthora infestans* (potato late blight) for 85 individual epidemics and estimated loss to within 5% for 78 of them and to within 10% for all.

The following study differs from the others in that loss prediction was the stated goal. Burleigh et al. (1972b) used data from 55 cultivar location-year combinations to construct a regression equation for the prediction of grain weight losses of wheat by leaf rust (*P. recondita tritici*). Crop loss was measured as the percentage reduction of grain weight from unsprayed plots as compared with fungicide treated neighbors. Independent variables used as predictors were disease severity (percentage of leaf area) estimates on the flag leaf or on all leaves per tiller recorded at boot (X_1, X_2), heading (X_3, X_4), early berry (X_5, X_6), and early dough (X_7, X_8) growth stages. A stepwise MLR program produced several equations of which one ($Y = 5.37 + 5.52X_2 - 0.33X_5 + 0.50X_7$) gave an R^2 of 0.79 and was used to estimate losses. The standard error of that equation was 9%. The equation fit fairly accurately moderate and severe levels of crop loss but overestimated low levels (Table 7.5). Of considerable interest is that fairly accurate predictions of crop loss could be made using predicted severity estimates. That is to say, loss estimates were made 30 days before ripeness of the crop. Forecasts were made with reasonable accuracy on varieties that possessed specific genes for resistance (Parker and Shawnee) and those that did not (Bison), which suggests that general predictive equations are possible.

Empirical Models

The aerobiology pathway is a significant factor in models of several plant disease systems, i.e., potato late blight, *Cercospora* leaf spot of sugar beets, *Cercospora* leaf spot of peanuts, rice blast, southern fusiform rust infection, downy mildew of lima beans, northern corn leaf blight, and early blight of celery. All these systems depend upon sporulation and aerial spore transport for their successful cycling from spore to spore.

Potato late blight

Potato late blight in Ireland was related successfully to synoptic weather regimes in the North Atlantic Ocean (Bourke 1957). These regimes were related to sporulation, spore dissemination, and subsequent impact or infection by the pathogen. A synoptic analog series for predicting outbreaks of this disease was developed for the north central states (Wallin and Riley 1960). Weather phenomena were related to the aerobiology pathway in the development of a method for forecasting late blight (Wallin 1962). Weather criteria constitute part of the basis for a computerized forecast of potato late blight in Pennsylvania (Krause et al. 1975).

Cercospora leaf spot of sugar beets

A weather model was developed for predicting favorable or unfavorable surface conditions for leaf spot development (Gullach and Wallin 1970). The model was based on the finding that longer dew periods result in greater spore production (Wallin and Loonan 1972), and that with longer dew periods at certain temperatures the spores produce more leaf infection (Wallin and Loonan 1971).

The flow pattern of the upper air at the 500-mbar level was found to be the key to the surface weather pattern influencing the development of the leaf spot fungus. Stable, long-wave conditions of the upper air provided more persistent surface conditions unfavorable for leaf spot development. A train of short waves produced favorable weather conditions for leaf spot development.

Table 7.5. Comparison of observed and predicted severity of *Puccinia recondita tritici* and damage on wheat.[a]

Location	Cultivar	Observed severity (%)				Loss estimated from observed severity (%)	Predicted severity (%)			Loss estimated from predicted severity (%)
		Boot[b]	EB[c]	ED[d]	Losses		Boot[b]	EB[c]	ED[d]	
Manhattan, Kans.	Bison	1	20	95	58	52	1	10	100	53
Do	Bison	1	25	95	55	49	1	4	66	38
Do	Guide	0	1	14	13	14	0	1	15	15
Do	Shawnee	0	1	1	14	8	0	1	7	11
Hutchinson, Kans.	Bison	1	1	16	2	15	1	1	4	10
Do	Parker	0	1	4	4	10	0	1	5	10
Altus, Okla.	Bison	1	24	24	5	13	0	12	10	9
Do	Triumph	1	1	20	4	17	1	1	12	13
Do	Guide	1	1	2	13	9	1	1	17	15

[a]From Burleigh et al. (1972). [b]Severity per tiller. [c]Severity on flag leaf at early berry. [d]Severity on flag leaf at early dough.

Cercospora leaf spot of peanuts

Weather variables influencing the sporulation and secondary infection of the peanut leaf spot fungus were documented and used in a model developed for forcasting the rate of increase in disease incidence (Jensen and Boyle 1966). The amount of secondary infection increased rapidly when diurnal periods of relative humidity ≥95% persisted for 10 h or more and when temperatures during the high humidity period were >21°C. This model has been employed successfully in the chemical control of the *Cercospora* leaf spot disease (Smith et al. 1974). A computer program for producing a worded daily leaf-spot spray advisory was developed (Parvin et al. 1974). The program was based upon the meteorological criteria.

Rice blast

The spread of the rice blast fungus was documented in field planting (Barksdale 1967). The spore load released into the air above the diseased crop was measured and related to disease increase, and a regression function was computed. The results indicated that this function might not be a useful model in predicting blast incidence but would predict forty to several hundred lesions per acre when a blast fungus spore was detected on a Rotorod sampler.

Fusiform rust of pine

The optimum criteria for spore production and subsequent infection of slash pine were determined (Davis and Snow 1968): (1) temperature above 16°C, (2) relative humidity at or near the saturation point for 9 h plus, (3) light surface winds and a surface-based inversion at night, and (4) showers in the afternoon or early evening. Weather models depicting those surface conditions were developed for predicting potential rust weather. High-pressure systems centered off the southeast coast of the United States were found to create the necessary circulation to move maritime tropical air over the Gulf states and produce the weather required for basidiospore production and pine infection. Both prognostic and synoptic charts should be used in appraising the rust hazard.

Downy mildew of lima beans

A method was developed for predicting the downy mildew disease of lima beans on the East Coast (Hyre 1957). Initial appearance of downy mildew was forecast after 8 consecutive mildew-favorable days and continuing mildew-favorable weather. A day was considered favorable for downy mildew when a 5-day moving mean temperature ending that day was less than 26°C with the minimum temperature ≥ 7°C coinciding with a 10-day total rainfall, ending that day, ≥ 3 cm. The processes in the life cycle of the fungus influenced by these criteria were sporulation, spore deposition, and secondary infection.

The criteria were employed as a basis for the development of a weather model for forecasting lima bean downy mildew (Scarpa and Raniere 1964). The downy mildew outbreaks were closely correlated with the prevalence of a persistent maritime air mass resulting in the accumulation of 40 h or more of dewpoints greater than 21°C. Based on these findings, when the dewpoint ≥21°C consecutively for 12 h and forecast to persist for at least another consecutive 28 h, warnings were issued.

Northern corn leaf blight

A model was developed in Florida for forecasting northern corn leaf blight (Berger 1970). The criteria were at least 7 h of humidities near 100% and temperatures above 15°C for significant sporulation. Spores formed at night were released in the morning as the humidity rapidly decreased. Nearly 50% of the spores were caught from 0800 to 1200. Eight or more blight-favorable hours required a regular fungicide to avoid serious losses. Numbers of airborne spores proved valuable in determining the actual daily blight threat during the season. Sheltered hygrothermographs in the field provided the weather data for the forecast model (Berger 1973a). The number of blight lesions were directly related to numbers of trap-monitored spores collected 7 days previously.

Early blight of celery

The aerobiology pathway was demonstrated in the development of criteria for predicting early blight increase in Florida (Berger 1973b). Disease incidence and weather during periods of spore formation and release influenced the number of blight spores trapped. Light rain in late afternoon or early morning lengthened the leaf wetness period and increased spore numbers. The number of spores trapped each day could be used to predict disease increase and thus could be extrapolated for spray application recommendations.

The above plant disease–weather models are examples of the role of aerobiology in plant disease epidemics and the subsequent forecasting of these diseases. The aerial movement or transport of spores is a key mechanism in the life cycle of most plant pathogens.

INSECT MODELS: AN ATMOSPHERIC DISPERSION MODEL FOR GYPSY MOTH LARVAE*

The gypsy moth (*Lymantria dispar* L.) is the cause of severe leaf defoliation to shade and orchard trees in the northeastern United States. Because it is rapidly spreading to the South and gradually to the West, however, it threatens to become a national problem. Spreading of the gypsy moth occurs in two ways: (1) transport of egg masses or larvae on vehicles and other forms of transportation and (2) airborne larval dispersal. The larvae emerge from overwintering eggs in late April or early May, climb trees, and are dispersed in the atmosphere, generally by hanging from leaves on silk threads that are broken by the force of the wind.

It is the second mode of dispersal that is of interest to us. Our primary objective is to develop a dispersal model whose use will allow accurate prediction of the concentration and location of succeeding generations from a specified parent population. As a result, we will achieve a prediction capability that will ultimately allow accurate forecasts of defoliation. A secondary objective is to identify critical factors in the dispersal process as shown by the model; by purposefully altering these factors with the imposition of controls in the field, it may be possible to successfully manage severe infestations without undue environment damage.

Atmospheric Dispersion Model

Diffusion model

An advecting Gaussian puff forms the basis for our atmospheric diffusion model describing larval transport (Mason 1975). McManus (1972) has found that the larvae have large settling velocities that are a function of both body weight and length of silk being trailed; a representative value for a newly hatched larva is 0.75 m/s. Such a value allows the settling process to be modeled explicitly by the inclusion of a settling term in the Gaussian puff kernel. Therefore we take as the appropriate Green's function for a point source the following:

---

*Conrad J. Mason and R. L. Edmonds

$$(2\pi)^{3/2}\sigma_x\sigma_y\sigma_z G(x,y,z,t,t') = \exp\left\langle -\left(\frac{1}{2}\right)\left\{\left[\frac{x - u(t - t')}{\sigma_x}\right]^2 + \left(\frac{y}{\sigma_y}\right)^2 + \left[\frac{z - v(t - t')}{\sigma_z}\right]^2\right\}\right\rangle \tag{7.19}$$

We assume the x-axis lies along the mean steady wind direction u; y and z are the crosswind and vertical axes, respectively; v is the settling velocity; and, like Turner (1964), that the dispersion coefficients are functions of the travel time $(t - t')$ and atmospheric stability. Consequently, the larval concentration at the point (x, y, z) and at time t owing to an instantaneous release $Q(t')$ at the origin at time t' is given by the convolution integral:

$$\chi(x,y,z,t) = \int_0^{t'} Q(t')G(x,y,z,t,t')\, dt' \tag{7.20}$$

The integration is performed numerically over 10-min intervals assuming that the release rate is constant over the same period. Procedures similar to those of Roberts et al. (1970) are used to extend the model's capabilities to treat varying winds and atmospheric stability while dispersion is occurring as well as to allow application to extended sources.

Source function model

We have developed a preliminary model to describe the larval activity that leads to their arrival at treetop dispersion sites based upon the laboratory and field studies of larval behavior by McManus (1972). The source function describing the larval release rate into the atmosphere is a function of time, relative humidity, and temperature because biological activity of the larvae is affected by the prevailing meteorological conditions. An egg mass inventory for a specific source function conducted during the overwintering stage establishes the spatial dependence of the source function.

The model consists of a closed set of ordinary differential equations relating the number of larvae hatched from a single egg mass to the number at the treetop dispersal sites. The system is solved using the continuous systems modeling program (CSMP) simulation language. Model details are given by Edmonds et al. (1973).

Larval reentrainment

After deposition has occurred, larvae do not necessarily start feeding. Instead, after a brief rest period, they climb to the treetop dispersal sites and participate again in the dispersal process. Since the larvae deposited at any point are indistinguishable from the larvae hatched at that point, the deposition process increases the release rate after a time interval $(t - t'')$ corresponding to the duration of the rest period. Consequently, the new release rate is given by

$$Q(t') = Q_o(t') + vAX(t' - t'') \tag{7.21}$$

where $Q_o(t')$ represents the release rate for the larvae hatched at the given point and the second term represents the influx of larvae deposited on a horizontal crown-level surface of area A at the same point.

Results

Representative model output is shown in Figure 7.24. Here, the number of larvae captured per square meter of vertical trap surface suspended at

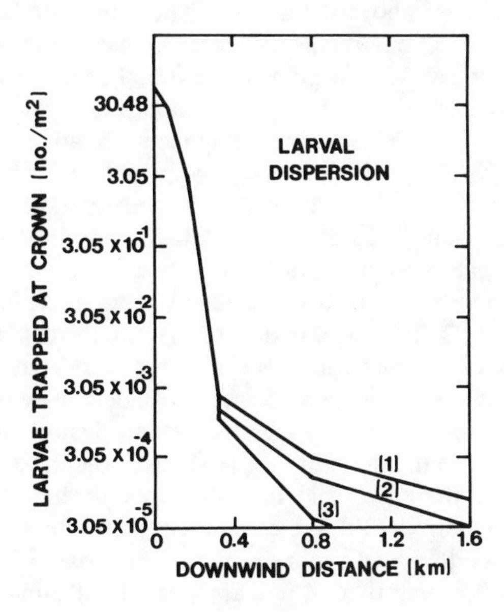

Figure 7.24. Number of larvae per square meter trapped at the crown as a function of downwind distance and atmospheric stability.

crown level is plotted versus downwind distance, assuming a constant 6.4 km/h wind for a period of 1 h, various atmospheric stability classes, an area source 0.16 km square, and a uniform release rate of one larva/min. Reentrainment is not included because of the short overall time interval.

The effect of the large settling velocity is clearly apparent: A change in larval areal density of approximately four decades occurs in a lateral distance of only about 0.32 km. Further, the effect of atmospheric stability on the dispersion process is not important up to this distance; only for longer range dispersal does stability influence the results. The above observations suggest the significant result that larval dispersion actually occurs as a series of short, successive hops. Field verification of the model has been initiated.

PALYNOLOGY MODELS*

Model Development

At the first workshop/conference of the IBP-Aerobiology Program (Manhattan, Kansas, January 1972), a committee[1] concerned with large particulates discussed the component elements of palynological systems and their functional relationships (Benninghoff and Edmonds 1972). At the time, the models discussed contained time dimensions of days, and space dimensions of about a cubic kilometer above the ground. The group produced a list of the primary biological and environmental factors that controlled atmospheric pollen concentrations at a given point in time and space. They also assessed the relative importance of each factor or component in the system, and estimated the state of knowledge of the components relative to the knowledge needed for quantitative prediction of airborne pollen concentrations. Palynology models aimed at paleoenvironmental reconstruction, which consists of much broader time and space dimensions (decades and 10^3 km^2), were discussed, but no formal recommendations were made.

During the second workshop/conference (Edmonds and Benninghoff 1973a) of the series (Boulder, Colorado, July 1972), the committee[2] discussed problems associated with modeling both the large- and small-dimension palynological systems. After discussion of specific models, they decided upon the pollen data needed for testing the suggested models. Data were to be collected from the field and from the literature for analyses at the third conference.

At the third conference, held in Ann Arbor during March 1973 (Edmonds and Benninghoff 1973b), the committee[3] split into small groups to test data against models, then met again during the final day of the conference to discuss the models. At that time, it was apparent that the meager data used to test the models of short time (days) and small distance (kilometer) dimensions

*A. M. Solomon and J. B. Harrington

were not adequate. It was obvious that this situation would remain unchanged until much more precise information was available.

The data for the models of longer time and space scales, however, could be applied to validation of a first-approximation model. The data were provided by H. Nichols and the group concentrated their efforts upon them. The results were written up by Harrington (Edmonds and Benninghoff 1973b:84–96) and they indicated that certain elements of atmospheric diffusion meteorology can be profitably applied to models of particle transport during multiyear periods and over distances of several thousand kilometers.

We will discuss these large-scale models with the addition of newly gathered data from the Sierra Nevada of California. We begin by reviewing the objectives of paleoenvironmental reconstruction by pollen analysis, the system variables upon which pollen analysis is based, and the models developed prior to the IBP-Aerobiology effort.

The palynological system

Pollen analysis is the use of fossil pollen grains to describe the prehistoric vegetation and plant distributions, and to describe indirectly the environmental parameters that controlled the plants. The pollen evidence is interpreted solely upon knowledge gleaned from studying relationships of modern pollen to the plants that produced it and of modern vegetation to the environmental parameters that control it.

Time

The time frame covers only the period during which modern pollen-plant-environment relationships have remained constant. This may be as short as 10^4 yr (Watts 1973). It more likely covers 10^5–10^6 yr, and 10^7 yr may not be an unreasonable period. No conclusive evidence and little suggestive evidence is yet available to indicate the actual length of this time period.

The time aspect of system variables is the major difference between pollen analytical phenomena and the other aerobiological phenomena discussed in this volume. The smallest increment of time studied may be 10^1 yr, but 10^2-yr increments during 10^4-yr periods represent the normal time scales of interest. Thus the objectives of pollen analysts may be compared with those of climatologists who integrate 10^1–10^2 yr of weather parameter values in order to examine patterns relating to longer time periods.

Data base

The system variables used in pollen analysis are also notably different from those used for discussion elsewhere in this volume. For all practical purposes, they are measured in ordinal rather than interval value units. Since the method of pollen analysis was first introduced (Von Post 1916), it has consisted of (1) extracting pollen grains from a sedimentary matrix, (2) identifying a given number of extracted pollen grains (commonly 200–1000), (3) calculating the relative frequency of each pollen taxon in the sample. Recent advances in the calculation of absolute pollen frequencies notwithstanding, the major share of pollen data used in future work will be derived from the relative pollen frequencies. Models that purport to describe or simulate pollen analytical systems must be designed for the broad time resolution and less precise data base that characterize pollen analysis.

Forcing functions

Important forcing functions of pollen analytical systems that control the scope and nature of models deserve mention. They include parameters of pollen sources and of pollen preservation environments, and differential dispersal of pollen grains.

Pollen sources vary in their productivity by a wide range (A. M. Solomon, chap. 2, this volume). Biotically pollinated plants produce little pollen and release even less. The partially insect-pollinated Norway maple (*Acer platanoides*) produces approximately 8×10^3 pollen grains/flower; a corresponding value for the wind-pollinated Scots pine (*Pinus sylvestris*) is about 1.6×10^5 pollen grains (Erdtman 1969). The insect-pollinated dog fennel (*Anthemis cotula*) and dandelion (*Taraxacum officionale*) emit pollen concentrations of $5.7–6.5 \times 10/m^3$ a meter away from the plants (Ogden et al. 1975), while the wind-pollinated tall ragweed (*Ambrosia trifida*) from the same plant family emits pollen concentrations of $7.7 \times 10^4/m^3$ at the same distance (Ogden and Lewis 1969). Thus pollen source strength per unit time (a pollinating season, or a year) is an important input.

The distance between the pollen sources and the deposition environment provides another important source of variation. The usual objectives of pollen analysis include understanding only the upland vegetation; the samples used are normally collected from lowlands, in lakes, and bogs. Thus the nearby pollen sources, which depend less upon climate than upon local edaphic conditions, are excluded by the investigator from the pollen types under consideration. Certain pollen types identifiable only at genus level may consist of species with both upland and lowland affinities. On the flat or rolling topography that typifies large portions of the North Temperate Zone where pollen analysis is usually applied, the distance that identifiable upland species grow from the sampling site, and the distribution patterns they assume are neither regular nor random. Indeed, wind pollination by necessity occurs among plants with clumped distributions (A. M. Solomon, chap. 2, this volume).

These problems are compounded by the differential dispersability of different pollen types. First, the vertical distance at which pollen is emitted varies considerably. Trees in closed-canopy forests discharge pollen only at the uppermost surface of the canopy. Trees in open situations emit pollen from all leafy surfaces of the tree. Herbaceous plants are generally less than 2 m high, so their pollen is released at that level. The wind environments in each situation are different and consequently the distance traveled by various portions of the total pollen emission varies accordingly.

Second, the various pollen types fall in the atmosphere at different speeds related to their volume and density. For example, spruce (*Picea excelsa*) and common ragweed (*Ambrosia elatior* [*artemisiifolia*]) have the same measured density (0.55); the spruce has approximately eight times the volume of ragweed and about six times the fall speed (Gregory 1973). Spruce and pine (*Pinus cembra*) pollen have similar dimensions and morphology, yet the former falls twice as rapidly as the latter (Gregory 1973). While pine pollen probably loses half its water (and one-third its specific gravity) upon entry into the atmosphere, spruce pollen loses very little.[4] Thus deposition rates must be considered separately for different pollen types.

The deposition environment from which pollen samples are recovered induces variability in pollen percentages. Lentic water of lakes or bogs is a normal pollen-sampling situation. It provides a stratigraphic, time-ordered sequence. Pollen grains are preserved best in the anaerobic-reducing conditions there. Pollen grains that sink rapidly, such as oak (*Quercus*) and birch (*Betula*) are less likely to be blown onto the shore by wind than, say, ragweed (*Ambrosia*) or pine (*Pinus*) pollen grains, which float for long periods of time (Davis and Brubaker 1973, Hopkins 1950). Once the pollen grains sink below the surface of the water, those that sink the most rapidly (e.g., oak) eventually become concentrated within the deepest parts of the water body, while other types, which remain suspended in the water (e.g., ragweed), are finally incorporated much more evenly into the lake sediment (Davis and Brubaker 1973). Other limnological processes further redistribute pollen grains (Davis 1973, Davis et al. 1971) before they finally become incorporated into stable sediments.

Although the differential preservation of pollen grains in aquatic environments has been examined closely (Sangster and Dale 1961, 1964, Havinga 1964, 1967, and others), it probably is unimportant with a few notable exceptions. The pollen grains of aspen (*Populus*), for example, disappear rapidly from lake sediments (Ritchie and Lichti-Federovich 1963). Pollen grains of all types are most rapidly corroded in shallow ponds, less so in lakes, and least in bogs (Sangster and Dale 1964). The foregoing points are summarized in the conceptual model presented in Figure 7.25.

Certain of these variables can be neutralized through selection of appropriate study regions, plants, and deposition sites. The models described below

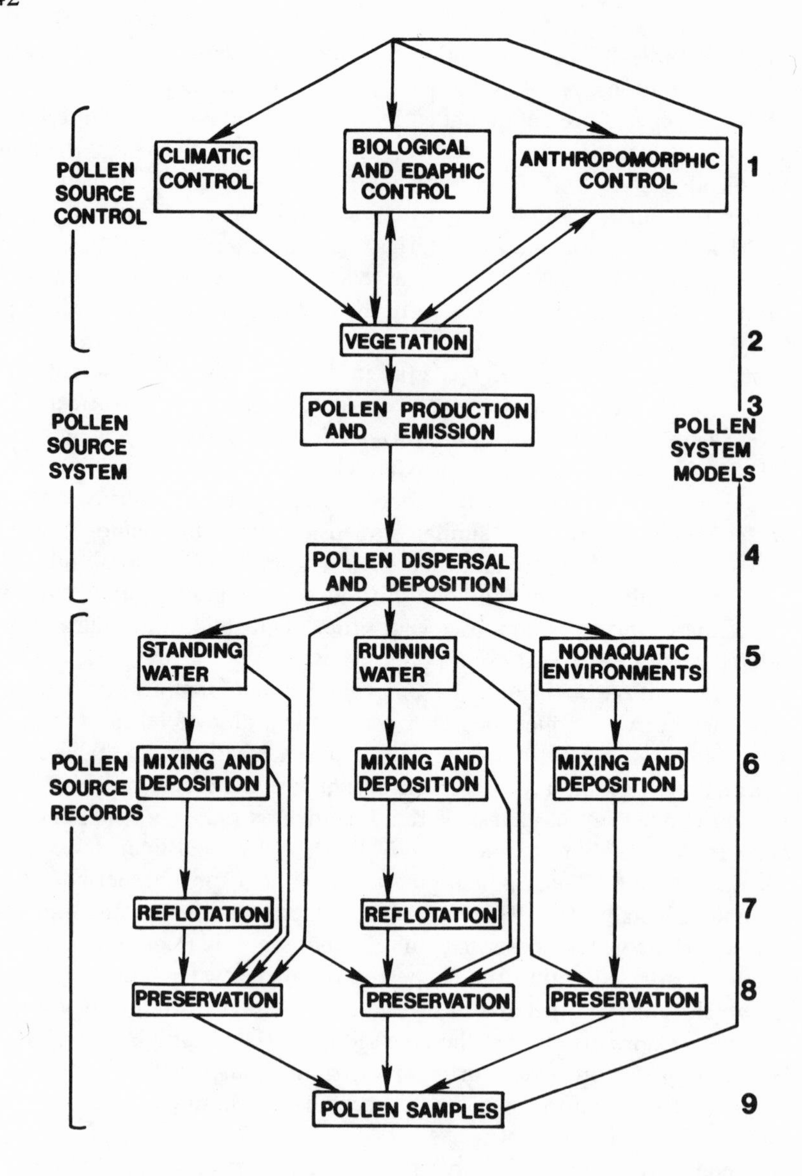

Figure 7.25. Conceptual model of palynological system and components.

pertain only to (1) lake environments of North America, (2) wind-pollinated plants that dominate upland vegetation, (3) natural plant communities (nonurban, nonagricultural) insofar as possible, and (4) time increments of 10^1–10^2 yr during the past 10^3–10^5 yr.

Major Non-IBP Models

Representation model

Since the formal inception of pollen analysis as a paleoecological tool, its adherents have been aware that major interpretive problems could be posed by genetically induced pollen productivity variations among the different species that contributed to the measured relative pollen frequencies of a given sample. In North America these difficulties were largely ignored until the appearance of a paper by M. B. Davis entitled, "On the theory of pollen analysis" (1963).

Earlier, Davis studied the relationship between proportions of tree species in a forest and the proportions of their pollen grains in the surface sediments of two lakes they surrounded (Davis and Goodlett 1960). Although studies of such relationships had been done in North America (Carroll 1943, Hansen 1949, Benninghoff 1960, Potter and Rowley 1960, and others), the conclusions drawn were new. Davis and Goodlett determined that the reconstruction of past vegetation from fossil pollen grains could not be valid until the following variables were quantified:

1. Pollen production differences between species as a function of genetic, environmental, and ontogenetic variations.
2. Pollen dispersal differences between species as a function of variations in topography, vegetation structure, yearly atmospheric conditions, and physical parameters of the pollen grains themselves.
3. Pollen deposition and preservation differences as a function of variations in the pollen grains themselves, and of the environment of deposition.
4. Pollen source distribution differences as a function of variations in local biotic and abiotic factors.

These conclusions led directly to the attempt (Davis 1963) at modeling the portion of the system represented by levels 2–9 in Figure 7.25. The aim was to assess the probable role the forcing functions represented by levels 3 and 4 (Figure 7.25) played in fossil pollen spectra. The model had the following constraints:

A. The vegetation consists of only three species, *a, b,* and *c.*

B. The plants of unspecified density are evenly distributed throughout a uniform infinite area (to avoid no. 4 above).

C. The percentages of pollen determined are a completely accurate sample of all the pollen delivered to any point in the area (to avoid no. 3 above).

D. If plants *a, b,* and *c* occur in the ratio of 1:1:1, their pollen productivity and dispersal differences produce a pollen ratio of 10:5:1, respectively (to control nos. 1 and 2 above).

The ratio of pollen frequency to plant frequency for each species was designated as its representation value, or *R*-value, R_a = species *a* pollen frequency/species *a* plant frequency. Although the *R*-values vary, by definition, the ratio of the *R*-values (R_a:R_b:R_c) is constant (10/1 = R_a; 5/1 = R_b; 1/1 = R_c, or 10:5:1). The objective of the model is to reconstruct vegetational frequencies ($V_{a,b,c}$) from pollen frequencies ($P_{a,b,c}$) with the *R*-value ratios (R_a:R_b:R_c) as parameters. The algebraic transfer functions given in Figure 7.26 were derived from the arithmetic operations demonstrated by Davis (1963, table 1).

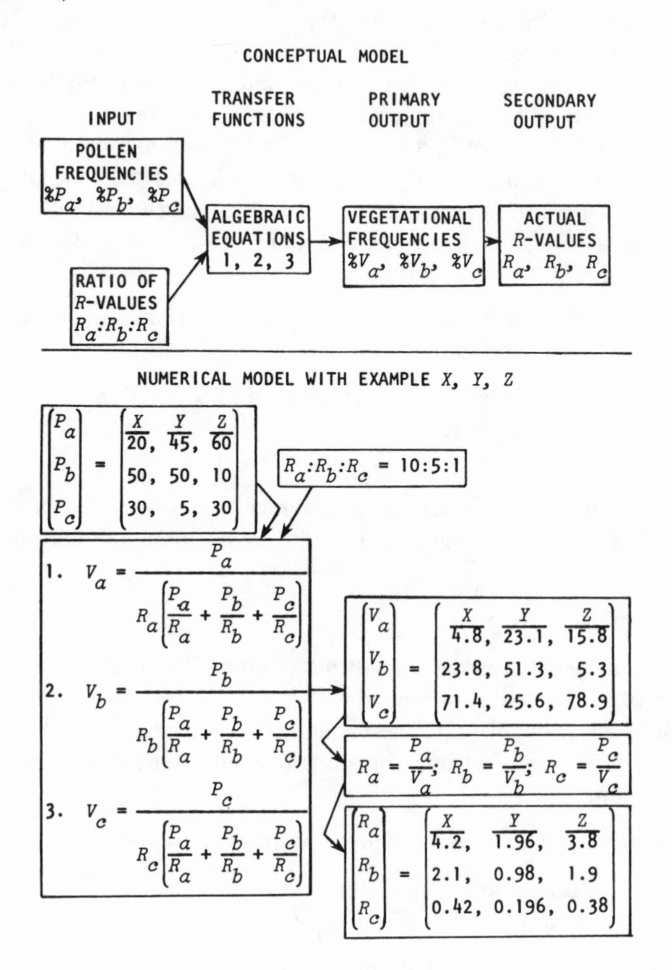

$$V_a = \frac{P_a}{R_a\left(\frac{P_a}{R_a}+\frac{P_b}{R_b}+\frac{P_c}{R_c}\right)} \quad (1)$$

$$V_b = \frac{P_b}{R_b\left(\frac{P_a}{R_a}+\frac{P_b}{R_b}+\frac{P_c}{R_c}\right)} \quad (2)$$

$$V_c = \frac{P_c}{R_c\left(\frac{P_a}{R_a}+\frac{P_b}{R_b}+\frac{P_c}{R_c}\right)} \quad (3)$$

$$R_a = \frac{P_a}{V_a};\ R_b = \frac{P_b}{V_b};\ R_c = \frac{P_c}{V_c}$$

Figure 7.26. *R*-value model.

There are several points the model brings to light. First, it is not intuitively obvious that the ratios of the *R*-values must indeed remain constant if they have been accurately defined. Were another species to replace species *a*, *b*, or *c*, however, the established representation relationship between the remaining two species would continue, despite pollen production and dispersal differences between the new species and the species it replaced. In addition, if species *a*, *b*, and *c* were suddenly invaded by other species that contributed to the pollen "spectrum," no change would occur in the original *R*-value ratio. This is demonstrated in Table 7.6.

Table 7.6. Hypothetical representation relationships of a five-species community.

Species	*R*-value ratio (*R*)	Community % pollen (*P*)	Corrected pollen values[a]	% veg[b] (*V*)	Actual[c] *R* values
a	10	15.0	1.5	3.31	4.53
b	5	4.0	0.8	1.77	2.265
c	1	30.0	30.0	66.24	0.453
d	3	30.0	10.0	22.06	1.359
e	7	21.0	3.0	6.62	3.171
		Σ100.0	Σ45.3	Σ100.0	

[a]Percent pollen/*R*-value ratio (*P*/*R*). [b]Corrected pollen value/*R*-value ratio × Σ corrected pollen value. [c]Percent pollen/percent vegetation (*P*/*V*).

Second, it is clear from the examples given in the model (Figure 7.26) that the actual *R*-values do indeed vary. When the plant that produces the most pollen (species *a*) replaces (all *R*-values decline) or is replaced by (all *R*-values increase) the plant that produces the least pollen (species *c*), the variability is most apparent (examples *X* and *Y*).

Third, the differences in overall pollen abundance achieved by various combinations of plant frequencies produces variance in pollen frequencies that may not be intuitively obvious. A comparison of examples *X* and *Y* and *Y* and *Z* in the model is illustrative. While species *b* doubles in the plant population at the expense of species *c* (*X* and *Y*), no change occurs in the pollen frequency of species *b*. Species *a* plants are partially replaced by species *c* (*Y* and *Z*), yet the pollen frequency of the declining species *a* continues to increase!

When Davis (1963) derived *R*-values from a lake in Vermont, and then applied the *R*-value model to objective interpretation of a core from the same lake, she concluded that there were very few pine trees some 8000 years ago during the pine pollen maximum that characterized the data. While this conclusion was eventually disproved (Davis 1967:226), Davis has pointed out that the important assumptions of the model (nos. B and C) rarely if ever exist in nature. Indeed, while the model seems to deal adequately with pollen analysis data from a theoretical standpoint, it has little validity from a practical one, for its assumptions are too far from reality.

The major problem evolves from the functional relationships between plant distributions and pollen dispersal potential. A producer of heavy, "short-range" pollen grains will induce high *R*-values if its distribution is clumped around the lake from which pollen frequencies are measured. Here, tree frequency in the total vegetation sample is low. The same plant growing abundantly in the vegetation, but not at the edge of the lake, produces low *R*-values because few of its pollen grains reach the lake.

A different situation occurs in plants that emit light, "long-range" pollen grains. They may not grow at all within the vegetation sampled, yet their pollen appears in the sediments, producing *R*-values that approach infinite values. Yet when the same plant grows abundantly in the vegetation, its pollen grains are so uniformly distributed that its *R*-values are reduced toward unity.

Because of these and other problems discussed by Janssen (1967) and Livingstone (1969), North American pollen analysts turned away from this kind of model for reconstructing past vegetation from pollen proportions.

The analog models

In an addendum to her discussion of the *R*-value model, Davis (1963) mentions that "(Martin and Gray 1962) have reconstructed past plant communities by matching the percentages of fossil pollen grains to percentages in surface samples from appropriate regions. . . . Although limited to instances where fossil and present pollen spectra are very similar both in composition and frequencies, the direct comparison method is extremely promising and represents a major advance over subjective interpretation."

The approach to interpretation of fossil pollen samples, in which modern analogs are sought to match fossil samples, has much to recommend it. One need not specify pollen percentages, for one studies the entire pollen spectrum and uses the measured variance in it as the pollen equivalent of a given plant community. One first defines vegetational units, then the variation in pollen spectra found within the defined vegetational units. Where the range of selected pollen frequencies in one unit does not overlap the range of the same pollen frequencies in a different unit, the units are considered distinguishable. The pollen spectrum of each is then matched with pollen spectra in fossil pollen sequences. In essence, this approach matches modern values in levels 2 and 9 (Figure 7.25) in the hope that processes represented in levels 3–9 can be ignored.

The approach demands only the assumption that similar pollen spectra (modern and fossil) are derived from similar vegetation. It is an exceedingly satisfying method because vegetation and the factors that controlled it can be reconstructed by actual research upon the modern vegetation analog.

J. G. Ogden, III (pers. commun.), tells the story of visiting a forest in Ohio for the first time with the unsettling feeling of having been there before. Months later he suddenly realized that the forest was one he had mentally reconstructed from fossil pollen spectra collected elsewhere!

What specific climatic parameters (winter snowfall, mean July temperature, growing season length, and so on) operated in southern Connecticut

9500–12,000 yr ago? Go to the point in northern Quebec (Figure 7.27) where modern pollen spectra resemble the southern Connecticut pollen assemblages of 9500–12,000 yr ago, and measure those parameters. While you are there, measure age distribution and species density of the forest trees as well!

McAndrews (1966) brought this approach to the attention of pollen analysts. His work in tracing the geography of the prairie-forest border in Minnesota during the past 12,000 yr included a qualitative dichotomous key (McAndrews 1966:52-53) to pollen assemblage zones based on the modern presettlement pollen spectra and associated presettlement plant communities.

Of the several pollen zones and subzones McAndrews was able to distinguish, two had no modern analogs, i.e., contained combinations of pollen taxa in proportions not found today. Subsequently this was shown to be a major problem for pollen spectra 8000 yr old and older (Davis 1967, 1969a). The vegetational relationships of these pollen spectra simply cannot be studied.

A second problem evident in the analog approach is that of variance in the data. While general plant-community types can be distinguished on the basis of analogs in which the high intercommunity variance in pollen percentages overcomes intracommunity variance, this is not true in smaller vegetational subdivisions (Davis 1967). Thus the resolution of details is quite limited.

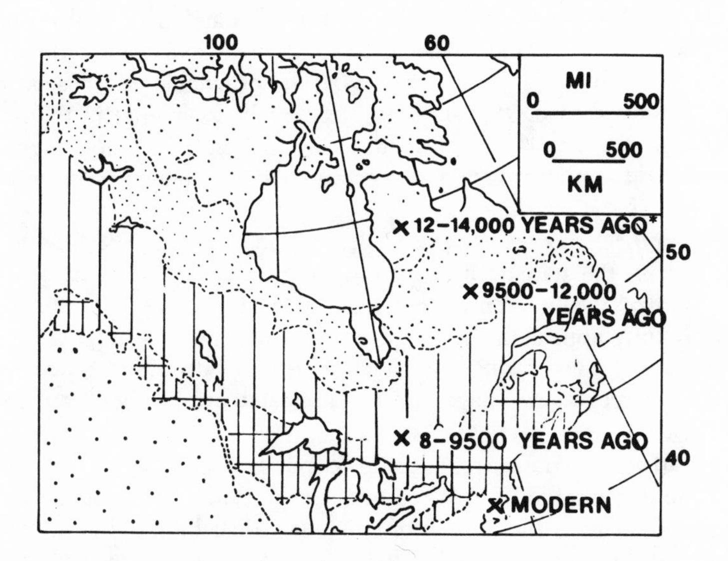

Figure 7.27. Localities (X) in Canada where surface pollen assemblages resemble fossil assemblages in southern New England. Age of analogous fossil material is indicated for each locality. *Note: This location in the tundra is extrapolated from the resemblance of fossil material to assemblages from northernmost Quebec (Bartley 1967). Surface samples are not yet available from the precise locality shown (from Davis 1969b).

Third, as Davis (1963) earlier pointed out, pollen spectra can be very similar, but result from quite different vegetation. This became a major problem, for example, when pollen analysts tried to distinguish the pollen spectrum of tundra vegetation from that produced by northern boreal forests. Was a tundra adjacent to North American ice sheets during full-glacial conditions as occurred in Europe, or did forests grow adjacent to the ice, as occurs in Canada and Alaska today? Little airborne pollen is produced on the modern tundra, so that pollen from boreal forests to the south constitute the major source of pollen in tundra sediments (Ritchie and Lichti-Federovich 1967). The pollen spectrum is therefore essentially the same in both the tundra and in the boreal forests, although the numbers of pollen grains are much reduced in the tundra.

Statistical approaches

A nonparametric statistical model

Attempts at quantifying the analog approach have been limited to the statistical treatment of analog data by Ogden (1969). Ogden's work utilized the nonparametric statistic, Spearman's rank correlation coefficient, r_S. By collecting a large number of pollen spectra from surface muds of lakes and bogs, Ogden produced an "analog" data bank for comparison with fossil pollen spectra. Those fossil samples with correlation coefficients above the critical value ($r_S > 0.80$ for $N = 23$ at the 0.90 level of significance based on a Z test) are considered to be the same pollen assemblage as that in the modern pollen analog. In general, Ogden assumes the highest correlation coefficients indicate the most similar pollen and vegetation. This is diagrammed in Figure 7.28.

This approach possesses several distinct advantages. The foremost advantage involves the statistic itself. By ranking the data, the investigator asks less precision of the inherently imprecise data. The most important pollen types consistently appear in the highest ranks, and the least important in the lowest ranks. Although some statistical noise is introduced when similar (usually low) percentages change ranks without statistically significant changes in percentage values, the r_S value is not strongly affected.

The statistic, of course, requires no particular population distribution for validity. The distribution of values a large set of sample percentages may assume is not likely to fit the normal or chi-square distribution, but in any case, the distributions have never been tested. The freedom from any statistical distribution probably adds validity to the model.

Overall, the approach is similarly free of assumptions. Although it presumes that similar pollen data are produced by similar vegetation and environmental parameters, little else is hypothesized. This advantage allows the investigator to relate pollen spectra objectively with a minimum of potentially unwarranted conclusions.

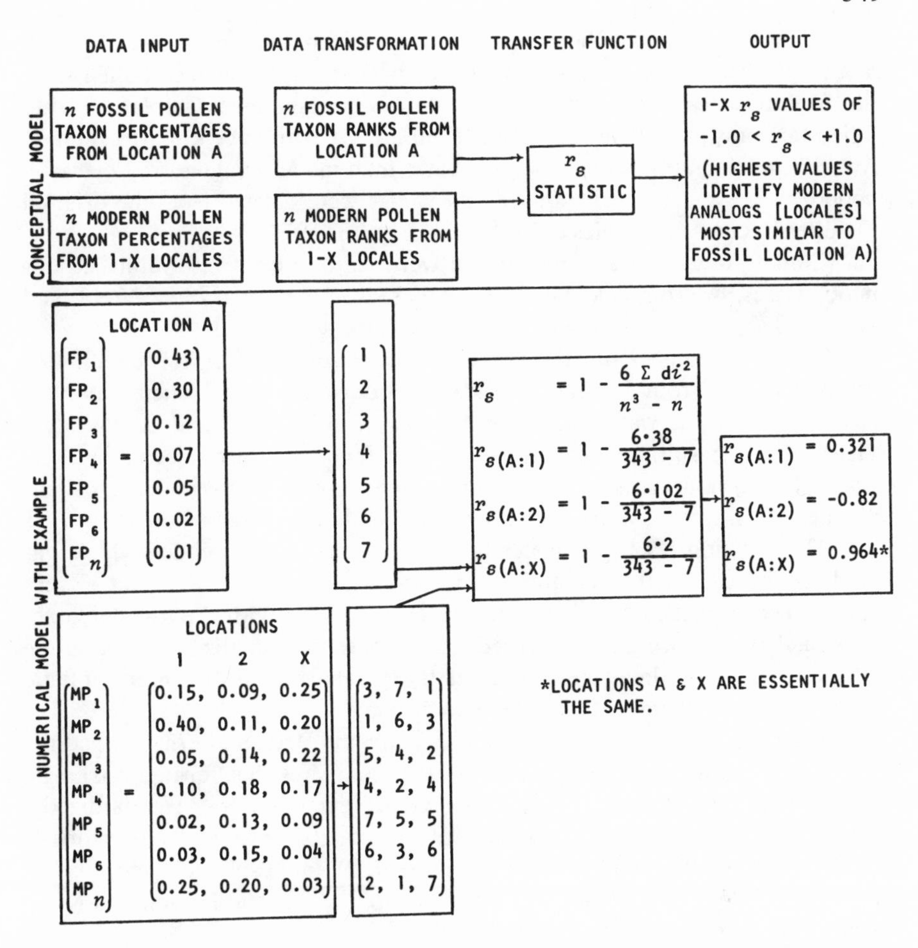

Figure 7.28. Statistical model of modern pollen analog approach.

The multivariate statistical model

The use of canonical correlation analysis to transform fossil pollen percentages into contemporaneous climatic parameter values originated with Bryson, Webb, and others at the University of Wisconsin's Center for Climatic Research. The method first appeared in doctoral dissertations by Cole (1969) and Webb (1971), then in published articles by Webb and Bryson (1972) and Bryson and Kutzbach (1974). The model is probably more similar to the ecological systems approach than is any other model in which relative pollen frequency data are used. The model consists of a statistical match between climatic parameters (level 1, Figure 7.25) and pollen values (level 9), excluding even vegetation (level 2).

The procedure consisted of deriving transfer functions from modern pollen and climate data, testing the transfer functions on an independent set of pollen and climate data, then applying the transfer functions to fossil pollen data to produce "fossil" climate data. For basic data, 96 surface pollen samples from 59 sites were used, 70% of which were lake mud samples. Twenty pollen taxa not present in all surface samples, six with very low percentages, five with maximum fossil values far greater than modern values, and one with modern values much greater than fossil values were deleted from the original 40 pollen taxa in all samples of Webb and Bryson (1972).

Climate data for each pollen site were constructed from published climatic atlases, or from calculated values based on measured ones (Webb and Bryson 1972).

To derive transfer functions, Webb and Bryson first transformed climate and pollen data from 34 sites into standard deviation units. Then canonical correlation analysis was applied to all pairs of pollen taxa and climate data sets. From the resulting correlations, transfer functions were selected by deletion of (1) all nonsignificant (low) correlations, and (2) all significant pollen-climate correlations that did not apply generally to the region studied.

The remaining set of transfer functions composed the model, which was validated by use of the remaining 25 sites. Pollen data at those sites were transformed into standard deviation units and treated with the transfer functions, and the resulting standardized values were denormalized into absolute values for climatic data. Deviations of these values from measured climatic values were then examined and biases were noted.

After testing the transfer functions, Webb and Bryson (1972) applied the model to three fossil pollen cores collected from sites that formed a triangle with legs of 241–322 km. Again, pollen percentages were "standardized," treated with transfer functions, and denormalized into absolute climatic values. The model is diagrammed in Figure 7.29 and Table 7.7, and, insofar as is possible, examples from Webb and Bryson (1972) and Bryson and Kutzbach (1974) are given.

The model is based upon certain important assumptions that should be briefly discussed. Webb and Bryson (1972) point out that results of the study "depend upon climate being the ultimate cause of the changes in the pollen records." In fact, this is probably not the case. Although climate may indeed control the composition of spatially stable plant associations, most of today's plant associations are more ephemeral than permanent. In two important papers, Davis (Davis et al. 1973, Davis 1976) points out that pollen and vegetational changes noted in eastern North America during the past 10,000 yr are due to variable migration rates, routes of major pollen-producing taxa, or both (biological control, level 1, Figure 7.25), rather than to climatic change as previously thought.

These migrations are still occurring today (Davis 1976) and their control is probably at least as much biological or edaphic (Brubaker 1975), as climatic. Correlations of climatic parameters that coincide with ephemeral vegetation associations are not the correlations of climate that cause permanent vegetation associations, as assumed by Webb and Bryson (1972) and Bryson

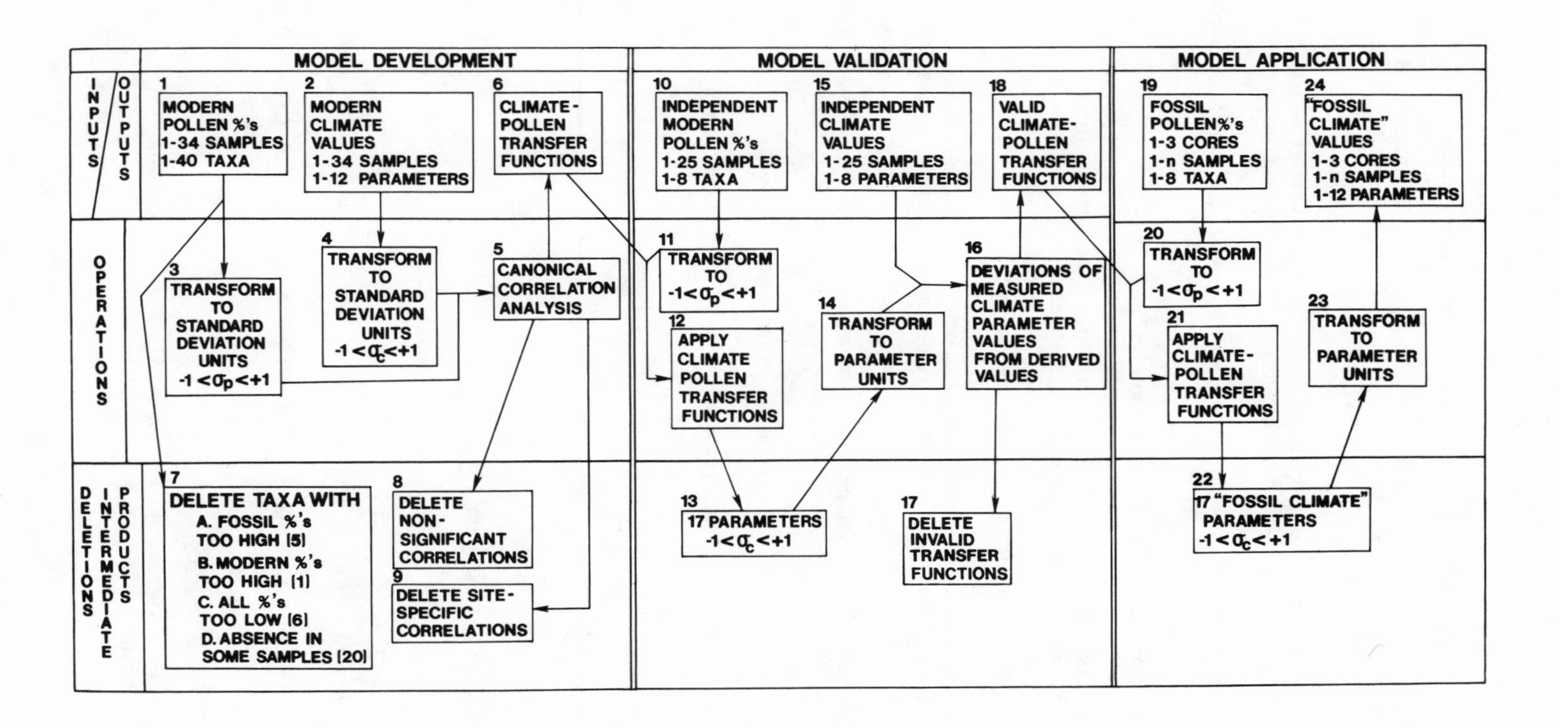

Figure 7.29. Multivariate statistical model: Conceptual model. (See Table 7.7.)

Table 7.7. Multivariate statistical model—Mathematical model.[a]

1	*Pinus, Larix, Ambrosia, Alnus, Picea, Juglans,* Gramineae, *Abies, Acer, Betula, Artemisia, Salix, Carya, Quercus, Fraxinus, Tilia; Ulmus, Ostrya, Corylus,* Chenopodiaceae, 20 other unnamed taxa; latitude
2	Precipitation minus potential evaporation; growing season precipitation; snowfall; hours of sunshine; degree days; growing season length; July mean temperature; maritme Tropical (mT), Return Polar (R), Pacific-south (Ps), Pacific-north (Pn), Arctic (A) air mass durations; mT, continental Tropical (cT), Ps, Pn, A July air mass frequencies
3-6	See Appendix in Webb and Bryson (1972).
7	(a) 20 unnamed taxa, absent in some samples; (b) *Alnus, Tilia, Abies, Salix, Artemisia,* Gramineae (all percentages too low); (c) *Ulmus, Ostrya-Carpinus, Larix, Fraxinus, Corylus* (fossil percentages too high); (d) *Ambrosia* (modern percentage too high)
8	Delete fourth to seventh canonical variates for climate, third to seventh canonical variates for air mass parameters (contribute too little information)
9	(a) Subdivide 58 samples into two sets, apply no. 1 (minus no. 7) above, no. 2 above, no. 6 above to each data set; (b) retain canonical variates common to both sets; (c) delete canonical variates unique to one data set or the other (coincidentally, same as no. 8 above)
10	*Pinus, Picea, Betula, Quercus, Acer, Carya, Juglans,* Chenopodiaceae; latitude
11	Same as no. 3
12	Same as no. 6
13	No. 2 in form of no. 4
14	Run no. 4, solving for $Y^*_{i,j}$
15	Same as no. 2
16	${}_nY^*_q - {}_nY^*_q = {}_nG_q = g_i$ where g_i are n-by-1 column vectors ($i = 1, \ldots, q$) and consist of deviations of a given variable at each of the 25 sites
17	"Snowfall," "growing season precipitation" in northern Wisconsin systematically underestimate true values
18	Same as no. 6
19	Figure 2—Webb and Bryson (1972—Fossil pollen diagram for Kirchner Marsh)
20	Same as no. 3

Table 7.7 (continued)

21	Same as in no. 6
22	Same as in no. 13
23	Same as in no. 14
24	Plot of average July temperatures from present to 13,500 years before present.

[a]From Webb and Bryson (1972).

and Kutzbach (1974). This problem is particularly apparent in the latter paper where pollen taxa primarily controlled by edaphic factors (*Larix, Salix,* some *Fraxinus,* and *Betula* species) and those controlled by anthropomorphic factors (*Ambrosia, Corylus,* Compositae) are mixed with the others, all as a measure of climate.

A second assumption of the model is "a constancy in the climatic response of the pollen types used" (Webb and Bryson 1972). This assumption is probably valid for the species involved, but certain important species were lumped into "pollen types" for which the assumption is not valid. For example, the white pine (*Pinus strobus*), which entered the area 7000 yr ago, was not distinguished from the red pine (*P. resinosa*) or jack pine (*P. banksiana*), which entered the area 10,500 yr ago (Wright 1968). Even today, the white pine does not occupy the northwestern jack pine region in which one-fourth of the Webb and Bryson sites are located. Not only do the three pine species respond differently to different edaphic and climatic parameters (Brubaker 1975), but white pine produces far more pollen than does either red or jack pine (L. B. Brubaker, unpubl. data, *fide* Davis et al. 1973). Thus constancy in the climatic response of the pine pollen occurs only when species composition of the genus remains constant. Again this is not the case, either spatially or temporally. Similar reasoning could be applied to indistinguishable species of birch and ash, which occur as species on both uplands (climatically controlled) and lowlands (edaphically controlled and adjacent to pollen sampling sites).

Despite these formidable problems involved with the model's implementation, the canonical correlation analysis approach represents a giant step beyond previous work in reconstructing climate or vegetation from pollen spectra. It clearly uses more of the information content of relative pollen frequencies than does any other method. It seems doubtful that any more precision would be produced by other stochastic methods when one considers the nature of the relative pollen frequency and the normal sampling errors inherent in it.

The IBP Models

The IBP-Aerobiology models differ from those discussed above in one major characteristic. The IBP models use transfer functions that simulate functional relationships between the environmental and physical variables (forcing functions), and the pollen deposition (system variables) in question. Because the output of the models is dependent upon the way in which the system actually functions, its results can be applied to regions and times different from those modeled. The models described above, on the other hand, depend at most upon statistical correlation transfer functions that correlate input and output of the modern system. They are stochastic models in the simplest sense, for they incorporate no knowledge of the functional relationships between forcing functions and system variables. Their results are applicable only to the region and time in which they were derived, the statements of Bryson and Kutzbach (1974) notwithstanding.

The models described below involve the subsystems concerned with pollen production, transport, and deposition (levels 3 and 4, Figure 7.25). They are based on an exceedingly simplified model of turbulent diffusion and deposition in the atmosphere. The models developed from discussions among participants of the palynology group at the third IBP-Aerobiology workshop (Edmonds and Benninghoff 1973b:84-96) as an attempt to describe pollen transport during decade-long periods. The models specifically apply to pollen spectra (relative pollen frequencies that compose a pollen sample) from nonforested plains or intermontane basins that occur near infinitely long forests (Arctic tundra) or linear, forested, mountain ranges (the basin and range Southwest).

As stated earlier (Edmonds and Benninghoff 1973b), the long-distance transport of pollen-size particles to a downwind point in space is a function of the depth of the turbulent boundary layer, the mean wind speed in the turbulent boundary layer, the settling speed of the pollen grains, the distance from the source to the downwind point (sample site), the strength of the mean vertical wind, and the magnitude of the vertical component of turbulence (whether it is sufficient to overcome gravitational settling).

The problem can be examined as follows. If the physical system is viewed as shown in Figure 7.30, then the flux of pollen grains through the x-facing sides of this cube are:

$$F(x) = u\chi \, \Delta y \, \Delta z \qquad (7.22)$$

where u is the horizontal wind speed and is the pollen concentration, and

$$F(x + \Delta x) = u\chi + \left[\frac{\partial (u\chi) \; \Delta x}{\partial x}\right] \Delta y \, \Delta z \qquad (7.23)$$

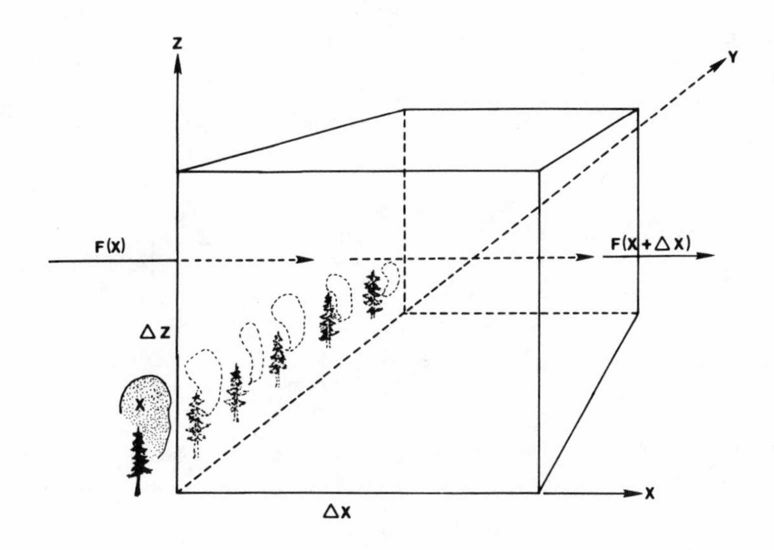

Figure 7.30. Box diagram of the atmosphere.

Then, the net flux into the cube from movement in the x direction is

$$F(x) - F(x + \Delta x) = -\left[\frac{\partial(u\chi)\ \Delta x\ \Delta y\ \Delta z}{\partial x}\right] \tag{7.24}$$

The net fluxes in the y and z directions, respectively, are found in a similar fashion to be

$$F(y) - F(y + \Delta y) = -\left[\frac{\partial(v\chi)\ \Delta x\ \Delta y\ \Delta z}{\partial y}\right] \tag{7.25}$$

$$F(z) - F(z + \Delta z) = -\left[\frac{\partial(w\chi)\ \Delta x\ \Delta y\ \Delta z}{\partial z}\right] \tag{7.26}$$

Now the total change of particles in the cube is equal to the net flux into the cube, or

$$\frac{\partial\chi\ \Delta x\ \Delta y\ \Delta z}{\partial t} = -\left[\frac{\partial(u\chi)}{\partial x} + \frac{\partial(v\chi)}{\partial y} + \frac{\partial(w\chi)}{\partial z}\right] \Delta x\ \Delta y\ \Delta z \tag{7.27}$$

This equation can be written in terms of mean values and deviations from the mean, such as $\chi = \bar{\chi} + \chi'$. The average value of Equation (7.27) is

$$\frac{\partial\bar{\chi}}{\partial t} = -\left[\frac{\partial(\bar{u}\bar{\chi} + \overline{u'\chi'})}{\partial x} + \frac{\partial(\bar{v}\bar{\chi} + \overline{v'\chi'})}{\partial y} + \frac{\partial(\bar{w}\bar{\chi} + \overline{w'\chi'})}{\partial z}\right] \tag{7.28}$$

Using the chain rule, the equation can be written

$$\frac{\partial\bar{\chi}}{\partial t} + \frac{\bar{u}\,\partial\bar{\chi}}{\partial x} + \frac{\bar{v}\,\partial\bar{\chi}}{\partial y} + \frac{\bar{w}\,\partial\bar{\chi}}{\partial z} \tag{7.29}$$

$$= -\left[\frac{(\overline{u'\chi'})}{\partial x} + \frac{(\overline{v'\chi'})}{\partial y} + \frac{(\overline{w'\chi'})}{\partial z}\right]$$

where

$$\bar{\chi}\left(\frac{\partial\bar{u}}{\partial x} + \frac{\partial\bar{v}}{\partial y} + \frac{\partial\bar{w}}{\partial z}\right) \equiv 0$$

because of the practical incompressibility of the air at subsonic wind speeds. The total derivative, i.e., the rate of change following the motion of a particle, is written

$$\frac{d\chi}{dt} = \frac{\partial\chi}{\partial t} + \frac{u\,\partial\chi}{\partial x} + \frac{v\,\partial\chi}{\partial y} + \frac{w\,\partial\chi}{\partial z} \tag{7.30}$$

Therefore the conservation of mass equation can be written as

$$\frac{d\bar{\chi}}{dt} = -\left[\frac{\partial(\overline{u'\chi'})}{\partial x} + \frac{\partial(\overline{v'\chi'})}{\partial y} + \frac{\partial(\overline{w'\chi'})}{\partial z}\right] \tag{7.31}$$

In the case of fluxes due to the mean wind speed, terms such as the last three in Equation (7.30) can be computed from a knowledge of the wind speed and either measured or computed values for the concentration gradient. These are advective or convective terms. The fluxes due to eddy motion such as $\overline{u'\chi'}$ in Equation (7.31), however, cannot be easily measured or computed. To close the equations, these diffusive terms must be approximated by assuming that the diffusive flux is equal to the product of a diffusivity and the gradient of the concentration. The simplest possible relations are

$$F'_x = -K_x\frac{\partial\bar{\chi}}{\partial x}; \quad F'_y = -K_y\frac{\partial\bar{\chi}}{\partial y}; \quad F'_z = -K_z\frac{\partial\bar{\chi}}{\partial z} \tag{7.32}$$

By substituting these terms for the eddy fluxes in Equation (7.31), we have

$$\frac{d\bar{\chi}}{dt} = \frac{\partial}{\partial x}\left(K_x\frac{\partial\bar{\chi}}{\partial x}\right) + \frac{\partial}{\partial y}\left(K_y\frac{\partial\bar{\chi}}{\partial y}\right) + \frac{\partial}{\partial z}\left(K_z\frac{\partial\bar{\chi}}{\partial z}\right) \tag{7.33}$$

where the eddy diffusivities K_x, K_y, and K_z are functions of x, y, z, t, and the dimensions of the diffusing cloud.

The major problem with this approach is that everything depends upon the definition of the average value. For example, if we average over the "infinite" time of interest, we take the chance that χ will be zero and moreover

$$\frac{d\bar{\chi}}{dt} \equiv 0 \quad \text{as is} \quad \frac{\partial\bar{\chi}}{\partial t} \tag{7.34}$$

If we average over a shorter time interval, then values of χ', u', and so on, depend very much on the time chosen. We hope to choose an averaging time such that the transports by the mean wind and those by the turbulent eddies have functional significance. It is apparent that no simple model can provide a general solution of the diffusion equation. Even the current Gaussian plume model is basically an empirical one, for despite the theories upon which it is based, certain terms ($\sigma_x, \sigma_y, \sigma_z$) are obtained empirically.

The model proposed here starts from the basic non-Fickian diffusion equation (Equation 7.29), which is simplified using the following assumptions or parameters of the situation:

1. The source is infinitely long, so that all variations in the y direction are zero. This eliminates the third and sixth terms.
2. Diffusion in the x direction is negligible compared with wind transport in the x direction. This eliminates the fifth term.
3. The pollen is contained in the turbulent boundary layer, where $\bar{w} \equiv 0$. In the fourth term, the fall speed of the pollen grain $-q$ is substituted for $\bar{w}$.

The final equation then becomes:

$$\frac{\partial\bar{\chi}}{\partial t} + u(z,t)\,\frac{\partial\bar{\chi}}{\partial x} = \frac{\partial}{\partial z}\left[K_z(z,t)\,\frac{\partial\bar{\chi}}{\partial z} + q\bar{\chi}\right] \tag{7.35}$$

This equation can be solved numerically if $u(z,t)$ is known or estimated, $K_z(z,t)$ is known or estimated, q is known, and the boundary conditions are known. The last involves (1) the pollen source strength as a function of time and position, (2) the pollen deposition rate at the lower boundary, and (3) the condition of zero diffusion at the upper boundary.

In the case of the Arctic pollen data to which the equation was applied (Edmonds and Benninghoff 1973b:84–96), the ground-level concentration was inferred from deposition measurements at various points downwind of the boreal forest pollen source, under the assumption that deposition equals the product of concentration and pollen fall speed integrated over the time during which the samples were operating. Without measurements either of source strength or of the concentration profile on the downwind edge of the source (K

and u could be estimated from obtainable meteorological data), little could be done with the equation. To avoid this basic problem, the equation was further simplified by allowing the pollen to be distributed uniformly in the mixing layer, thereby eliminating diffusion. After estimating a mixing depth and wind speed, and taking into account the differences in day and night diffusion, the relative change of deposition with distance was calculated from the resulting integration:

$$N(t) = N(0) \exp - \frac{q}{H}t_1 \left[1 - \frac{q}{H}(t - t_1)\right] \tag{7.36}$$

where N is deposition of pollen grains, q the fall speed of pollen, and H the mixing height. Time is measured from the time of emission; t_1 is the time of sunset and t is the time at which deposition takes place.

Considering the number of estimated quantities in the equation, the results (Figure 7.31) are not inconsistent with measured values. Refinements probably will not yield greater accuracy because the problem at this point is still exceedingly complex and contains too many unknown parameters. Clearly, until simultaneous pollen deposition and concentration measurements, along with the necessary meteorological variables, are made, good agreement between theory and observation will not be possible.

Nowhere is this more apparent than in the pollen data we collected in intermontane valleys near linear, isolated mountain ranges (Solomon 1975, A. M. Solomon and E. B. Leopold, manuscript in preparation). Here, the deposition of tree pollen downwind of the source was expected to follow the relation:

$$N(X) = N(0) \exp -\left(\frac{q}{uH}X\right) \tag{7.37}$$

where $N(X)$ is the deposition at distance X downwind, $N(0)$ is the deposition at the pollen emission source, q is the fall speed of the particle, u is the mean wind speed, and H is the height of the tree population above the valley floor.

Exponential equations (written in linear form [$\ln y = \ln a + bx$] and solved as linear regression) were derived from pollen deposition in rain gages at six downwind sites covering a distance of 25 km downwind from the Sierra Nevada sites (Figure 7.32). For yellow pine (*Pinus*–Diploxylon) and piñon pine (*Pinus*–Haploxylon), respectively, the equations

$$N(X) = N(0) \exp (-0.111X); \quad N(0) = 491.3 \tag{7.38}$$

and

$$N(X) = N(0) \exp (-0.134X); \quad N(0) = 1003.8 \tag{7.39}$$

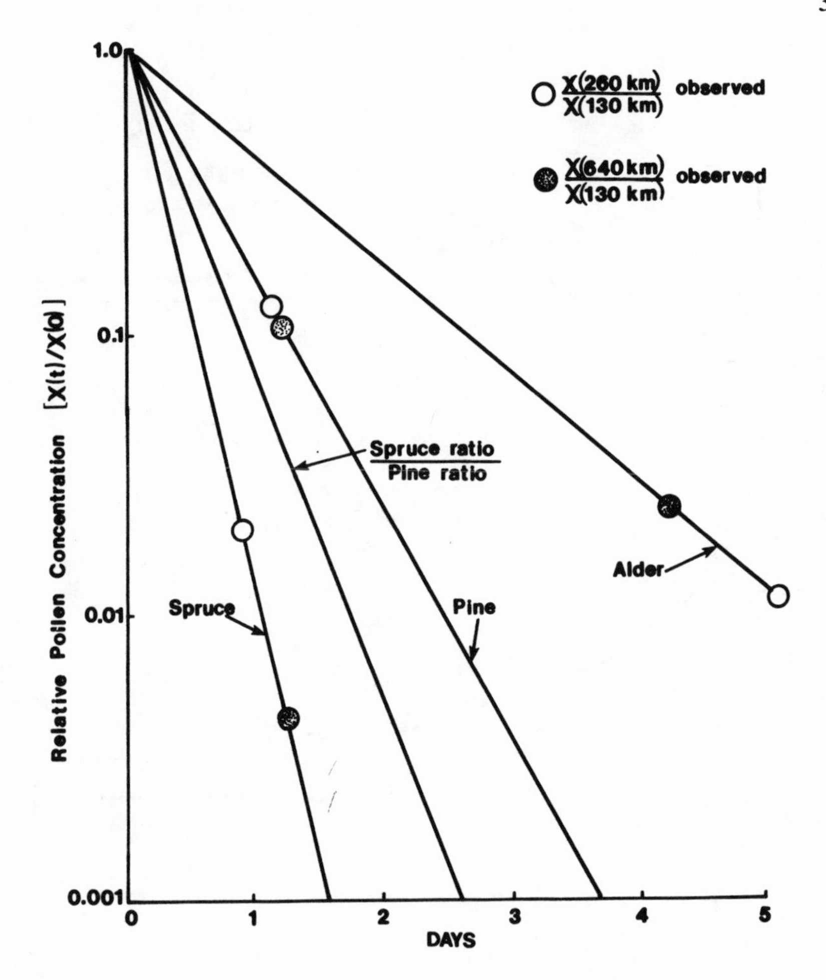

Figure 7.31. Rate of decrease of relative spruce, pine, and alder pollen concentrations with travel time in days. The time of travel has been adjusted to make the observed values fit an equation adapted from Equation (7.36), in which daylight continues for 24 hours and the boundary layer has a depth H equal to 1700 m (from Edmonds and Benninghoff 1978b).

provide close fit with the data ($r = -0.95$, $r^2 = 0.90$; $r = -0.95$, $r^2 = 0.91$, respectively). If one uses the parameters of the situation, however, where q = 2–5 cm/s (Gregory 1973), H = 1300 m, and u = 5–10 m/s, the equation provides values of q/uH between 4 and 8×10^{-8}, instead of the 1×10^{-1} measured for that term. Clearly, the model has no value without measurements of the real processes that are bringing the pollen back to earth. In this case, pollen is apparently moved upslope under the daytime valley winds that prevail at ground level. It is then returned back (eastward) across the valley with prevailing westerly winds, which are first encountered at the Sierra crest.

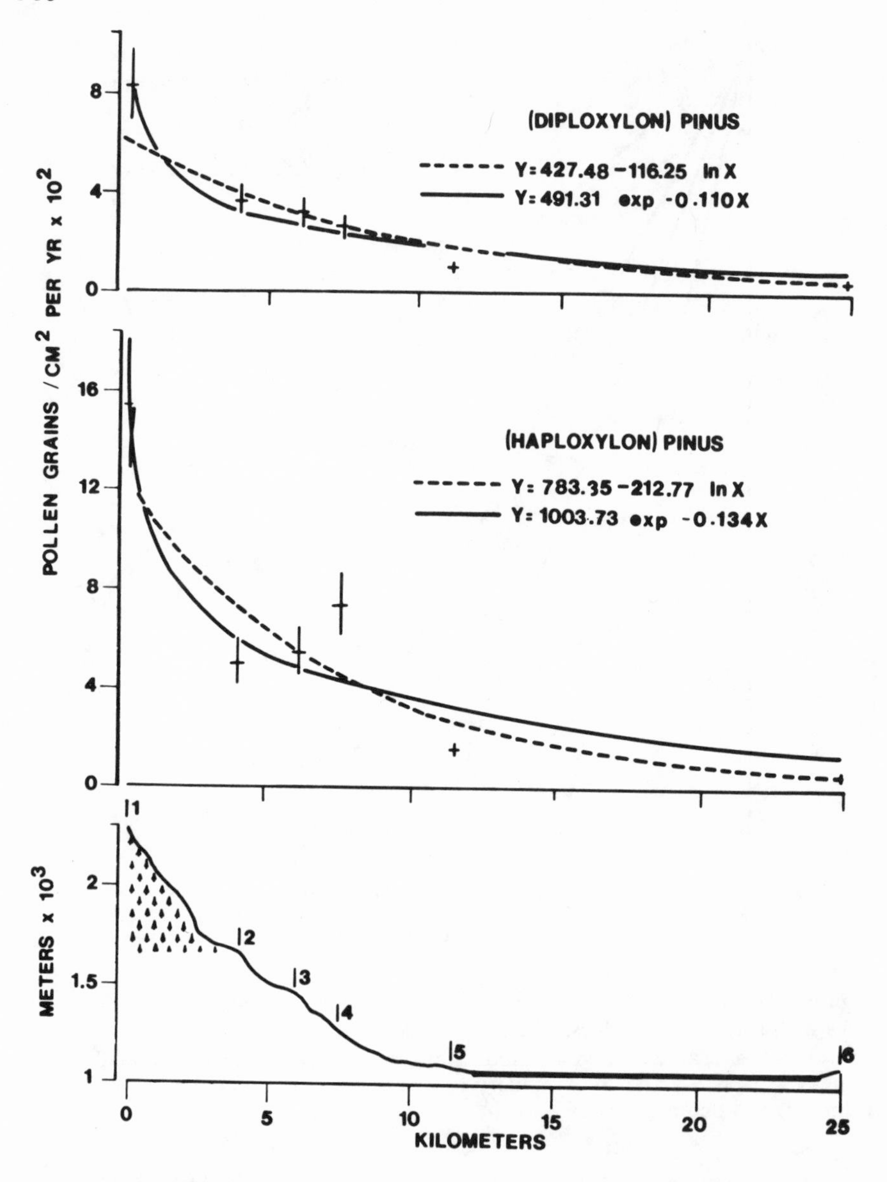

Figure 7.32. Pine pollen, Owens Lake; 95% confidence interval; {†, measured APF value.

During upslope transport, much of the pollen measured at source plants is deposited. The measurement of $N(0)$ then should have been collected at the crest where it may be several orders of magnitude lower than on the lower slope. On the other hand, we are somewhat pleased that the derived equation at least consists of the form suggested by meteorological considerations. It reinforces our hope that the systems approach, applied to functional relationships among forcing functions and system variables, contains the potential for valid solution of palynological environmental reconstruction.

AIRBORNE MICROBIAL MODELS*

In order to understand, predict, and, if desired, manage the loadings of atmospheric microbes in a system as complex as the atmosphere, the precision and organization characteristics of the modeling approach is a useful and perhaps essential tool. One such model, shown in Figure 7.33, depends upon

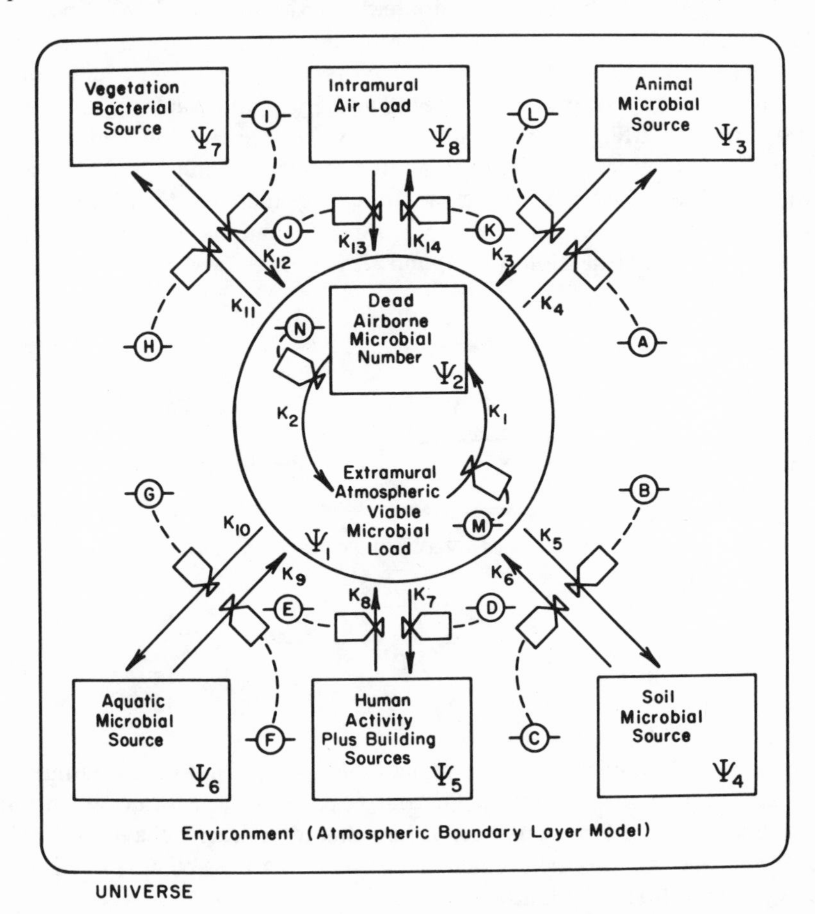

Figure 7.33. Systems diagram of the viable microbial input/output loads to the atmosphere. For key, see Forrester (1961).

*B. Lighthart

knowledge of the state variables (e.g., loadings) and rates of organism "flows" between state variables. Our knowledge of these elements is limited to the state variables in Table 2.1 (chap. 2, this volume).

Models of environmental effects of survival of airborne bacteria have been constructed in various forms that may be categorized as (a) biologically modified determinate models and (b) biological process models. Biological process models have been developed for entire ecosystems during the U.S. International Biological Program. The model shown in Figure 7.33 is one such model in which the extramural atmospheric load of viable microorganisms is described as the integration of dynamic processes of injection, airborne survival, and deposition. Injection of microorganisms into the atmosphere may be from many sources (Gregory and Monteith 1967, Gregory 1973, Spendlove 1974, Hers and Winkler 1973), and at rates depending upon many variables. Airborne survival is a function of many factors, both atmospheric and biological, while deposition of particles larger than 0.5 μm in diameter is largely due to gravitational settling and impaction. The system presented in Figure 7.33 is a set of state variables (Ψ_1–Ψ_8) representing an array of microbial species (or physiological potentials) inputting or outputting microbes to the extramural atmosphere at rates designated K_1–K_{14}. State equations for the generalized system are given by Equations (7.40) and (7.41).

rate of change of state variable Ψ_n (7.40)

$$= \frac{d(\Psi_n)}{dt} = \sum_{i=1}^{8} K_{2i}\Psi_{i+1} \quad \sum_{i=0}^{7} K_{2i+1}\Psi_i$$

value of state variable Ψ_n at time t (7.41)

$$= \int_{t_o}^{t} \frac{d(\Psi_n)}{dt} \Psi_{n_t} + \Psi_{n_{t_o}}$$

where Ψ_n = the nth microbial state variable, K_i = the ith input or output rate to a state variable, and t = time. Of course, each of the K_n rates is an integrated function of environmental conditions that affect each behavioral feature of each microbe in a specific predictable manner. See Caswell et al. (1972) for a detailed exposition of systems equations.

To use the systems model presented, one may either generate the functions necessary to describe the sets of relations within each cell by aerosolizing under known environmental conditions, or attempt to lump or integrate groups of cells with a relatively homogeneous set of cellular relations. Because of the copius data necessary to run each model, particularly the dispersal model, one would probably start work with the "lumped" case. In this regard, Spendlove (1975) has applied both empirical data and theoretical models to study penetration of extramural aerosols into structures of various types. His study showed that the typical residential dwelling, as well as many public buildings, provides no protection to the inhabitants from potentially hazardous extramural aerosols. Further, it may be possible to relate indoor to outdoor airborne bacteria loads using the model of Shair and Heiter (1974) and Hales et al. (1974).

The relationship of meteorological conditions to injection rates of microbes from various sources and their subsequent survival in the atmosphere must be understood. Further, if spatial predictability of viable cells is to be achieved, the biological model must be incorporated into a realistic, atmospheric, turbulent diffusion model. The importance of this wedding of models is illustrated by a recent experiment of A. F. Frisch (personal communication). He observed that radar-reflective chaff statically released (simulating a tree branch) approximately 1 m above the sunheated ground rose convectively in a bubble of hot air to at least 1 km.

Finally, unifying models based on characteristics of the diurnal structure of the atmospheric boundary layer at local sites have been discussed, but have not been implemented (B. Lighthart and A. F. Frisch, unpublished data).

The information necessary to prepare a realistic version of the model presented (Figure 7.33) would include quantitative and qualitative data and atmospheric injection rates for specific human and nonhuman sources. Little information is available about the quantitative nature of microorganisms from natural sources and relatively little is known of the quality of these sources as they affect the airborne load. Further, and more significantly, little is known about the quantity of the natural airborne load that would give a firm clue to the major sources of aeroplanktonic bacteria at any geographical location at any time. Little is known of the all-important injection rates from either human or nonhuman sources. Also, the effects of atmospheric conditions that affect airborne survival must be understood better. For instance, the "open air factor," which arises from urban air containing mixtures of olefins and ozone, is known to be deleterious to the airborne stability of certain bacteria and viruses (Druett 1973a,b, Druett and May 1968). Likewise the exacerbating effect of air ions on influenza virus infections has been reported (Krueger and Reed 1972). In contrast, M. T. Hatch (personal communication) has observed that certain sulfur dioxide concentrations actually enhance airborne influenza survival. What is apparent is that data are accumulating on airborne microbes, although not as rapidly as one might wish. It should be understood, of course, that collection of suitable data from various sources generally requires considerable resources in personnel and equipment. These resources are frequently beyond the reach of most investigators.

FOOTNOTES

[1]James B. Harrington, Wayne M. Wendland (meteorology), Jane Gray, Harvey Nichols, Siewert Nilsson, Allen M. Solomon, Thompson Webb, III (palynology), R. I. Gara, Michael L. McManus (entomology).

[2]Gray, Harrington, Nichols, Nilsson, Solomon, and Webb from the previous group; W. S. Benninghoff, Donald Walker (palynology), George F. Estabrook, Ronald N. Kikert (biomstry) were added to the group.

[3]Benninghoff, Estabrook, Harrington, Nichols, Solomon.

[4]Both pollen types have "bladders" attached at opposite ends of the grains, collectively providing about half the volume of the grain. Bladders of fresh spruce pollen from atmospheric collections and from bulk samples are always water filled, while those of pine are consistently dry and contain only air. The rapid drying of ragweed pollen grains, reducing their mass but not their volume, was carefully studied by Harrington and Metzger (1963).

LITERATURE CITED

Barksdale, T. H. 1967. Spread of rice blast in small fields. Plant Dis. Rep. 51:243–247.

Benninghoff, W. S. 1960. Pollen spectra from bryophytic polsters, Inverness Mud Lake bog, Cheboygan County, Michigan. Pap. Mich. Acad. Sci. Arts Lett. 65:41–60.

Benninghoff, W. S., and R. L. Edmonds. 1972. Ecological systems approaches to aerobiology. I. Identification of component elements and their functional relationships. US/IBP Aerobiol. Program Handb. 2. Univ. Michigan, Ann Arbor. 158 p.

Berger, R. D. 1970. Forecasting *Helminthosporium turcicum* attacks in Florida sweet corn. Phytopathology 60:1284. (Abstr.)

Berger, R. D. 1973a. *Helminthosporium turcicum* lesion numbers related to numbers of trapped spores and fungicide sprays. Phytopathology 63:930–933.

Berger, R. D. 1973b. Early blight of celery: Analysis of disease spread in Florida. Phytopathology 63:1161–1165.

Blakeman, J. P., and A. K. Fraser. 1969. The significance of temperature during sporulation on the biology of pycnidiospores of *Mycosphaerella ligulicola*. Ann. Appl. Biol. 63:295–301.

Blakeman, H. P., and G. Hadley. 1968. The pattern of asexual sporulation in *Mycosphaerella ligulicola*. Trans. Br. Mycol. Soc. 51:643–651.

Bourke, P. M. A. 1957. The use of synoptic weather maps in potato blight epidemiology. Tech. Note 23, Irish Meteorol. Serv., Dublin. 35 p.

Boyce, J. S. 1961. Forest pathology. McGraw-Hill, New York.

Brubaker, L. B. 1975. Postglacial forest patterns associated with till and outwash in north central Upper Michigan. Quat. Res. 5:499–528.

Bryson, R. A., and J. E. Kutzbach. 1974. On the analysis of pollen-climate canonical transfer functions. Quat. Res. 4:162–174.

Burleigh, J. R., M. G. Eversmeyer, and A. P. Roelfs. 1972a. Development of linear equations for predicting wheat leaf rust. Phytopathology 62:947–953.

Burleigh, J. R., A. P. Roelfs, and M. G. Eversmeyer. 1972b. Estimating damage to wheat caused by *Puccinia recondita tritici*. Phytopathology 62:944–946.

Burleigh, J. R., R. W. Romig, and A. P. Roelfs. 1969. Characterization of wheat rust epidemics by numbers of uredia and numbers of urediospores. Phytopathology 59:1229–1237.

Butt, D. J., and D. J. Royle. Multiple regression analysis in the epidemiology of plant diseases. Pages 78–114 *in* J. Krantz, ed. Epidemics of plant diseases—Mathematical analysis and modeling. Springer-Verlag, New York.

Carroll, G. 1943. The use of bryophytic polsters and mats in the study of Recent pollen deposition. Am. J. Bot. 30:361–366.

Caswell, H., H. E. Koenig, J. A. Resh, and Q. E. Ross. 1972. An introduction to systems science for ecologists. Pages 3–78 *in* B. C. Patten, ed. Systems analysis and simulation ecology, vol. II. Academic Press, New York.

Civerolo, E. L. 1975. Quantitative aspects of pathogenesis of *Xanthomonas pruni* in peach leaves. Phytopathology 65:258–264.

Cole, H. 1969. Objective reconstruction of the paleoclimatic record through application of eigenvectors of present-day pollen spectra and climate to late Quaternary pollen stratigraphy. Ph.D. dissertation, Univ. Wisconsin.

Croke, E. J., J. E. Carson, D. F. Gatz, H. Moses, A. S. Kennedy, J. A. Gregory, J. J. Roberts, K. Croke, J. Anderson, D. Parsons, J. Ash, J. Norco, and R. P. Carter. 1968. Chicago air pollution system model. Third Q. Progr. Rep., Argonne Natl. Lab., ANL/ES-CC-003.

Davis, M. B. 1963. On the theory of pollen analysis. Am. J. Sci. 261:897–912.

Davis, M. B. 1967. Pollen accumulation rates at Rogers Lake, Connecticut, during late- and postglacial time. Rev. Palaeobot. Palynol. 2:219–230.

Davis, M. B. 1969a. Palynology and environmental history during the Quaternary period. Am. Sci. 57:317–332.

Davis, M. B. 1969b. Climatic changes in southern Connecticut recorded by pollen deposition at Rogers Lake. Ecology 50:409–422.

Davis, M. B. 1973. Redeposition of pollen grains in lake sediment. Limnol. Oceanogr. 18:44–52.

Davis, M. B. 1974. Holocene migrations of ecotones: Continuing northward migration of tree species throughout the Holocene caused changes in community composition and migrations of ecotones. Am. Quat. Assoc., 3rd Bienn. Meet. Abstr. p. 18–21.

Davis, M. B. 1976. Pleistocene biogeography of temperate deciduous forests. Geosci. Man 13:13–26.

Davis, M. B., and L. B. Brubaker. 1973. Differential sedimentation of pollen grains in lakes. Limnol. Oceanogr. 18:634–646.

Davis, M. B., L. B. Brubaker, and J. M. Beiswenger. 1971. Pollen grains in lake sediments: Pollen percentages in surface sediments from southern Michigan. Quat. Res. 1:450–467.

Davis, M. B., L. B. Brubaker, and T. Webb, III. 1973. Calibration of absolute pollen influx. Pages 9–27 *in* H. J. B. Birks and R. G. West, eds. Quaternary plant ecology. John Wiley & Sons, New York.

Davis, M. B., and J. C. Goodlett. 1960. Comparison of the present vegetation with pollen spectra in surface samples from Brownington Pond, Vermont. Ecology 41:346–357.

Davis, R. T., and G. A. Snow. 1968. Weather systems related to fusiform rust infection. Plant Dis. Rep. 52:419–422.

Dirks, V. A., and R. W. Romig. 1970. Linear models applied to variation in numbers of cereal rust urediospores. Phytopathology 60:246–251.

Druett, H. A. 1973a. Effect on the viability of microorganisms in aerosols of the rapid rarefaction of the surrounding air. Pages 90–94 *in* J. F. P. Hers and K. C. Winkler, eds. Airborne transmission and airborne infection. Oosthoek Publ. Co., Utrecht, The Netherlands.

Druett, H. A. 1973b. The open air factor. Pages 141–149 *in* J. F. P. Hers and K. C. Winkler, eds. Airborne transmission and airborne infection. Oosthoek Publ. Co., Utrecht, The Netherlands.

Druett, H. A., and K. R. May. 1968. Unstable germicidal pollutants in rural air. Nature (Lond.) 220:395.

Edmonds, R. L. 1972. Modeling of aerobiology systems. Pages 12–17 *in* W. S. Benninghoff and R. L. Edmonds, eds. Ecological systems approaches to aerobiology. I. Identification of component elements and their functional relationships. US/IBP Aerobiol. Program Handb. 2. Univ. Michigan, Ann Arbor.

Edmonds, R. L. 1973. Prediction of spread of *Fomes annosus* in forests. Pages 77–92 *in* R. L. Edmonds and W. S. Benninghoff, eds. Ecological systems approaches to aerobiology. II. Development, demonstration, and evaluation of models. US/IBP Aerobiol. Program Handb. 3. (NTIS no. AP-USIBP-H-73-3.) Univ. Michigan, Ann Arbor.

Edmonds, R. L., and W. S. Benninghoff, eds. 1973a. Ecological systems approaches to aerobiology. II. Development, demonstration, and evaluation of models. US/IBP Aerobiol. Program Handb. 3. (NTIS no. AP-USIBP-H-73-3.) Univ. Michigan, Ann Arbor. 186 p.

Edmonds, R. L., and W. S. Benninghoff, eds. 1973b. Ecological systems approaches to aerobiology. III. Further model evelopments. US/IBP Aerobiol. Program Handb. 4. (NTIS no. AP-USIBP-H-73-4.) Univ. Michigan, Ann Arbor. 118 p.

Edmonds, R. L., and C. H. Driver. 1974. Dispersion and deposition of spores of *Fomes annosus* and fluorescent particles. Phytopathology 64:1313–1321.

Edmonds, R. L., I. Von Lindern, C. J. Mason, M. L. McManus, and W. Wallner. 1973. Report of entomology group. Page 57 *in* R. L. Edmonds and W. S. Benninghoff, eds. Ecological systems approaches to aerobiology. III. Further model developments. US/IBP Aerobiol. Program Handb. 4. (NTIS no. AP-USBIP-H-73-4.) Univ. Michigan, Ann Arbor.

Erdtman, G. 1969. Handbook of palynology. Hafner Press, New York. 486 p.

Eversmeyer, M. G., and J. R. Burleigh. 1970. A method of predicting epidemic development of wheat leaf rust. Phytopathology 60:805–811.

Eversmeyer, M. G., J. R. Burleigh, and A. P. Roelfs. 1973. Equations for predicting wheat stem rust development. Phytopathology 63:348–351.

Eyal, Z., and O. Ziv. 1974. The relationship between epidemics of *Septoria* leaf blotch and yield losses in spring wheat. Phytopathology 64:1385–1389.

Forrester, J. W. 1961. Industrial dynamics. M.I.T. Press, Cambridge, Mass. 464 p.

Friedlander, S. K., and J. H. Seinfeld. 1969. A dynamic model of photochemical smog. Environ. Sci. Technol. 3:1175.

Gregory, P. H. 1968. Interpreting plant disease dispersal gradients. Annu. Rev. Phytopathol. 6:189–212.

Gregory, P. H. 1973. The microbiology of the atmosphere, 2nd ed. Halsted Press Div., John Wiley & Sons, New York. 377 p.

Gregory, P. H., E. J. Guthrie, and M. E. Bunce. 1959. Experiments on splash dispersal of fungus spores. J. Gen. Microbiol. 20:328–354.

Gregory, P. H., and J. L. Monteith, eds. 1967. Airborne microbes. Cambridge Univ. Press, Cambridge, England. 385 p.

Gullach, C. B., and J. R. Wallin. 1970. A suggested relationship between sugar beet leaf spot and upper-air flow patterns. Int. J. Biometeorol. 14:349–355.

Hales, C. H., A. M. Rollinson, and F. H. Shair. 1974. Experimental verification of linear combination model for relating indoor-outdoor pollutant concentrations. Environ. Sci. Technol. 8:452–453.

Hanna, S. R. 1971. A simple method of calculating dispersion from urban area sources. J. Air Pollut. Control Assoc. 21:775.

Hansen, H. P. 1949. Pollen content of moss polsters in relation to forest composition. Am. Midl. Nat. 42:473–479.

Harrington, J. B., Jr., and K. Metzger. 1963. Ragweed pollen density. Am. J. Bot. 50:532–539.

Havinga, A. J. 1964. Investigation into the differential corrosion susceptibility of pollen and spores. Pollen Spores 6:621–635.

Havinga, A. J. 1967. Palynology and pollen presentation. Rev. Palaeobot. Palynol. 2:81–98.

Hecht, T. A., and J. H. Seinfeld. 1972. Development and validation of a generalized mechanism for photochemical smog. Environ. Sci. Technol. 6:47.

Hers, J. F. P., and K. C. Winkler, eds. 1973. Airborne transmission and airborne infection. Oosthoek Publ. Co., Utrecht, The Netherlands. 610 p.

Hopkins, J. S. 1950. Differential flotation and deposition of coniferous and deciduous tree pollen. Ecology 31:633–641.

Hyre, R. A. 1957. Forecasting downy mildew of lima bean. Plant Dis. Rep. 41: 7–9.

Ingold, C. T. 1971. Fungal spores—Their liberation and dispersal. Clarendon Press, Oxford, England. 302 p.

James, W. C. 1974. Assessment of plant diseases and losses. Annu. Rev. Phytopathol. 12:27–48.

James, W. C., and C. S. Shih. 1973. Relationship between incidence and severity of powdery mildew and leaf rust on winter wheat. Phytopathology 63:183–187.

James, W. C., C. S. Shih, W. A. Hodgson, and L. C. Callbeck. 1972. The quantitative relationship between late blight of potato and loss of tuber yield. Phytopathology 62:92–96.

Janssen, C. R. 1967. A comparison between the Recent regional pollen rain and the sub-Recent vegetation in four major vegetation types in Minnesota (U.S.A.). Rev. Palaeobot. 2:331–342.

Jenkyn, J. F., and A. Bainbridge. 1974. Disease gradients and small plot experiments on barley mildew. Ann. Appl. Biol. 76:269–279.

Jensen, R. E., and L. W. Boyle. 1966. A technique for forecasting leafspot on peanuts. Plant Dis. Rep. 50:810–814.

Johnson, W. B. 1972. The status of air quality simulation modeling. Proc. Interagency Conf. Environ. Lawrence-Livermore Lab., Livermore, Calif.

Kerr, A., and W. R. F. Rodrigo. 1967a. Epidemiology of tea blister blight (*Exobasidium vexans*). III. Spore deposition and disease prediction. Trans. Br. Mycol. Soc. 50:49–55.

Kerr, A., and W. R. F. Rodrigo. 1967b. Epidemiology of tea blister blight (*Exobasidium vexans*). IV. Disease forecasting. Trans. Br. Mycol. Soc. 50:609–614.

Kerr, A., and N. Shanmuganathan. 1966. Epidemiology of tea blister blight (*Exobasidium vexans*). I. Sporulation. Trans. Br. Mycol. Soc. 49:139–145.

Kincaid, R. R., F. G. Martin, N. Gammon, Jr., H. L. Breland, and W. L. Pritchett. 1970. Multiple regression of tobacco black shank, root knot and coarse root indexes on soil pH, potassium, calcium and magnesium. Phytopathology 60: 1513–1516.

Kranz, J. 1970. Schätzklassen für krankheitsbofall. Phytopathol. Z. 69:131–139.

Kranz, J. 1974. The role and scope of mathematical analysis and modeling in epidemiology. Pages 7–54 *in* J. Kranz, ed. Epidemics of plant diseases—Mathematical analysis and modeling. Springer-Verlag, New York.

Kranz, J., M. Mogk, and A. Stumpf. 1973. EPIVEN—Ein simulator für apfelschorf. Z. Pflanzenkr. 80:181–187.

Krause, R. A., L. B. Massie, and R. A. Hyre. 1975. Blitecast: A computerized forecast of potato late blight. Plant Dis. Rep. 59:95–98.

Krueger, A. P., and E. J. Reed. 1972. Effect of air ion environment on influenza in the mouse. Int. J. Biometeorol. 16:323–327.

Lamb, R. G., and M. Neiburger. 1971. An interim version of a generalized urban air pollution model. Atmos. Environ. 5:239.

Larsen, R. I. 1971. A mathematical model for relating air quality measurements to air quality standards. Environ. Prot. Agency (U.S.) Publ. AP Ser. AP-89. U.S. Gov. Print. Off., Washington, D.C.

Leonard, K. J. 1969. Factors affecting rates of stem rust increase in mixed plants of susceptible and resistant oat varieties. Phytopathology 59:1845–1850.

Livingstone, D. A. 1969. Communities of the past. Pages 83–104 *in* K. N. H. Greenidge, ed. Essays in plant geography and ecology. N.S. Mus., Halifax.

McAndrews, J. H. 1966. Postglacial history of prairie, savanna, and forest in northwestern Minnesota. Mem. Torrey Bot. Club. 22:1–72.

McCoy, R. E. 1971. Epidemiology of chrysanthemum *Ascochyta* blight. Ph.D. thesis, Cornell Univ., Ithaca, N.Y. 177 p.

McCoy, R. E. 1973. Ballistics of *Mycosphaerella ligulicola* ascospore discharge. Phytopathology 63:793–794.

McCoy, R. E., and A. W. Dimock. 1972. Relationship of temperature and humidity to development of *Mycosphaerella* lesions on chrysanthemum. Phytopathology 62:1195–1196.

McCoy, R. E., and A. W. Dimock. 1973. Environmental factors regulating ascospore discharge by *Mycosphaerella ligulicola*. Phytopathology 63:586–589.

McCoy, R. E., R. K. Horst, and A. W. Dimock. 1972. Environmental factors regulating sexual and asexual reproduction by *Mycosphaerella ligulicola*. Phytopathology 62:1188–1195.

MacCracken, M. C., T. V. Crawford, K. R. Peterson, and J. B. Knox. 1971. Development of a multi-box air pollution model and initial verification for the San Francisco Bay area. Lawrence-Livermore Lab. Rep. UCRL-73348, Livermore, Calif.

McManus, M. L. 1973. A dispersal model for larvae of the gypsy moth. Page 129 *in* R. L. Edmonds and W. S. Benninghoff, eds. Ecological systems approaches to aerobiology. II. Development, demonstration, and evaluation of models. US/IBP Aerobiol. Program Handb. 3. (NTIS no. AP-USIBP-H-73-3.) Univ. Michigan, Ann Arbor.

Martin, D. O. 1971. An urban diffusion model for estimating long-term average values of air quality. J. Air Pollut. Control Assoc. 21:16.

Martin, P. S., and J. Gray. 1962. Pollen analysis and the Cenozoic. Science 137:103–111.

Mason, C. J. 1975. A model to predict the dispersal of gypsy moth larvae. Rep. 012094-1-F, Univ. Michigan, Ann Arbor.

Mason, C. J., and R. Edmonds. 1974. An atmospheric dispersion model for gypsy moth larvae. Page 251 *in* Proc. AMS/WMO Symp. Atmos. Diffus. Air Pollut., September 1974, Santa Barbara, Calif.

Mason, C. J., and M. L. McManus. 1975. Verification of an atmospheric dispersion model for gypsy moth larvae. EOS 56:995.

Massie, L. B. 1973. Modeling and simulation of southern corn leaf blight disease caused by race T of *Helminthosporium maydis* NISIK and MIYAKE. Ph.D. thesis, Pennsylvania State Univ., University Park. 84 p.

Massie, L. B., R. R. Nelson, and G. Tung. 1973. Regression equations for predicting sporulation of an isolate of race T of *Helminthosporium maydis* on a susceptible male-sterile corn hybrid. Plant Dis. Rep. 57:730–734.

Milford, S. N., G. C. McCoyd, L. Aronowitz, J. H. Scanlon, and C. Simon. 1971. Developing a practical dispersion model for an air quality region. J. Air Pollut. Control Assoc. 21:549.

Mogk, M. 1973. Untersuchungen zur Epidemiologie von *Colletotrichum coffeanum* Noack sensu Hindorf in Kenia: Eine Analyse der Wirt-Parasit-Umwelt-Beziehungen. Ph.D. thesis, Univ. Giessen, West Germany.

Moses, H. 1970. Tabulation techniques. *In* A. C. Stern, ed. Proc. Symp. Mult.-Source Urban Diffus. Models. Environ. Prot. Agency (U.S.) Publ. AP Ser. AP-86. U.S. Gov. Print. Off., Washington, D.C.

Mundy, E. J. 1973. The effect of yellow rust and its control on the yield of Joss Cambier winter wheat. Plant Pathol. 22:171–176.

Ogden, E. C., and D. M. Lewis. 1969. Concentrations of airborne pollen near the source plants. Paper presented at 11th Int. Bot. Congr., Seattle, Wash. 5 p. (Mimeo.)

Ogden, E. C., G. S. Raynor, and J. V. Hayes. 1975. Travels of airborne pollen. EPA-650/3-75-003. NTIS, Springfield, Va. 100 p.

Ogden, J. G., III. 1969. Correlation of contemporary and late Pleistocene pollen records in the reconstruction of postglacial environments in northeastern North America. Mitt. Int. Verein. Limnol. 17:64–77.

Ono, K. 1965. Principles, methods and organization of blast disease forecasting. Pages 173–194 *in* The rice blast disease. (Proc. Symp., July 1963.) Int. Rice Res. Inst., Los Baños, Laguna, Philippines.

Parvin, D. W., Jr., D. H. Smith, and F. L. Crosby. 1974. Development and evaluation of a computerized forecasting method for *Cercospora* leafspot of peanuts. Phytopathology 64:385–388.

Pasquill, R. 1962. Atmospheric diffusion. Van Nostrand, London. 297 p.

Potter, L. D., and J. Rowley. 1960. Pollen rain and vegetation, San Augustin Plains, New Mexico. Bot. Gaz. 122:1–25.

Ritchie, J. C., and S. Lichti-Federovich. 1963. Contemporary pollen spectra in central Canada. 1. Atmospheric samples at Winnipeg, Manitoba. Pollen Spores 5: 95–114.

Ritchie, J. C., and S. Lichti-Federovich. 1967. Pollen dispersal phenomena in arctic-subarctic Canada. Rev. Palaeobot. Palynol. 3:255–266.

Roberts, J. J., E. S. Croke, and A. S. Kennedy. 1970. An urban atmosphere dispersion model. Pages 6.1–6.72 *in* A. C. Stern, ed. Proc. Symp. Mult.-Source Urban Diffus. Models. Environ. Prot. Agency (U.S.) Publ. AP Ser. AP-86. U.S. Gov. Print. Off., Washington, D.C.

Roelfs, A. P. 1972. Gradients in horizontal dispersal of cereal rust urediospores. Phytopathology 62:70–76.

Romig, R. W., and L. Calpouzos. 1970. The relationship between stem rust and loss in yield of spring wheat. Phytopathology 60:1801–1805.

Romig, R. W., and V. A. Dirks. 1966. Evaluation of generalized curves for number of cereal rust urediospores trapped on slides. Phytopathology 56:1376–1380.

Rowe, R. C., and R. L. Powelson. 1973. Epidemiology of *Cercosporella* foot-rot of wheat: Disease spread. Phytopathology 63:984–988.

Royle, D. J. 1973. Quantitative relationships between infection by the hop downy mildew pathogen, *Pseudoperonospora humuli,* and weather and inoculum factors. Ann. Appl. Biol. 73:19–30.

Royle, D. J., and G. E. Thomas. 1972. Analysis of relationships between weather factors and concentrations of airborne sporangia of *Pseudoperonospora humuli*. Trans. Br. Mycol. Soc. 58:79–89.

Sangster, A. G., and H. M. Dale. 1961. A preliminary study of differential pollen grain preservation. Can. J. Bot. 39:35–43.

Sangster, A. G., and H. M. Dale. 1964. Pollen grain preservation: Underrepresented species in fossil spectra. Can. J. Bot. 42:437–440.

Scarpa, M. J., and L. C. Raniere. 1964. The use of consecutive hourly dewpoints in forecasting downy mildew of lima bean. Plant Dis. Rep. 48:77–81.

Schmidt, R. A., R. E. Goddard, and C. A. Hollis. 1974. Incidence and distribution of fusiform rust in slash pine plantations in Florida and Georgia. Univ. Fla. Agric. Exp. Stn. Tech. Bull. 763. 21 p.

Schmitt, C. G., C. H. Kingsolver, and J. F. Underwood. 1959. Epidemiology of stem rust of wheat. I. Wheat stem rust development from inoculation foci of different concentration and spatial arrangement. Plant Dis. Rep. 43:601–606.

Schrödter, H. 1964. Methodisches zur Bearbeitung phytometeoropathologischer Untersuchungen, dargestellt am Beispiel der Temperaturrelation. Phytopathol. Z. 53:154–166.

Schrödter, H., and J. Ullrich. 1965. Untersuchungen zur Biometeorologie und Epidemiologie von *Phytophthora infestans* (Mont.) de By. auf mathematischstatisticher Grundlage. Phytopathol. Z. 54:87–103.

Schrödter, H., and J. Ullrich. 1967. Eine mathematischstatistische Lösung des Problems der Prognose von Epidemien mit helfe Meteorologischer parameter dargestellt am Beispiel der kartoffelkrautfaule (*Phytophthora infestans*). Agric. Meteorol. 4:119–135.

Shair, F. H., and K. L. Heiter. 1974. Theoretical model for relating indoor pollutant concentrations to those outside. Environ. Sci. Technol. 8:444–451.

Shrum, R. 1975. Simulation of wheat stripe rust using EPIDEMIC, a flexible plant disease simulator. Pa. State Univ., Agric. Exp. Stn. Prog. Rep. 347. University Park, Pa.

Slade, D. H., ed. 1968. Meteorology and atomic energy. U.S.A.E.C., Div. Tech. Inf. Publ. TID 24190. 445 p.

Smith, D. H., F. L. Crosby, and W. J. Ethredge. 1974. Disease forecasting facilitates chemical control of *Cercospora* leafspot of peanuts. Plant Dis. Rep. 58: 666–668.

Sokal, R. R., and F. J. Rohlf. 1969. Biometry. W. H. Freeman, San Francisco. 776 p.

Solomon, A. M. 1975. A rational model for interpretation of pollen samples from arid regions. *In* Abstr. 7th Annu. Meet. Am. Assoc. Strat. Palynol., Houston, Tex.

Spendlove, J. C. 1974. Industrial, agricultural, and municipal microbial aerosol problems. Dev. Ind. Microbiol. 15:20–27.

Spendlove, J. C. 1975. Penetration of structures by microbial aerosols. Dev. Ind. Microbiol. 16:427–436.

Start, G. E., C. R. Dickson, and L. L. Wendell. 1975. Diffusion in a canyon within rough mountainous terrain. J. Appl. Meteorol. 14:333.

Turner, D. B. 1964. A diffusion model for an urban area. J. Appl. Meteorol. 3: 83.

Turner, D. B. 1969. Workbook of atmospheric dispersion estimates. US-DHEW Public Health Serv. Publ. 999-AP-26. 84 p.

Underwood, J. F., C. H. Kingsolver, C. E. Peet, and K. R. Bromfield. 1959. Epidemiology of stem rust of wheat. III. Measurements of increase and spread. Plant Dis. Rep. 43:1154–1159.

Van der Plank, J. E. 1963. Plant diseases: Epidemics and control. Academic Press, New York. 349 p.

Van der Plank, J. E. 1975. Principles of plant infection. Academic Press, New York. 210 p.

Von Post, L. 1916. Om Skogsträdpollen i sydsvenska torfmosslagerföljder. Geol. Foren Stockholm Forh. 38:384–390, 392–393.

Wade, N. 1972. A message from corn blight: The dangers of uniformity. Science 177:678–679.

Waggoner, P. E. 1968. Weather and the rise and fall of fungi. *In* W. P. Lowry, ed. Biometeorology. Oregon State Univ. Press, Corvallis.

Waggoner, P. E. 1974. Simulation of epidemics. Pages 137–160 *in* J. Kranz, ed. Epidemics of plant diseases—Mathematical analysis and modeling. Springer-Verlag, New York.

Waggoner, P. E., and J. G. Horsfall. 1969. EPIDEM. Conn. Agric. Exp. Stn. (New Haven) Bull. 698. 80 p.

Waggoner, P. E., J. G. Horsfall, and R. J. Lukens. 1972. EPIMAY. Conn. Agric. Exp. Stn. (New Haven) Bull. 729. 84 p.

Wallin, J. R. 1962. Summary of recent progress in predicting late blight epidemics in the United States and Canada. Am. Potato J. 39:306–312.

Wallin, J. R., and D. V. Loonan. 1971. Effect of leaf wetness duration and air temperature on *Cercospora beticola* infection of sugar beet. Phytopathology 61: 546–549.

Wallin, J. R., and D. V. Loonan. 1972. The increase of *Cercospora* leaf spot in sugar beets and periodicity of spore release. Phytopathology 62:570–572.

Wallin, J. R., and J. A. Riley. 1960. Weather map analysis—An aid in forecasting potato late blight. Plant Dis. Rep. 44:227–234.

Wallin, J. R., and P. E. Waggoner. 1950. The influence of climate on the development and spread of *Phytophthora infestans* in artificially inoculated potato plots. Plant Dis. Rep. Suppl. 190:19–33.

Watts, W. A. 1973. Rates of change and stability in vegetation in the persepctive of long periods of time. Pages 195–206 *in* H. J. B. Birks and R. G. West, eds. Quaternary plant ecology. John Wiley & Sons, New York.

Webb, T., III. 1971. The late- and postglacial sequence of climatic events in Wisconsin and east-central Minnesota: Quantitative estimates derived from fossil pollen spectra by multivariate statistical analysis. Ph.D. dissertation, Univ. Wisconsin.

Webb, T., and R. A. Bryson. 1972. Late- and postglacial climatic change in the northern Midwest, U.S.A.: Quantitative estimates derived from fossil pollen spectra by multivariate statistical analysis. Quat. Res. 2:70–115.

Wright, H. E. 1968. The roles of pine and spruce in the forest history of Minnesota and adjacent areas. Ecology 49:937–955.

8. THE CHALLENGES OF AEROBIOLOGY

R. L. Edmonds and W. S. Benninghoff

MONITORING PROGRAMS IN OUTDOOR ENVIRONMENTS

The US/IBP-Aerobiology Program has set the stage for future integrated aerobiological studies and there is now more than ever a conviction that aerobiology is fast growing in importance. There is an urgent need for more research. Despite the work that has already been accomplished, we are still largely ignorant of the biological content of the atmosphere and its interactions with terrestrial organisms.

The need for a global environmental monitoring network was advanced in 1968 and a commission of global monitoring of the environment was established under SCOPE (Special Committee on Problems of the Environment) of the International Council of Scientific Unions (ICSU). This culminated in a report on global environmental monitoring (SCOPE 1, 1971), which was submitted to the United Nations Conference on the Human Environment in Stockholm in 1972. The concept of monitoring and early warning systems is not a new one. Many countries have traditionally operated such systems and many of these national activities are coordinated internationally by organizations such as WMO, WHO, and FAO. The efforts, however, are uneven on a worldwide scale, particularly those that deal with man-induced phenomena. SCOPE 1 aimed at a system involving measurements and observations directed toward a description of the environment and its change. Such a system will have two basic functions:

1. It will establish existing natural base-line values of the bioenvironmental state against which contaminated states can be compared in relation to effect levels on biota and man.
2. It will provide the basic information for the early detection and global extent of bioenvironmental changes and their causes.

In addition to these primary functions, the system must be supported by, and supportive to, a continuing research program directed to the causes and effects of bioenvironmental change.

Interest steadily increases in programs for measuring biological and health phenomena, especially in relation to changes in physical parameters of the environment, such as chemical composition of the atmosphere. Biologists have been extremely slow to respond to the needs and opportunities presented. A primary factor responsible for this is that individuals cannot respond effectively; only groups of scientists organized for action can respond to this need. Mr. Donald H. Pack, leader of the National Oceanic and Atmospheric Administration's (NOAA) Project on Geophysical Monitoring for Climatic Change, has repeatedly urged that aerobiologists get under way with a design for a monitoring program so that we can begin to see what the problems and potentialities are in such an approach. It is not proposed that the aerobiology system monitor gaseous or particulate pollution, as that will be covered by other groups. The interactions of biological material with these pollutants is of interest, however.

The aerobiology monitoring system should consist of three levels, urban, regional, and global (base line). Such a monitoring system would:

1. Provide knowledge that would assist in the understanding, control, and forecasting of airborne human, animal, and plant diseases.

2. Assist in the detection of major changes in the general composition and productivity of natural vegetation, microflora, and microfauna.

3. Provide a data base that would aid the development of mathematical simulation models of aerobiological interactions that can predict the results expected from varying types and degrees of remedial actions and can guide further research.

4. Contribute to the better management of renewable natural resources such as forest and crop systems, through evidences of effects of management on, for example, pollen and pollinators, insect pests, and diseases.

Pilot monitoring programs are needed, particularly at base-line stations, to identify what should be monitored on a long-term basis. It is neither feasible nor useful to monitor all groups of organisms. Some priority urban and regional scale problems in the United States have been identified by the aerobiology program. Table 8.1 is a suggested monitoring scheme that considers urban, regional, and global problems. The list is not exhaustive. Figure 8.1 shows the geographical distribution of some of these regional problems, including two air pollution areas where interactions of air pollutants and biological material are known to occur. There is yet a need to determine the optimum number of sampling sites and instruments and this can only be done with a pilot program of perhaps five years. The sites will need to tie in with other networks for measuring chemical changes and the like, since much of the interacting data, particularly meteorological data will be obtained in this way. There should be no restriction on the number of sites as it may be necessary to have more aerobiological stations. Data from the chemical monitoring stations can be used to assess, for example, the impact of air pollutants on airborne microorganisms.

Table 8.1. Suggested pilot aerobiology monitoring scheme for the United States.

Scale	Field of aerobiology	Organisms to be monitored (by section of country)		
		East	Midsection	West
Urban	human allergology	ragweed, tree pollen, molds	grass, crop pollen, molds, algae	bur sage and grass pollen
	human disease[a]	bacteria from sewage treatment, insect vectors	same as East	same as East
Regional	palynology	tree, grass, shrub pollen	same as East	same as East
	entomology	gypsy moth, mosquitoes	gypsy moth, aphids, locusts, mosquitoes	Douglas-fir tussock moth, mosquitoes, bark beetles
	plant pathology	fusiform rust, *Fusarium, Fomes annosus*	wheat rusts, corn blight, *Fusarium*	white pine blister rust, *Fusarium, Fomes annosus*
	animal disease	IBR, foot-and-mouth disease	IBR, equine encephalitis, foot-and-mouth disease	equine encephalitis
Global	Pilot program at special sites such as Mauna Loa, Point Barrow, and American Samoa to identify what should be monitored. A suggestion is *Cladosporium* spp., which are distributed worldwide.			

[a]Most airborne human diseases are thought to be transmitted primarily indoors, and it is not known what levels of potentially pathogenic organisms occur outdoors.

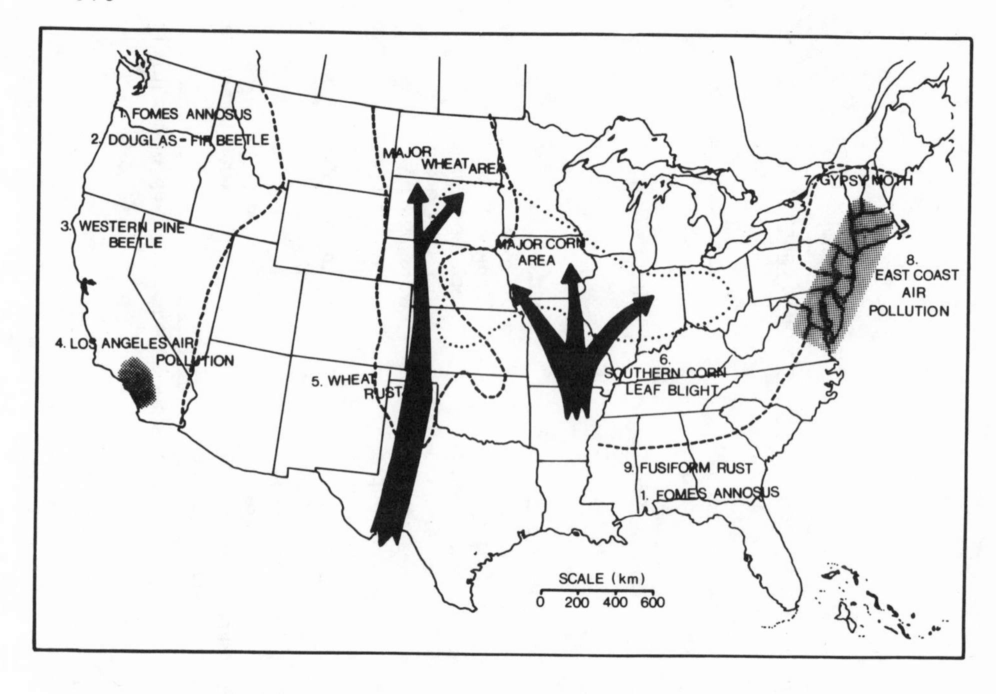

Figure 8.1. Some regional aerobiological problems in the United States.

Specific Monitoring Objectives with Particular Reference to Base-Line Stations

1. Measure, at intervals determined to be appropriate, the total organic content (dead and alive) of the air at different sites around the world (e.g., Mauna Loa, Point Barrow, American Samoa).
2. At appropriate intervals and with a specific variety of culture media and other tests for viability, sample the atmospheric biota, calculate the viability status of organism groups specifically sampled, and estimate the total living biota for a unit volume of the sampled air mass.
3. From collections obtained through projects 1 and 2, inventory the organisms that are reasonably easy to identify; group them into classes such as bacteria, fungi, pollen, and so on; and calculate or estimate the numbers per unit volume of air. Repeat this for several diurnal cycles in each season of the year.
4. Select potential or tentative "sentinel" or "indicator" organisms from those whose spores tend to be caught in air samplers during a given biological or meteorological event.
5. Seek to identify the source areas of airborne biological material and the

atmospheric transport pathway to the collection site.

To achieve the above will be a major undertaking. It could be made more practical by first initiating a pilot study of periodic intensive sampling (perhaps for a 2-wk period several times a year). Such samples would represent a data base for subsequent intercorrelations and identification of suitable "indicator" components if they exist. Laboratory facilities and identification centers will be needed to achieve this task. Modern taxonomic approaches should be used and archiving of collected material should be done so that it can be examined at a later date. Although aerobiological samples exhibit enormous temporal and spatial variability, this should not discourage attempts at assessing global and regional variations. Powerful statistical techniques now exist for extracting trends and cycles from "noisy" data.

INTRAMURAL AEROBIOLOGY

The contamination of home and industrial environments with microbiological material or other pollutants is a major health problem, and the methods by which this particulate matter gets distributed, deposited, and absorbed by recipient persons certainly is a problem that needs further investigation. For example, home heating and air-conditioning systems can be serious sources of aeroallergens (dusts, molds). The relationships between outdoor and indoor levels and types of materials should be further investigated. Such studies should be linked to outdoor urban monitoring networks with respect to sources of organisms and air pollutants, both inside and outside of buildings.

SAMPLING INSTRUMENTS

Several hundred instruments have been developed for sampling particles in the atmosphere. The most common principles used are those of sedimentation or impaction and the material is either identified by direct visual identification under a microscope or by culturing. Unfortunately, there is no one sampler yet developed that can adequately sample all the sizes and types of material present in the atmosphere.

We are recommending the following four instruments, which are shown in Figure 8.2, so that some degree of standardization can be obtained.

The aerobiology program has been recommending the following four instruments, which are shown in Figure 8.2, so that some degree of standardization can be obtained.

A. The Tauber trap—a simple and cheap sedimentation sampler for pollen and spores, which can be used in remote locations where power cannot be obtained. Its disadvantage is that it is not a volumetric sampler and provides only semiquantitative data by visual microscopic identification. It is

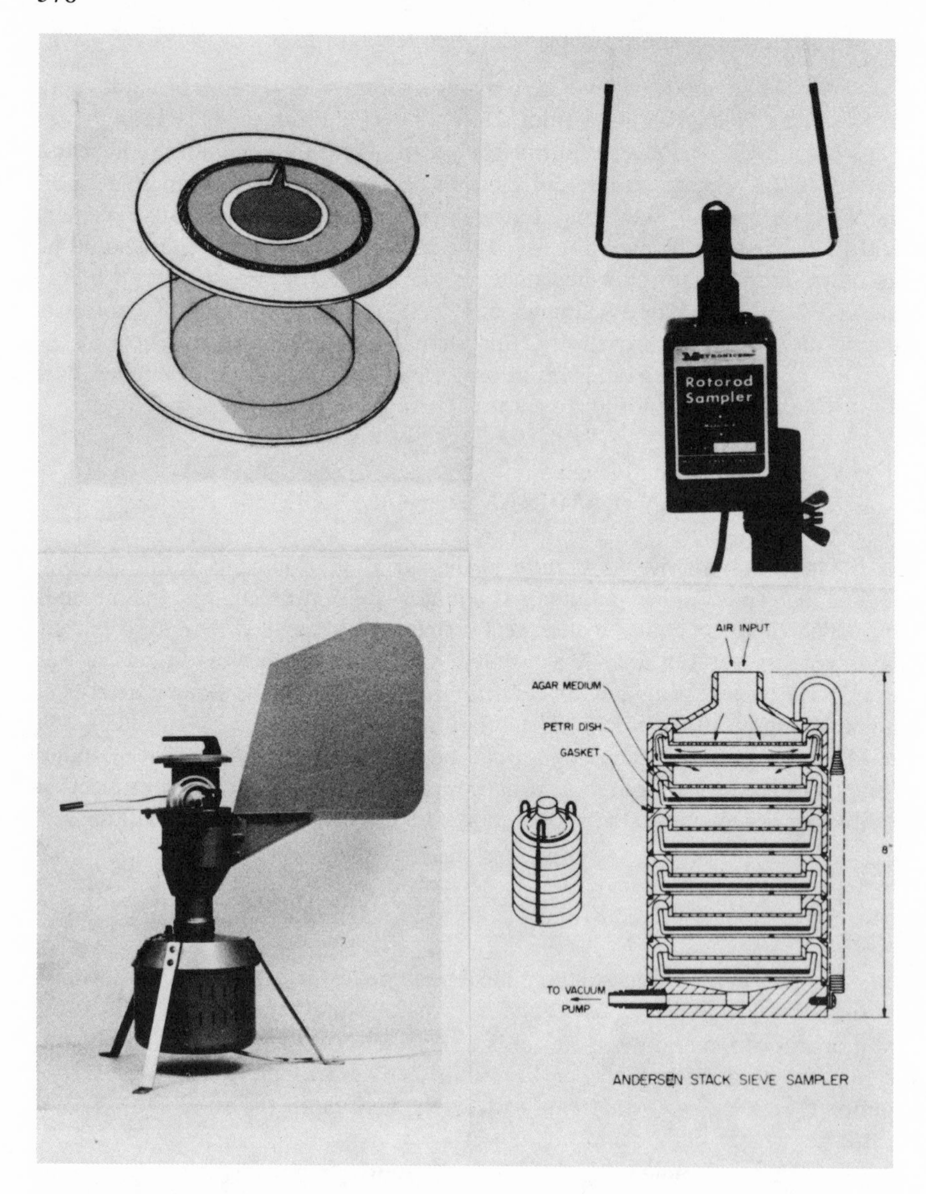

Figure 8.2. Instruments recommended for sampling airborne microorganisms: (A) Tauber trap, (B) Rotorod sampler, (C) Burkard version of Hirst trap, (D) Andersen sampler.

useful for showing trends.

B. The Rotorod sampler—a rotating-arm, portable impaction sampler for pollen and spores; AC and DC versions are available. It is a volumetric sampler and is highly efficient, largely independent of wind speed. Its efficiency, however, is very dependent on particle size and that must be taken into account. It is generally recommended for short-period sampling of up to 2 h. Identification is generally by visual microscopic identification, but cultural techniques can be used.

C. The Burkard version of the Hirst trap—a suction-type, nonportable, impaction sampler for pollen and spores. It is a volumetric sampler vaned into the wind; the material is impacted on a moving tape. The sampling period is 7 days. Identification is by visual microscopic inspection. Hourly and daily trends can be investigated.

D. The Andersen sampler—a suction-type impaction sampler that can be used for both direct visual identification and culturing. Particle sizes are separated by impaction on the various stages. It samples bacteria as well as pollen and spores.

Specialized instruments are needed for sampling insects (radar has been used for tracking swarms) and other larger material and smaller materials such as viruses, although viruses are generally carried on "rafts" of organic material that can be sampled in an Andersen sampler.

The needs for more trained specialists to identify these collections are great. It is a tedious job to identify the materials and there is a need for improved manuals for precise microscopic identification of specific particle types, especially fragments of algae and insects.

A whole new set of collecting devices is needed, particularly to examine the viability of particles. Rapid assessment is needed if we are to use monitoring systems as early warning devices for epidemics.

Advances in the field of remote sensing should be incorporated into aerobiological studies particularly with regard to assessment of impact.

DATA STORAGE

There is at present no means for keeping account of acquired data of current efforts in aerobiology other than that by (1) the Overseas Centre for Pest Control in London on locust and similar insect plagues in Africa, the Middle East, and Southern Asia, and by (2) the WHO Vector Biology Control Unit in Geneva on mosquitoes and the human diseases they carry. The International Association for Aerobiology can be used to exchange information but cannot, of itself, provide for data storage. To learn of the directions and amounts of change, or to discover anomalies, we must have the proper records made and kept in an accessible manner. In 1970 and 1971 it appeared that the SAROAD (Storage and Retrieval of Atmospheric Data) System of the National Air Pollution Control Administration and its successor in the EPA would handle aerobiological data for the United States, but this system is no longer available for such uses. In intergovernmental agencies pressure is growing for centralized data storage of this kind (witness the World Data Centers' geophysical data resulting from the 1956 IGY and subsequent periods). The National Academy of Sciences/National Research Council Committee on Aerobiology is currently studying the question of storage of aerobiological data.

INDEX